METHODS IN PHARMACOLOGY AND TOXICOLOGY

Series Editor
Y. James Kang
Department of Pharmacology and Toxicology
University of Louisville
Louisville, KY, USA

For further volumes:
http://www.springer.com/series/7653

Methods in Pharmacology and Toxicology publishes cutting-edge techniques, including methods, protocols, and other hands-on guidance and context, in all areas of pharmacological and toxicological research. Each book in the series offers time-tested laboratory protocols and expert navigation necessary to aid toxicologists and pharmaceutical scientists in laboratory testing and beyond. With an emphasis on details and practicality, *Methods in Pharmacology and Toxicology* focuses on topics with wide-ranging implications on human health in order to provide investigators with highly useful compendiums of key strategies and approaches to successful research in their respective areas of study and practice.

Methods for Novel Psychoactive Substance Analysis

Edited by

Marta Concheiro

Department of Sciences, John Jay College of Criminal Justice, City University of New York, New York, NY, USA

Karl B. Scheidweiler

Scientific Affairs, Toxicology R&D, Abbot/Immunalysis, Pomona, CA, USA

Humana Press

Editors
Marta Concheiro
Department of Sciences, John Jay
College of Criminal Justice
City University of New York
New York, NY, USA

Karl B. Scheidweiler
Scientific Affairs, Toxicology R&D
Abbot/Immunalysis
Pomona, CA, USA

ISSN 1557-2153 ISSN 1940-6053 (electronic)
Methods in Pharmacology and Toxicology
ISBN 978-1-0716-2607-8 ISBN 978-1-0716-2605-4 (eBook)
https://doi.org/10.1007/978-1-0716-2605-4

This Humana imprint is published by the registered company Springer Science+Business Media, LLC, part of Springer Nature.
The registered company address is: 1 New York Plaza, New York, NY 10004, U.S.A.

Preface

The goal of this handbook, entitled *Methods for Psychoactive Substance Analysis*, from the series Methods in Pharmacology and Toxicology, is to provide an overview, but at the same time, a detailed information on current advances in analytical methods and strategies employed for monitoring and discovering novel psychoactive substances (NPS) in clinical and forensic laboratories.

NPS are a chemically heterogeneous group of substances that mimic the effects of traditional drugs of abuse, such as cannabinoids, amphetamines, cocaine, or heroin, but are not internationally controlled. The main classes of NPS in terms of prevalence include synthetic cannabinoids, synthetic cathinones, synthetic opioids, and designer or synthetic benzodiazepines. NPS are constantly emerging to circumvent legislation and drug testing efforts requiring robust, sensitive, and specific analytical methodology to yield clinically/forensically relevant results, while being flexible for incorporating new NPS to keep pace with this dynamic drug class. Most NPS compounds do not cross-react with existing immunoassay screening reagents, and development/validation of new immunoassay reagents lags behind emerging waves of NPS introduced by clandestine laboratories. Delays while NPS reference standards become available also complicate the development of confirmatory methods, making up-to-date NPS analysis a challenge.

Selecting the appropriate sample matrix and analytical testing approach(es) are key for laboratories faced with NPS drug testing. Clinical and forensic laboratories are pursuing novel analytical strategies to monitor this constantly changing group of substances in classic (blood and urine) and alternative biological samples. NPS analysis in urine is especially challenging due to the lack of research documenting NPS metabolic routes and availability of NPS metabolite reference standards. In recent years, there is heightened interest in alternative biological matrices for NPS testing. Oral fluid has been investigated to detect recent drug exposure in driving under the influence cases. There is also interest in NPS testing with hair, to document past exposure and history of NPS consumption, and meconium, to determine drug exposure during pregnancy. Therefore, there is much clinical and forensic interest in robust, sensitive, selective, and flexible analytical methodologies for NPS testing of a wide variety of biological matrices to accurately document NPS exposure.

Forensic and clinical toxicologists, pharmacologists, and chemists in academic and research settings, as well as in private laboratories, will benefit from this book. The book describes the state-of-the-art methodologies for a wide range of novel psychoactive substances in different biological samples, employing screening and confirmatory techniques, and new metabolites identification. This book also targets undergraduate and graduate students interested in analytical methodologies in toxicology.

New York, NY, USA — *Marta Concheiro*
Pomona, CA, USA — *Karl B. Scheidweiler*

Contents

Contributors

JEAN-CLAUDE ALVAREZ • *Pharmacology and Toxicology, Paris-Saclay University (Versailles Saint-Quentin-En-Yvelines University), Inserm U-1173, Raymond Poincaré hospital, Assistance Publique-Hôpitaux de Paris (AP-HP), Garches, France*

PIETRO BRUNETTI • *Department of Excellence of Biomedical Science and Public Health, University "Politecnica delle Marche", Ancona, Italy*

FRANCESCO PAOLO BUSARDÒ • *Department of Excellence of Biomedical Science and Public Health, University "Politecnica delle Marche", Ancona, Italy*

CLAUDIA CARELLI • *Department of Public Health, Experimental and Forensic Medicine, University of Pavia, Pavia, Italy*

MARTA CONCHEIRO • *Department of Sciences, John Jay College of Criminal Justice, City University of New York, New York, NY, USA*

ANA DE-CASTRO-RÍOS • *Toxicology Service, Institute of Forensic Sciences, Universidade de Santiago de Compostela, Santiago de Compostela, Spain*

VINCENT DI FAZIO • *Laboratory of Toxicology, National Institute of Criminalistics and Criminology (NICC), Brussels, Belgium*

ANNAGIULIA DI TRANA • *Department of Excellence of Biomedical Science and Public Health, University "Politecnica delle Marche", Ancona, Italy*

JEAN-MICHEL GAULIER • *Unité Fonctionnelle de Toxicologie, CHU Lille, Lille, France*

SILVIA GRAZIANO • *National Centre on Addiction and Doping, National Institute of Health, Rome, Italy*

HENRIK GRÉEN • *Department of Biomedical and Clinical Sciences, Linköping University, Linköping, Sweden; Department of Forensic Genetics and Forensic Toxicology, National Board of Forensic Medicine, Linköping, Sweden*

ROBERT KRONSTRAND • *Department of Forensic Genetics and Forensic Toxicology, National Board of Forensic Medicine, Linköping, Sweden*

ALEX J. KROTULSKI • *NPS Discovery, Center for Forensic Science Research and Education, Fredric Rieders Family Foundation, Willow Grove, PA, USA*

NUNZIA LA MAIDA • *Department of Excellence of Biomedical Science and Public Health, University "Politecnica delle Marche", Ancona, Italy*

ISLAM AMINE LARABI • *Pharmacology and Toxicology, Paris-Saclay University (Versailles Saint-Quentin-En-Yvelines University), Inserm U-1173, Raymond Poincaré hospital, Assistance Publique-Hôpitaux de Paris (AP-HP), Garches, France*

ELENA LENDOIRO • *Toxicology Service, Institute of Forensic Sciences, Universidade de Santiago de Compostela, Santiago de Compostela, Spain*

ALFREDO FABRIZIO LO FARO • *Department of Excellence of Biomedical Science and Public Health, University "Politecnica delle Marche", Ancona, Italy*

SARA MALACA • *Department of Excellence of Biomedical Science and Public Health, University "Politecnica delle Marche", Ancona, Italy*

GIULIO MANNOCCHI • *International School of Advance Studies (ISAS), University of Camerino, Camerino, Italy*

MATTEO MORETTI • *Department of Public Health, Experimental and Forensic Medicine, University of Pavia, Pavia, Italy*

LUCA MORINI • *Department of Public Health, Experimental and Forensic Medicine, University of Pavia, Pavia, Italy*

SIMONA PICHINI • *National Centre on Addiction and Doping, National Institute of Health, Rome, Italy*

LUKE N. RODDA • *Forensic Laboratory Division, Office of Chief Medical Examiner, City and County of San Francisco, San Francisco, CA, USA*

NELE SAMYN • *Laboratory of Toxicology, National Institute of Criminalistics and Criminology (NICC), Brussels, Belgium*

KAREN S. SCOTT • *Department of Pathology, University of Alabama at Birmingham, Birmingham, AL, USA*

STEVEN TOWLER • *Department of Sciences, John Jay College of Criminal Justice, City University of New York, New York, NY, USA*

STEPHANIE R. VANLYSEBETTENS • *Department of Pathology, University of Alabama at Birmingham, Birmingham, AL, USA; Forensic Science, Arcadia University, Glenside, PA, USA*

MARIA ROSARIA VARÌ • *National Centre on Addiction and Doping, National Institute of Health, Rome, Italy*

SVANTE VIKINGSSON • *Center for Forensic Sciences, RTI International, Research Triangle Park, Durham, NC, USA*

SARAH M. R. WILLE • *Laboratory of Toxicology, National Institute of Criminalistics and Criminology (NICC), Brussels, Belgium*

Chapter 1

Target Analysis of Synthetic Cannabinoids in Blood and Urine

Alex J. Krotulski

Abstract

The synthetic cannabinoid subclass of NPS remains one of the largest and the most complex in terms of chemistry and behavior. Several considerations must be understood prior to development of target analysis for this drug class. Unlike other NPS subclasses, synthetic cannabinoids and their metabolites are neutral to acidic analytes. To compound this distinction, these analytes behave differently in blood and urine samples, requiring additional attention prior to development and implementation. As a result, synthetic cannabinoids require special extraction and analytical methods for their characterization. This chapter describes all aspects associated with target analysis of synthetic cannabinoids, including sample preparation methods and instrumental platforms. Types of target analysis discussed include liquid chromatography tandem mass spectrometry (LC-MS/MS) [the most commonly employed], gas chromatography mass spectrometry (GC-MS), immunoassay, and high-resolution mass spectrometry (HRMS).

Key words Synthetic Cannabinoids, LC-MS/MS, GC-MS, Immunoassay, LC-HRMS, Blood, Urine

1 Introduction

Target analysis for synthetic cannabinoids can be one of the most challenging aspects associated with new psychoactive substance (NPS) testing. To begin, synthetic cannabinoids vary greatly in their chemistry and pharmacokinetic nature from other NPS subclasses. This does not allow laboratories to apply that same broad-based testing methodologies or sample preparation protocols developed for other subclasses to the synthetic cannabinoid subclass. From sample collection to data interpretation, including all of the analytical aspects in between, synthetic cannabinoids require specialized treatment and experienced handling. This chapter will focus on important aspects associated with the target analysis of synthetic cannabinoids in blood and urine, focusing specifically on sample preparation, analytical methods, and final considerations.

Marta Concheiro and Karl B. Scheidweiler (eds.), *Methods for Novel Psychoactive Substance Analysis*,
Methods in Pharmacology and Toxicology, https://doi.org/10.1007/978-1-0716-2605-4_1,

Synthetic cannabinoids were introduced into the recreational drug market around 2008 when scientists began documenting the influx of these new drugs on plant-like material or other substrates entering the United States and Europe [1]. The first synthetic cannabinoids were substances pirated from academic and pharmaceutical research studies, namely, by Hebrew University (Jerusalem, Israel), John W. Huffman (Clemson University, SC), Alexandros Makriyannis (Northeastern University, MA), and Pfizer (New York City, NY) [2]. This process of stealing legitimate research findings and using the information as blueprints to develop the next "legal high" is a common practice among clandestine chemists and laboratories to this day. Since the emergence of the first generation of synthetic cannabinoids back in 2008, there have been more than 200 new synthetic cannabinoids synthetized and introduced into the recreational drug supply [3–5]. As time has evolved, so too has the structure of these drugs. Most synthetic cannabinoids retain common features, often referred to as the head, core, tail, and linked group (Fig. 1), although some exceptions exist. Understanding this historical perspective is essential to laying the groundwork for the development and implementation of target methods. While it will not be covered here in detail, changes to the synthetic cannabinoid structure also lead to changes in cannabinoid receptor activation and binding, which has previously resulted in large difference in potency and toxicity (e.g., HU-210 or JWH-018 vs. 5F-MDMB-PINACA or MMB-FUBINACA).

Synthetic cannabinoids can be subcategorized based on (generally) their core and linked group region [6]. Common subclassifications include cyclohexylphenols (e.g., CP 47,497), naphthoylindoles (e.g., JWH-018, AM-1120), benzoylindoles (e.g., RCS-4), phenylacetylindoles (e.g., JWH-250), alkylindoles (e.g., UR-144, XLR-11), indole carboxylates (e.g., PB-22, NM-2201), indole carboxamides (e.g., 5F-APP-PICA, 5F-MMB-PICA), and indazole carboxamides (e.g., 5F-MDMB-PICA, MMB-FUBINACA), the latter two of which are most commonly seen in the late 2010s and early 2020s. Each of these subclassifications contains additional subclassifications as well, and the number of structural variations seems endless. However, the important information to understand and remember is that these changes in subclassification (and ultimately substructure) lead to changes in chemistry and therefore behavior of the drugs during analysis. Early-generation subclassifications (i.e., cyclohexylphenols, naphthoylindoles, benzoylindoles) are more nonpolar in chemistry, while later generations (i.e., indole and indazole carboxamides) are more neutral in their behavior – although assessment of overall structure with varying head and tail groups is necessary for ultimate chemical behavior.

The metabolism of synthetic cannabinoids is another important aspect to review prior to discussing analytical methods. Unlike

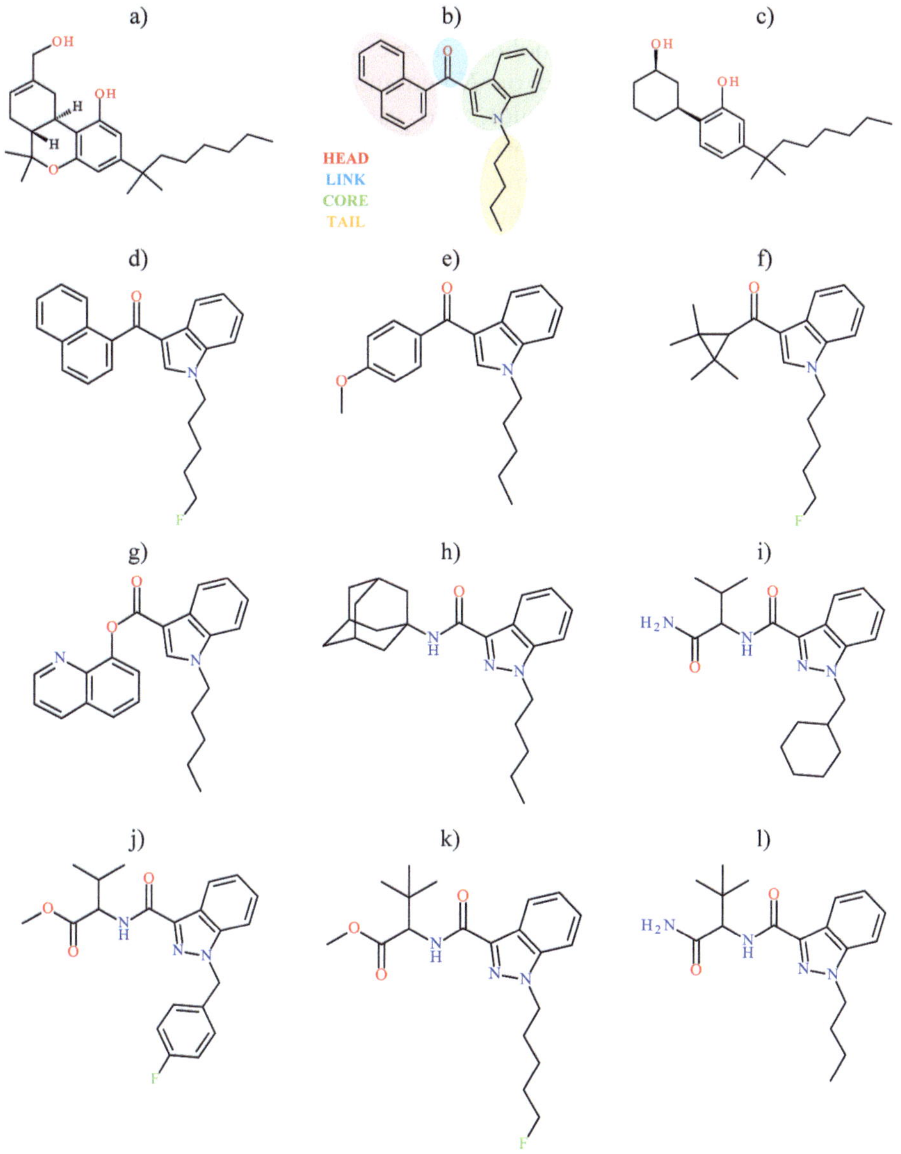

Fig. 1 Structures of synthetic cannabinoid spanning several generations: (**a**) HU-210, (**b**) JWH-018, (**c**) CP 47,497, (**d**) AM-2201, (**e**) RCS-4, (f) XLR-11, (**g**) PB-22, (**h**) AKB-48, (**i**) AB-CHMINACA, (**j**) MMB-FUBINACA, (aka FUB-AMB) (**k**) 5F-MDMB-PINACA (aka 5F-ADB), and (**l**) ADB-BINACA (aka ADB-BUTINACA)

other subclasses of NPS, synthetic cannabinoids are extensively metabolized in the human body, resulting in little to no parent drug remaining in urine (and sometimes other biological) specimens. This requires laboratories to extensively test for synthetic cannabinoid metabolites, which in turn requires research studies to confirm the major and minor metabolites produced after synthetic cannabinoid consumption. Similar to the conversion of tetrahydrocannabinol (THC) to its metabolites (e.g., THC-COOH), synthetic cannabinoids are generally converted to more polar and "acid" metabolites. These acid metabolites, commonly named butanoic acid or *N*-pentanoic acid (based on different location of the molecule), also vary widely in chemistry and behavior from the parent drug, complicating analytical processes further. The most common metabolic conversions to produce major human metabolites include ester (or amide) hydrolysis and hydroxylation followed by carboxylation. For example, JWH-018 is metabolized to produce JWH-018 *N*-pentanoic acid, and 5F-MDMB-PINACA is metabolites to produce 5F-MDMB-PINACA 3,3-dimethylbutanoic acid (*see* Fig. 2) [7, 8]. These more polar and acidic metabolites can require separate and special samples preparation procedures and analytical methods, not to mention the potential for secondary metabolites (e.g., glucuronides) to be produced.

A final phenomenon to discuss about synthetic cannabinoids in their documented instability, as this affects considerations for development of appropriate targeted analyses. Research reports show that some synthetic cannabinoids (especially those with ester head groups) can be highly unstable at common refrigerated conditions [9]. Fortunately, the breakdown process (e.g., ester hydrolysis) is similar to that of metabolism, and the characterized metabolites can be used to track or monitor instability. For example, 5F-MDMB-PINACA is highly unstable and breaks down into 5F-MDMB-PINACA 3,3-dimethylbutanoic acid when stored in blood under room temperature and refrigerated conditions. Unfortunately, most laboratories are not looking for the metabolite/breakdown product in blood, so positive cases could be missed. Ultimately, prior research work sheds light on the need for analytical methods to include or consider metabolites in more than just urine specimens.

As one can now understand, target analysis of synthetic cannabinoids in blood and urine is a complex and unique universe, varying quite extensively from other NPS subclasses. Not only does a scientist and laboratory need to consider the evolving nature of the synthetic cannabinoid subclass, but they also need to factor in aspects related to biological processes – both are tough tasks that require education, training, expertise, and constant innovative thinking.

2 Target Methods

"Target analysis" in forensic toxicology workflows can be described as three distinct points in the analytical process: (1) sample preparation and/or extraction, (2) instrumental analysis, and (3) data processing. Each can be designed to target specific drugs or drug features (also known as data/information dependent), or,

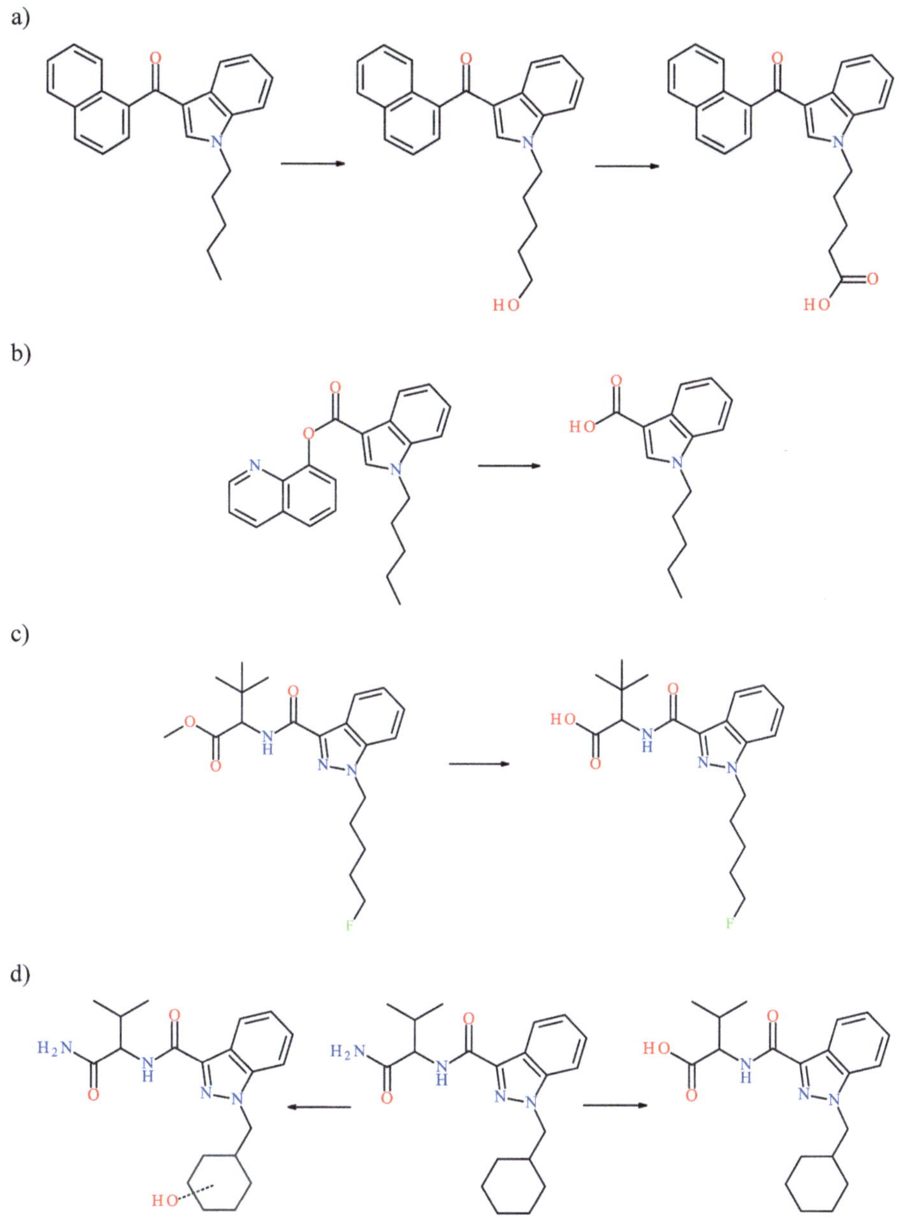

Fig. 2 Common metabolic routes for synthetic cannabinoid: (**a**) Hydroxylation followed by carboxylation of JWH-018, (**b**) ester hydrolysis of PB-22, (**c**) ester hydrolysis of 5F-MDMB-PINACA, and d) amide hydrolysis of AB-CHMINACA (right) and hydroxylation (right)

alternatively, each can be designed to encompass many drugs and drug-related molecules over a large spectrum (also known as "non-targeted" or data/information independent). Figure 3 breaks down how targeted and non-targeted aspects fall under these analytical categories. This chapter will focus specifically on targeted instrumental analysis methods but will briefly discuss the other two tangential areas.

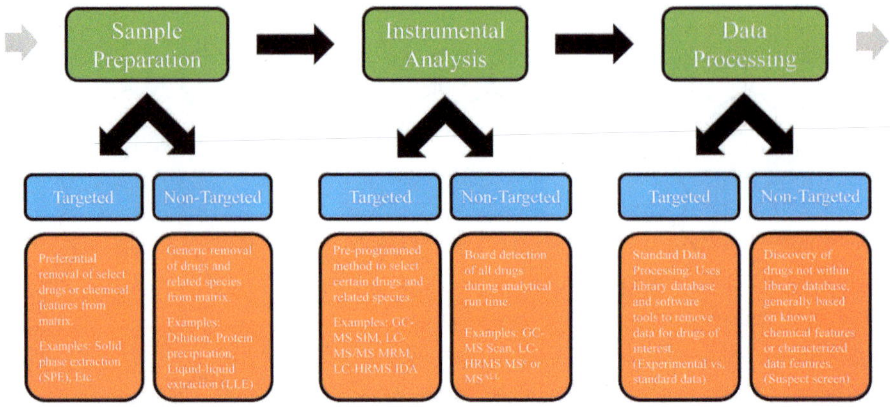

Fig. 3 Distinction between processes in the overall analytical method and descriptions of how each can encompass targeted and non-targeted aspects

In forensic toxicology, target analysis can be used for both drug screening and confirmation, although, in this modern era where high-resolution mass spectrometry (HRMS) is available, drug screening has moved more toward non-targeted approaches. Target analysis of synthetic cannabinoids is most commonly employed via liquid chromatography tandem-mass spectrometry (LC-MS/MS), as this instrument provides high levels of specificity, sensitivity, and flexibility. Less commonly now due to the ever-changing landscape of drug structure and increased potency, immunoassay-based testing and gas chromatography mass spectrometry (GC-MS) are rarely used, respectively.

2.1 Sample Preparation

Before discussing how a laboratory uses instrumental analysis to detect synthetic cannabinoids, a short examination of sample preparation techniques is warranted. There are three general manners in which synthetic cannabinoids are removed from biological material for analysis: (1) liquid-liquid extraction (LLE), (2) solid-phase extraction (SPE), and (3) protein precipitation for blood or simple dilution for urine. LLE procedures can be effective for simple yet broad removal of drugs from matrix; however, singular pH and universal solvents can lead to missed synthetic cannabinoids of interest or dirty extracts with high levels of background, respectively. LLE is commonly used for non-targeted extraction approaches, which can pair well with targeted analysis. SPE generally generates a cleaner extract; however, development of SPE procedures can take more time and be more complex than LLE. SPE can also cater to differing chemistries, if needed. Protein precipitation and dilution preparation methods are often the simplest and cheapest; however, issues can arise as they result in "dirty" samples for injection into an instrument. Issues with extraction selectivity are not a concern with these two though. Several published sample preparation and extraction methods exist for synthetic cannabinoids, some of which vary based on the generation of drugs analyzed and their chemistries.

When targeting synthetic cannabinoid parent drugs, LLE procedures generally involve the use of a buffered solution around pH 9–10 (or sometimes less) to render the synthetic cannabinoids unionized for migration to an organic solvent [10–14]. The composition and strength of the buffer can vary greatly depending on the laboratory developing the method, which is not uncommon when discussing LLE. Since synthetic cannabinoids are chemically nonpolar, the organic solvent will consist of a singular or mixture of highly nonpolar solvents (e.g., hexane) with more middle of the road polarities (e.g., ethyl acetate, MTBE). When targeting synthetic cannabinoid metabolites, the same LLE procedures cannot be used since the metabolites are generally more acidic and more polar. The pH adjustment of the samples is acidic, unionizing the metabolites for migration to the organic layer. But similarly, the organic layer is generally a mixture of highly nonpolar and slightly polar solvents.

Due to their targeted nature, there are many SPE procedures that are used to extract synthetic cannabinoids from matrices, all of which vary greatly based on the intended utility [10, 15–22]. Reverse-phase SPE and ion exchange SPE are common, as well as mixed mode cartridges which employ combinations of the two.

2.2 Quantitation

As with quantitation of any drug or NPS, the potency of the drug often dictates the necessary dynamic range for assessment of drug concentration. This is the case for synthetic cannabinoids as well, although recent generations have higher potencies (compared to older generations) resulting in lower quantitative values in human biological samples. Quantitative analysis for synthetic cannabinoids in blood and urine generally ranges from 0.1 to 10 ng/mL, 0.5 to 50 ng/mL, or 1 to 100 ng/mL; however, this is highly dependent on laboratory procedures and policies, as well as scientist experience and input.

A common approach to quantitation is to use 0.5 mL of biological samples and spike calibrators between 0.5 and 50 ng/mL and controls accordingly. Table 1 shows an example preparation for quantitative analysis of synthetic cannabinoids.

2.3 LC-MS/MS (LC-QQQ-MS)

As previously mentioned, LC-MS/MS is the most commonly used analytical instrument for the detection of synthetic cannabinoids in blood and urine. Multiple reaction monitoring (MRM) is most commonly used to accurately detect the transition of precursor ion to product ions, similar to the way this instrument is used for drug testing in other areas of forensic toxicology. This technique is broken into its subparts for discussion: liquid chromatography (separation), tandem quadrupole mass spectrometry (detection), and data processing.

Table 1
Preparation of calibrators and controls for quantitative assessment of synthetic cannabinoids in 0.5 mL blood

Sample ID	Working calibrator/ control (ng/μL)	Spike volume (μL)	Final concentration (ng/mL)
Standard 1	1	25	50
Standard 2	1	10	20
Standard 3	1	5	10
Standard 4	0.1	25	5
Standard 5	0.1	10	2
Standard 6	0.1	5	1
Standard 7	0.01	25	0.5
Control low	0.01	75	1.5
Control mid	1	4	8
Control high	1	20	40

2.3.1 Liquid Chromatography

The nonpolar nature of synthetic cannabinoid parent drugs can require specialized chromatographic separation methods. Typically, reverse-phase chromatography is used in forensic toxicology testing, which includes the use of polar mobile phase and nonpolar stationary phase. Since synthetic cannabinoid parent drugs are often highly nonpolar in chemical behavior, this means special attention must be placed on the development of methods to remove (elute) these nonpolar analytes from the nonpolar stationary phase and in a reasonable amount of time. Since THC is also highly nonpolar (and more nonpolar than most synthetic cannabinoids), this drug can be used for comparison when assessing some of the most extreme chromatographic behavior (i.e., the latest elution time). Elution of synthetic cannabinoids in reverse-phase chromatographic systems can require high percent organic mobile phase composition, use of organic mobile phases that are less polar, and/or increased hold time at high organic composition. This process is "trial and error" during the method development stage. Remember, other NPS subclasses behave differently in these chemical environments so what is good for synthetic cannabinoids may not be good for all NPS and vice versa.

As with all aspects of synthetic cannabinoid testing, chromatographic separation is not so simple. Synthetic cannabinoid metabolites can be vastly different in their chemical behavior from the parent drug, largely due to changes in functional group configuration and therefore overall molecule polarity. These metabolites, containing hydroxy and carboxy groups, are attracted more to the polar mobile phase than the parent drug, leading to shorter time spent interacting with the nonpolar stationary phase and ultimately a shorter elution time. If a laboratory is interested in detecting both parent drug and metabolites in a single chromatographic method, this means attention must be placed on both the late eluting nonpolar parent drugs and the early to mid-eluting more polar metabolites. This can be challenging depending on the chromatography system used and the overall setup (i.e., column, mobile phase, run time).

Tables 2 and 3 show two chromatographic separations which were successfully developed and employed for the detection of

Table 2
Two liquid chromatography gradients developed by Krotulski et al. for the detection of synthetic cannabinoids

Time (min) – Method 1	Time (min) – Method 2	%A	%B
0	0	95	5
1	0.5	95	5
10	4	5	95
13	6	5	95
13.5	6.1	95	5
15.5	7	95	5

Mobile phase A is ammonium formate (A, 10mM, pH 3 with formic acid), and mobile phase B is 0.1% formic acid in methanol/acetonitrile (B, 50:50 v:v)

Table 3
A liquid chromatography gradient developed by Rigdon et al. for the detection of synthetic cannabinoids

Time (min)	%A	%B
0	55	45
2	55	45
6	15	85
6.1	5	95
7	15	85
7.1	55	45
8.5	55	45

Mobile phase A is 0.05% acetic acid in water, and mobile phase B is 0.05% acetic acid in acetonitrile

synthetic cannabinoids. Table 2 details a more standard broad-based screening approach developed by Krotulski et al. [10, 23], beginning at high percent aqueous and transitioning over time to high percent organic. The analytical column used was a Phenomenex® Kinetex C18 analytical column (50 mm × 3.0 mm, 2.6 μm). The total run time was 15.5 min. Table 2 details a more specialized chromatographic separation method for JWH-018 and its metabolites, developed by Ridgon et al. [24]. Contrarily to the first method described, this method employs a mobile phase gradient with more emphasis on the organic phase. It should be noted that while this method may work for the chromatographic separation of other synthetic cannabinoids, it may not be ideal or appropriate for all.

2.3.2 Tandem Quadrupole Mass Spectrometry

Target analysis for synthetic cannabinoids requires development of target analytical methods that are designed to detect certain masses associated with the drugs of interest. This first required the acquisition of this information for later verification. Drug standards for synthetic cannabinoids are infused – a process by which the drug is directly injected – into the mass spectrometer to evaluate their behavior. First, the precursor ion (or monoisotopic mass calculated based on a drug's molecular formula) is discovered via a simple MS1 scan in the first quadrupole (Q1). Second, this precursor ion is placed through a series of tests to determine optimal conditions (e.g., voltages, gas flow) for its detection. Third, the precursor ion is fragmented (Q2), and the resulting product ions are discovered through MS2 scan in Q3. During this process, additional optimal conditions can be tested. Finally, all of this information is compiled together for experimentation as a developed analytical method.

The weakest point in the core synthetic cannabinoid structure is generally the linked group – an important factor that drives the manner in which these drugs fragment. The mass spectrum for most synthetic cannabinoids consists of a large base peak and handful of smaller ions (*see* Fig. 4 for APP-BINACA). The base peak correlated with this fragmentation produces an ion which retains the oxygen atom of the linked group (to form an acylium ion), the core, and the tail. Another common fragment ion results from the removal of the tail from this ion and can be spotted as a 144 or 145 ion, for indole and indazole cores, respectively. This information is important to understand as the MRM transitions selected for target analysis generally arise from these molecular cleavages.

Since synthetic cannabinoids can be found at low concentrations (<5 ng/mL), some additional mass spectrometer optimization may be required. However, there are many different instrument vendors with different hardware, software, and names for certain gas flows and voltages that makes it difficult to explain within the short focus of this chapter. In addition, rapid

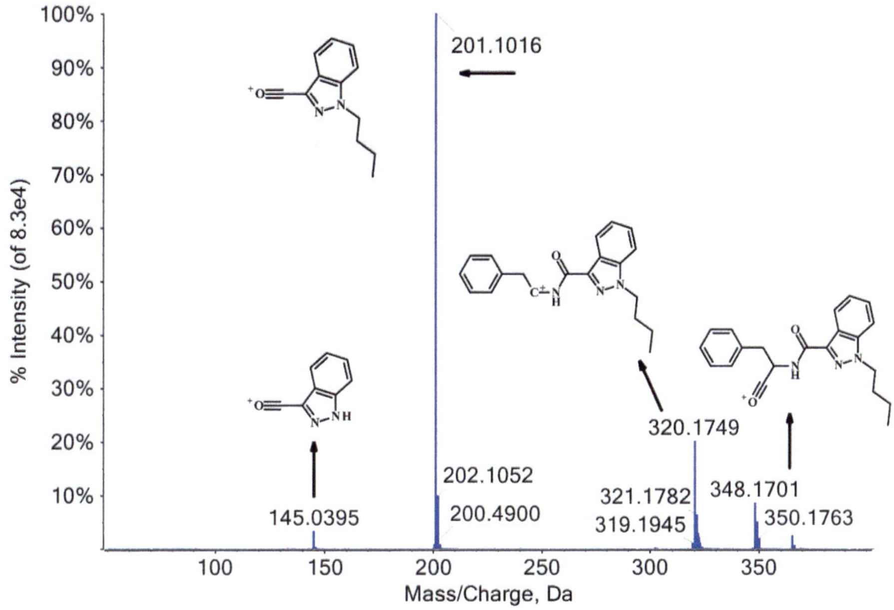

Fig. 4 Fragmentation pattern of APP-BINACA showing the major fragment ions produced

advancements in technologies and added hardware would make this discussion quickly obsolete. It is always recommended that users of analytical instrumentation become familiar with all of these aspects prior to method development.

Table 4 provides a list of previously commonly encountered synthetic cannabinoids and their mass spectrometer set points. This includes name, precursor ion, and product ions. Target analysis generally includes at least two transitions from precursor to product ion.

2.3.3 Data Analysis

Data analysis for target testing of synthetic cannabinoids is generally quite simple and follows that of typically target methods in forensic toxicology. Chromatographic behavior is evaluated based on retention time (RT) and peak shape. There are no set guidelines for these criteria, but generally the RT from an authentic sample should be very close to the RT of the calibrators or controls and peak shape should be Gaussian [25]. Deviations in RT for any chromatographic system will be dependent on the technology used (i.e., UPLC vs HPLC) and the overall run time. Currently in modern forensic toxicology with the introduction of UPLC, RT should not deviate more than 0.05–0.1 min across a run or batch. However, as methods increase in run time (>15 min), laboratories may see more variability.

An additional chromatographic feature that should be monitored is ion ratio – the ratio of peak area for precursor-product ion transition #1 to #2. Inherently, this value will be dependent on the

Table 4
Ions that could be selected for MRM transitions of commonly encountered synthetic cannabinoids

Name	Precursorion (Q1)	Production (Q2)
JWH-018	342.2	155.0 214.1
AM-2201	360.2	155.0 232.1
XLR-11	330.2	125.1 232.1
PB-22	359.2	214.1 144.0
AKB-48	366.3	135.1 107.1
AB-CHMINACA	357.2	241.1 145.0
MMB-FUBINACA	384.2	253.1 109.0
5F-MDMB-PINACA	378.2	233.1 145.0
ADB-BINACA	331.2	201.1 145.0

intensity of the product ions chosen and mass spectrometer set-points. Most guidance would suggest that ion ratio be between 20% and 30% of target (i.e., calibrators, controls). Ion ratio remains an important piece of data to access with using triple quadrupole systems as this value can help distinguish the drug of interest for interfering analytes.

With LC-MS/MS target analysis, no mass spectral data itself is generally evaluated, although the addition of MS2 scans (via quadrupole or ion trap) may be of interest. In those cases, library comparisons can include their own criteria and provide higher levels of certainty for positive identification.

2.4 GC-MS

With the advancements of modern forensic toxicology, GC-MS analysis is more rarely used, especially for the detection of NPS and synthetic cannabinoids. GC-MS generally does not provide the needed analytical sensitivity for the detection of low concentration synthetic cannabinoid parent compounds, and the nature of the instrument setup usually requires use of derivatization for synthetic cannabinoid metabolites with polar functional groups containing active hydrogens.

Table 5
GC-MS method for the generic detection of synthetic cannabinoids (and other NPS) as described by Krotulski et al.

GC-MS method parameters	Setpoints
Instrument	Agilent5975 Series GC/MSD System
Column	Zebron™ Inferno™ ZB-35H (15 m × 250 μm × 0.25 μm)
Carrier gas	Helium (flow: 1 mL/min)
Temperatures	Injection port: 265 °C Transfer line: 300 °C MS source: 230 °C MS quad: 150 °C
Oven program	60 °C for 0.5 min, 35 °C/min to 340 °C for 6.5 min
Injection parameters	Injection type: splitless Injection volume: 1 μL
MS parameters	Mass scan range: 40–550 m/z Threshold: 250

Gas chromatography is generally implemented using an inert carrier gas (e.g., helium) and nonpolar stationary phase (e.g., hydrocarbon-bonded silica). We now know that synthetic cannabinoids are nonpolar in chemical behavior, so this chromatographic process results in more interaction with the stationary phase and therefore longer retention times. An example gas chromatographic method is shown in Table 5 which was used for the detection of 4F-MDMB-BINACA and other synthetic cannabinoids [26]. Like all analytical systems, there are many approaches to chromatographically separating substances of interest, and these methods will be highly dependent on instrument, need, and laboratory preferences.

Single quadrupole mass spectrometer (as with GC-MS) target analysis of synthetic cannabinoids can be developed using either scan or selected ion monitoring (SIM) operations – whereby all masses are acquired over the mass range or only select ions, respectively. Scan acquisition allows for library searching while SIM data is evaluated by assessment of ion ratios. For synthetic cannabinoids, the same fragment ions seen with LC-MS/MS [ESI+] above are often seen with GC-MS [EI] as well (which is sometimes not the case with other drugs). Those similar ions are likely useful for setting up a SIM method but should be evaluated more appropriately in the context of the method and its intended use. After acquisition, data is processed in a similar manner to above but within instrument or software different perspectives.

2.5 Immunoassay

Immunoassay is rarely used for the target analysis of synthetic cannabinoids due to low utility and the absence of cross reactivity with traditional kits [27]. The use of immunoassay as a tool for NPS detections is contingent on this idea of cross reactivity – or getting a positive result for a drug that the assay was not designed for. With early generations (e.g., JWH-018), immunoassays were developed to assist with target screening [28, 29], whether in clinical or forensic settings, but the development of these assays quickly fell out of cycle and favorability due to development time being longer than life cycle of the drug on the market or at least too long in time to capture rising cases and peak positivity. Immunoassays function on the development of antibodies that react to the drug as their antigen. The structural diversity among the synthetic cannabinoids class affects how generated antibodies can react to new structural changes. This is not a new phenomenon for NPS, but rather it is a phenomenon that is useful to some other subclasses. For example, new synthetic opioids do not cross react with traditional opiate (morphine) kits, while some new benzodiazepines show good cross reactivity with traditional benzodiazepine (temazepam) kits. For these reasons, target analysis of synthetic cannabinoids is dominated by the use of mass spectrometry.

2.6 LC-HRMS (Targeted Testing)

The focus of this chapter is on target analysis of synthetic cannabinoids, and screening methods are described elsewhere (Chap. 11). However, it is important to understand that the use of high-resolution mass spectrometry (HRMS) is often via target analysis, whether during their instrumental acquisition or data processing. Common data-dependent acquisition (DDA) (also known as information-dependent acquisition [IDA]) modes are designed as targeted analyses based on certain criteria. These criteria can include m/z, retention time, and/or intensity. In short, the instrument will only acquire certain elements of data (i.e., MS/MS spectra) for analytes that meet the criteria during the run time. Krotulski et al. described this information in more detail, specifically comparing targeted and non-targeted acquisition modes [23].

An additional level of complexity is how data is processed for HRMS analyses. Even if the data is acquired in a non-targeted fashion, data processing of HRMS datafiles is almost exclusively accomplished via target analysis (some exceptions certainly apply). Laboratories process data in a targeted fashion due to the overwhelming amount of data collected during an HRMS analysis. If you think back to LC-MS/MS, the instrument is only acquiring data for what you tell it to. With HRMS, the instrument may be acquiring thousands to millions of data points, so targeted data process is often the only reasonable means to interpreting these complex massive datafiles. Most commonly, target data processing for synthetic cannabinoids is accomplished via tolerance around set criteria. These criteria include mass (from formula), retention time,

Table 6
Standard criteria for processing and evaluation of HRMS data

Criteria	Pass	Additional review	Fail
Mass error (ppm)	<5	<10	>10
Retention time error (min)	<0.25	<0.35	>0.35
Isotope ratio (% difference)	<30	<100	>100
Library score	>70	>50	<50
Signal-to-noise ratio	>10	N/A	<10
Peak intensity (counts)	>800	N/A	<800

Table 7
Information for commonly encountered synthetic cannabinoids that can be used for HRMS target analysis [10]

Name	Formula	Exact mass (Da)	Retention time (min)
JWH-018	$C_{24}H_{23}NO$	342.1852	10.39
AM-2201	$C_{24}H_{22}FNO$	360.1758	9.66
XLR-11	$C_{21}H_{28}FNO$	330.2228	10.06
PB-22	$C_{23}H_{22}N_2O_2$	359.1754	10.12
AKB-48	$C_{23}H_{31}N_3O$	366.2540	11.33
AB-CHMINACA	$C_{20}H_{28}N_4O_2$	357.2285	9.39
MMB-FUBINACA	$C_{21}H_{22}FN_3O_3$	384.1718	9.42
5F-MDMB-PINACA	$C_{20}H_{28}FN_3O_3$	378.2188	9.62
ADB-BINACA	$C_{18}H_{26}N_4O_2$	331.2129	8.94

isotope distribution, and/or library score. Krotulski et al. described these criteria in more detail elsewhere [10, 23]. Criteria used for targeted data processing are shown in Table 6.

As mentioned, using HRMS, synthetic cannabinoids are often targeted via their mass, which is generated based on the drug's chemical formula. The chemical formula of a drug can be converted to what's called its "exact mass" – a mass with four (or more) decimals places that accounts for mass defect and ultimately becomes a more unique identifier. Afterall, this premise is what affords the utility of HRMS in general. Table 7 shows a list of commonly encountered synthetic cannabinoids to include their formula and accurate mass.

3 Considerations

Target analysis of synthetic cannabinoids remains one of the most challenging aspects of NPS detection. Scientists and laboratories must remain aware of the challenges associated with this type of testing and be prepared to develop manners to remedy them. When developing a method for synthetic cannabinoids, one must first determine their scope of analysis, i.e., what synthetic cannabinoids are you testing for. This includes the distinction between parent drugs and metabolites; however, sometimes that question comes following the determination of matrix type. When developing any toxicology method, one must determine what matrix will be assessed. Generally, the scientist will select their primary matrix of interest (e.g., blood) and then determine if supplemental or secondary matrices need to be added (e.g., urine, tissue). As one can now understand from this chapter, here is where the decision of parent drugs and/or/vs. metabolites may come into play (do not forget the assessment of stability). Once a scope of analysis is set and the matrix is selected, the next step is to determine the analytical platform. As mentioned, this is most commonly LC-MS/MS (aka LC-QQQ-MS). An additional step to consider is expected concentration of the synthetic cannabinoids (and their metabolites) in question, although information may or may not be available depending on the novelty of the drug itself.

An additional consideration that should be explored is the need for future updating to the method developed. We know that specific NPS are not here to stay, generally coming and going within the span of only a year or so. As laboratories develop their methods, one should consider whether or not the method can or will need to be updated and/or expanded in the future. Generally, the process of expanding an LC-MS/MS method for a new analyte of interest is not too complicated. Once mass spectrometer setpoints are discovered via drug infusion, a series of simple experiments can be completed to assess the fitness of the drug for the already existing method (examples for experiments might include assessments of recovery, matrix effects, interferences, etc.). It is hard to predict how often a target method should be updated for synthetic cannabinoids (it is drug, scheduling, and region dependent); however, it is suggested that laboratories evaluate the drug market and consider an update at least every 3 months. Otherwise, laboratories are increasingly at risk of missing synthetic cannabinoids not within their target scope of analysis but present in the toxicology casework.

Implementation of a target method will generally take between 1 month and 6 months (or more), often dependent on laboratory policies. With the literature base growing for synthetic cannabinoids, laboratories can rely on this information for the purpose of expediting method development. Most new and emerging synthetic cannabinoids are detectable via previously developed extraction approaches, given that their chemistry is not too dissimilar. Previously developed chromatographic separation can also serve useful, although the question always comes down to positional isomers and drugs that share the same, similar, or close masses. This roadmap can be paired alongside quick in-laboratory experiments to get the method up and running. Method development generally takes a few days to a few weeks to complete. Method validation generally takes at least 1 week to complete, if not more. The final aspect is processing all of the data and assessing the performance of the method before implementation with authentic samples, which can take days to weeks to months depending on many factors.

Target analysis of synthetic cannabinoids may seem like a routine approach to testing; however, innovative thinking can allow laboratories to get ahead of ever-changing drug trends. Although target testing does not allow the analyst to see what they are not looking for, inclusion of "suspect" masses or transitions can help determine what other synthetic cannabinoids could be out there. This could help decide where the laboratory may want to spend more time with future developments or refinements. For example, if a laboratory knows a methyl ester synthetic cannabinoid exists, they might want to add the same mass spectrometer setpoints for the ethyl ester variant by simply changing the precursor ion mass by adding 14 amu. If a peak appears within this new mass channel during testing, the laboratory may want to investigate the emergence of a new analogue in that specific sample.

For many laboratories, target analysis of synthetic cannabinoids (whether by LC-MS/MS, HRMS, or even GC-MS) is a vital tool for answering important toxicology-based questions concerning drug use and effects. The testing is often more complex and less routine than other NPS classes, and special considerations are generally encouraged. No one knows what is next on the horizon for synthetic cannabinoids, and, because of this, laboratories must remain vigilant and aware of recreational drug markets and their emerging trends. NPS Discovery [www.npsdiscovery.org], a program at the Center for Forensic Science Research and Education (CFSRE), serves as a international open-access drug early warning system and includes online databases with analytical information about emerging NPS. This program allows scientist to remain at the forefront of both target and non-target analysis. A target analysis for synthetic cannabinoids is often ineffective and outdated within a year or two, and methods developed more than 5 years ago might be forgotten and certainly inadequate. However, the science that is published remains a key blueprint for potential future assessment of synthetic cannabinoids. In summary, if you think your target analysis for synthetic cannabinoids is good enough, it is likely not good at all – keep thinking about ways in which you can better the assay via expansion or refinement.

References

1. Mason M (2013) Lethal highs: how CBP's laboratories are protecting America from designer drugs. Frontline Mag 6:11–16
2. Seely KA, Lapoint J, Moran JH, Fattore L (2012) Spice drugs are more than harmless herbal blends: a review of the pharmacology and toxicology of synthetic cannabinoids. Prog Neuro-Psychopharmacol Biol Psychiatry 39:234–243. https://doi.org/10.1016/j.pnpbp.2012.04.017
3. EMCDDA home page | www.emcdda.europa.eu. http://www.emcdda.europa.eu/. Accessed 30 June 2017
4. National Forensic Laboratory Information System (NFLIS). https://www.deadiversion.usdoj.gov/nflis/
5. NPS Discovery (2019). Cent Forensic Sci Res Educ. https://www.npsdiscovery.org/. Accessed 11 Jan 2019
6. Abbate V, Schwenk M, Presley BC, Uchiyama N (2018) The ongoing challenge of novel psychoactive drugs of abuse. Part I. Synthetic cannabinoids (IUPAC Technical Report). Pure Appl Chem 90:1255–1282. https://doi.org/10.1515/pac-2017-0605
7. Rajasekaran M, Brents LK, Franks LN et al (2013) Human metabolites of synthetic cannabinoids JWH-018 and JWH-073 bind with high affinity and act as potent agonists at cannabinoid type-2 receptors. Toxicol Appl Pharmacol 269:100–108. https://doi.org/10.1016/j.taap.2013.03.012
8. Yeter O, Ozturk YE (2019) Metabolic profiling of synthetic cannabinoid 5F-ADB by human liver microsome incubations and urine samples using high-resolution mass spectrometry. Drug Test Anal. https://doi.org/10.1002/dta.2566
9. Krotulski AJ, Bishop-Freeman SC, Mohr ALA, Logan BK Evaluation of synthetic cannabinoid metabolites in human blood in the absence of parent compounds – a stability assessment. J Anal Toxicol. https://doi.org/10.1093/jat/bkaa054
10. Krotulski AJ, Mohr ALA, Logan BK (2020) Emerging synthetic cannabinoids: development and validation of a novel liquid chromatography quadrupole time-of-flight mass spectrometry assay for real-time detection. J Anal Toxicol 44:207–217. https://doi.org/10.1093/jat/bkz084
11. Knittel JL, Holler JM, Chmiel JD et al (2016) Analysis of parent synthetic cannabinoids in blood and urinary metabolites by liquid chromatography tandem mass spectrometry. J Anal Toxicol 40:173–186. https://doi.org/10.1093/jat/bkv137
12. Mogler L, Franz F, Rentsch D et al (2018) Detection of the recently emerged synthetic cannabinoid 5F-MDMB-PICA in 'legal high' products and human urine samples: detection of 5F-MDMB-PICA in herbal blends and human urine. Drug Test Anal 10:196–205. https://doi.org/10.1002/dta.2201
13. Giorgetti A, Mogler L, Halter S et al (2020) Four cases of death involving the novel synthetic cannabinoid 5F-Cumyl-PEGACLONE. Forensic Toxicol 38:314–326. https://doi.org/10.1007/s11419-019-00514-w
14. Tynon M, Homan J, Kacinko S et al (2017) Rapid and sensitive screening and confirmation of thirty-four aminocarbonyl/carboxamide (NACA) and arylindole synthetic cannabinoid drugs in human whole blood. Drug Test Anal 9:924–934. https://doi.org/10.1002/dta.2096
15. Gaunitz F, Kieliba T, Thevis M, Mercer-Chalmers-Bender K (2020) Solid-phase extraction–liquid chromatography–tandem mass spectrometry method for the qualitative analysis of 61 synthetic cannabinoid metabolites in urine. Drug Test Anal 12:27–40. https://doi.org/10.1002/dta.2680
16. Chimalakonda KC, Moran CL, Kennedy PD et al (2011) Solid-phase extraction and quantitative measurement of omega and omega-1 metabolites of JWH-018 and JWH-073 in human urine. Anal Chem 83:6381–6388. https://doi.org/10.1021/ac201377m
17. Kleis J-N, Hess C, Germerott T, Roehrich J Sensitive screening of synthetic cannabinoids using liquid chromatography quadrupole time-of-flight mass spectrometry after solid phase extraction. Drug Test Anal. https://doi.org/10.1002/dta.3052
18. Sobolevsky T, Prasolov I, Rodchenkov G (2012) Detection of urinary metabolites of AM-2201 and UR-144, two novel synthetic cannabinoids. Drug Test Anal 4:745–753. https://doi.org/10.1002/dta.1418
19. Sundström M, Pelander A, Angerer V et al (2013) A high-sensitivity ultra-high performance liquid chromatography/high-resolution time-of-flight mass spectrometry (UHPLC-HR-TOFMS) method for screening synthetic cannabinoids and other drugs of abuse in urine. Anal Bioanal Chem 405:8463–8474. https://doi.org/10.1007/s00216-013-7272-8

20. Hasegawa K, Wurita A, Minakata K et al (2014) Identification and quantitation of 5-fluoro-ADB, one of the most dangerous synthetic cannabinoids, in the stomach contents and solid tissues of a human cadaver and in some herbal products. Forensic Toxicol 33:112–121. https://doi.org/10.1007/s11419-014-0259-0
21. Castaneto MS, Wohlfarth A, Desrosiers NA et al (2015) Synthetic cannabinoids pharmacokinetics and detection methods in biological matrices. Drug Metab Rev 47:124–174. https://doi.org/10.3109/03602532.2015.1029635
22. Kleis J, Germerott T, Halter S et al (2020) The synthetic cannabinoid 5F-MDMB-PICA: a case series. Forensic Sci Int 314:110410. https://doi.org/10.1016/j.forsciint.2020.110410
23. Krotulski AJ, Varnum SJ, Logan BK (2020) Sample mining and data mining: combined real-time and retrospective approaches for the identification of emerging novel psychoactive substances. J Forensic Sci 65:550–562. https://doi.org/10.1111/1556-4029.14184
24. Rigdon A. LC/MS/MS analysis of metabolites of synthetic cannabinoids JWH-018 and JWH-073 in Urine 1:9
25. European Commission (2002) 2002/657/EC: Commission Decision of 12 August 2002 implementing Council Directive 96/23/EC concerning the performance of analytical methods and the interpretation of results. Off J Eur Commun
26. Krotulski AJ, Mohr ALA, Kacinko SL et al (2019) 4F-MDMB-BINACA: a new synthetic cannabinoid widely implicated in forensic casework. J Forensic Sci. https://doi.org/10.1111/1556-4029.14101
27. Kronstrand R, Brinkhagen L, Birath-Karlsson C et al (2014) LC-QTOF-MS as a superior strategy to immunoassay for the comprehensive analysis of synthetic cannabinoids in urine. Anal Bioanal Chem 406:3599–3609. https://doi.org/10.1007/s00216-013-7574-x
28. Arntson A, Ofsa B, Lancaster D et al (2013) Validation of a novel immunoassay for the detection of synthetic cannabinoids and metabolites in urine specimens. J Anal Toxicol. https://doi.org/10.1093/jat/bkt024
29. Rodrigues WC, Catbagan P, Rana S et al (2013) Detection of synthetic cannabinoids in oral fluid using ELISA and LC-MS-MS. J Anal Toxicol 37:526–533. https://doi.org/10.1093/jat/bkt067

Chapter 2

Target Analysis of Synthetic Cathinones in Blood and Urine

Elena Lendoiro and Ana de-Castro-Ríos

Abstract

Synthetic cathinones are new psychoactive substances which act as central nervous system stimulants with mixed sympathomimetic activity. In the last decades, they entered in the drug market as substitutes of traditional drugs and made necessary the development of analytical methods for their determination in different biological samples. Blood and urine are the two main biological matrices in clinical and forensic toxicology for testing and evaluation of drug consumption. So the most common analytical approaches for target analysis of synthetic cathinones in blood and urine are described, including cleanup methodologies for purification and concentration of the analytes (protein precipitation, liquid-liquid extraction, and solid-phase extraction) and instrumental analysis for their identification and quantification (LC-MS/MS and GC-MS). Moreover, analytical challenges such as identification of enantiomers and stability issues during transport, storage, handling, and analytical conditions are discussed. Finally, an overview of blood concentrations found in real clinical and forensic cases is included.

Key words Synthetic cathinones, Blood, Urine, LC-MS/MS, GC-MS, Liquid-liquid extraction, Solid-phase extraction

1 Introduction

Synthetic cathinones are the second largest group of new psychoactive substances (NPS) monitored by the European Monitoring Centre for Drugs and Drug Addition (EMCDDA) and by the United Nations Office on Drugs and Crime (UNODC) [1, 2]. Synthetic cathinones are sympathomimetic drugs chemically related to cathinone, a natural alkaloid presents in the leaves of the plant *Catha edulis* (Khat). Khat grows in the region of northeast Africa and the Arabian Peninsula [3], where the leaves are traditionally chewed for their euphoric and stimulant effects [4]. Cathinone structure, like amphetamine, consists of a linear chain of three carbons with a phenyl group attached to the first carbon and an amino group attached to the second, but in cathinone an additional carbonyl group at the β carbon is present (*see* Fig. 1). This modification increases the polarity of the molecule and reduces its ability

Marta Concheiro and Karl B. Scheidweiler (eds.), *Methods for Novel Psychoactive Substance Analysis*,
Methods in Pharmacology and Toxicology, https://doi.org/10.1007/978-1-0716-2605-4_2,

Fig. 1 Chemical structures of amphetamine and cathinone

to cross the blood-brain barrier and, therefore, its potency [5, 6]. Cathinone acts as a central nervous system stimulant with mixed sympathomimetic activity [5].

Multiple cathinone derivatives can be synthesized by adding different substituents to the core skeleton of the cathinone molecule, like to the α-carbon, to the N-terminus and/or to the aromatic ring [7, 8] (*see* Fig. 2). This facilitates the entering of new synthetic cathinones into the drug market, and currently 138 synthetic cathinones are monitored by the EMCDDA [1]. Some of these derivatives had not been recently synthetized (e.g., mephedrone was developed in 1929, MDPV in 1969, and α-pyrrolidinohexiophenone (α-PHP) and α-pyrrolidinopentiophenone (α-PVP) in the 1960s [5, 9]), but they were recently introduced in the recreational drug market [1] as substitutes of traditional drugs with similar structures and effects like cocaine and amphetamines [10]. Some synthetic cathinones have been considered as therapeutic drug candidates for the treatment of different diseases, but nowadays, only bupropion is authorized for medical use (antidepressant and smoking cessation aid) in the USA and Europe [11, 12].

Synthetic cathinones consumption produces stimulating, euphoric, and empathogenic effects but also other serious side effects, such as mydriasis, tachycardia, hypertension, hyperthermia, seizures, rhabdomyolysis, renal and hepatic failure, and even death [5, 13]. Moreover, psychological side effects like agitation, hallucinations, erratic and delusional behavior, and psychosis have also been described [14]. These stimulant effects are associated with an inhibition of the monoamine transporters and an increase of the monoamine neurotransmitters (dopamine, serotonin, and norepinephrine) in the synaptic gap [5, 15]. Specifically, synthetic cathinones can be classified as cocaine-MDMA mixed-type cathinones (which inhibit the uptake of dopamine and norepinephrine in a nonselective manner), methamphetamine-like-type cathinones (which inhibit catecholamine uptake and release dopamine), and pyrovalerone-type cathinones (which block the selective uptake of catecholamines but do not release the substrate) [15].

The administration of these compounds is mainly by snorting, injection, smoking, or oral consumption, and the effects are observed from 30–45 min to 1–3 h after intake [16, 17]. However, some side effects can last for several hours and until days [16]. The unintentionally consumption of these substances has been

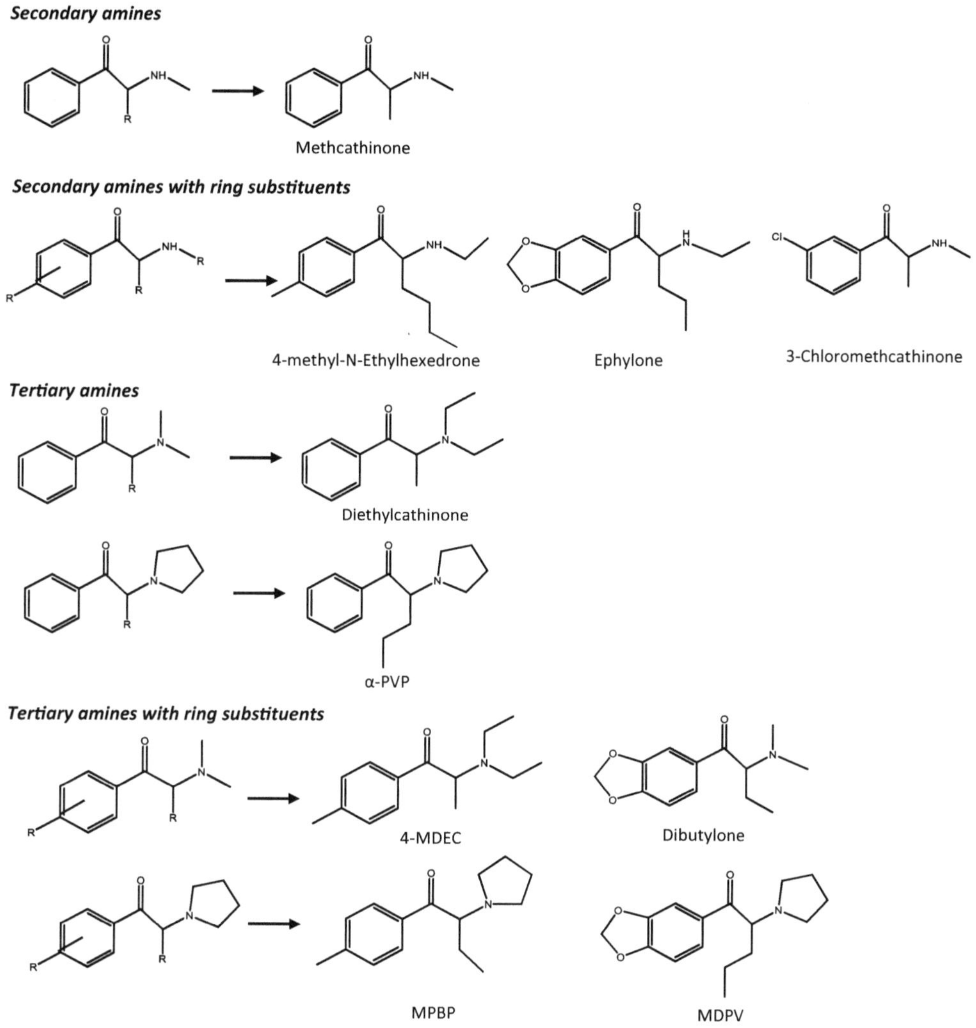

Fig. 2 Chemical structures of synthetic cathinone derivatives, including the most common substituents and examples of each group

described because they are frequently present as adulterants of other illegal drugs of abuse.

The detection and quantification of NPS, and specifically of synthetic cathinones, present several challenges from the analytical point of view, such as the high number of compounds that have to be monitored, the constant introduction of new compounds in the drug market, the scarce analytical information about them, the lack of commercial reference standards and deuterated analogs, the high number of structurally related isomers and analogs, and the absence of information about their metabolic profile, among others. This chapter summarizes the main analytical approaches for the determination of synthetic cathinones in blood and urine, the two main biological matrices in clinical and forensic toxicology for testing and evaluation of drug consumption.

2 Analytical Techniques

The first published analytical methods for the determination of synthetic cathinones in blood and/or urine were developed for one compound [4, 18] or a limited number of compounds [9, 16, 19–24], mainly for the analysis of samples from a specific forensic case. Although the development of methodologies for the analysis of a limited number of compounds and its application in authentic cases is still useful [14, 15, 23, 25–34], there are also some published methods for the simultaneous determination of a wide number of compounds [3, 6, 7, 17, 35–46]. Due to the rapid evolution of the NPS market and the polydrug profile of many users, multianalyte methods are more useful in toxicology laboratories.

Even though the quantification of drug concentrations in urine has a limited interpretative value compared to blood [10], several quantitative methods in urine have been published [11, 17, 19, 20, 22, 25–27, 33–35, 43, 44, 47–49]. It is known that synthetic cathinones are metabolized by phase I transformations such as N-dealkylation, reduction, hydroxylation, oxidation, and demethylenation [35], but many of these metabolites have not yet been characterized or commercial standards are not available. For this reason, most analytical methods in urine include only the detection of the parent drug [4, 16, 20, 22, 25–27, 33–35]. This strategy has proved to be useful for routine analysis because these compounds are generally consumed in high doses and the amount of parent drug present in urine samples is enough for their detection [35, 43].

For blood analysis, serum or plasma is commonly used in clinic cases from living individuals. The concentrations in serum and plasma appear to be similar although some differences could take place for the use of different additives in plasma/blood collection tubes that make a difference in their stability [13]. However, in postmortem cases, whole blood is the matrix of choice, and variations in drug concentrations depending on the site of collection can be detected due to different factors such as the condition of the body, the degree of trauma, the phenomenon of postmortem redistribution, the postmortem interval, the ambient temperature, and the degree of decomposition of the body [50].

The analytical procedure for the target analysis of synthetic cathinones in blood and/or urine is performed in two steps. First, a cleanup step for purification by removal of interferences that can affect the detection of the analytes or the performance of the instrumentation, for concentration of the analytes from the biological matrix and for solvent switching into an analytically more compatible solvent. Second, the instrumental analysis for the confirmation and quantification of the analytes. The quality of the results depends on the correct performance of these steps.

2.1 Cleanup

Several sample preparation techniques have been described for the target analysis of synthetic cathinones in blood and/or urine, including protein precipitation [21, 29, 33, 39, 40], liquid-liquid extraction [9, 10, 14, 16, 17, 19, 25, 27, 28, 30–32, 41, 46, 51], solid-phase extraction [3, 5, 11, 12, 18, 20, 23, 34, 36–38, 43, 52–56], solid-phase microextraction by packed sorbent [24, 57], dispersive solid-phase extraction [15, 22], salting-out assisted liquid-liquid extraction [58], molecularly imprinted polymer [59, 60], or a combination of techniques [35]. The selection of the optimum cleanup technique depends on the number and the physicochemical properties of the analytes, the analytical requirements of instrumental analysis, the complexity of the biological matrix, or the need for high throughput.

The minimum sample cleanup treatment described for the analysis of blood samples was the use of an organic solvent for protein precipitation [21, 29, 33, 39, 40]. However, as urine samples are less complex, there are some methods that analyze these compounds directly by dilute and shoot [8, 42, 45, 61, 62]. The dilute-and-shoot procedure is advantageous in practical application by diminishing the process of sample preparation but is an unrealistic procedure for the analysis of more complex samples or a large number of compounds. Generally, low volumes of urine (50–200 μL) are diluted with a mixture of aqueous and organic solvents in proportion 1:4–1:20 and centrifuged at high speed [8, 42, 45, 61, 62]. In some cases, before injection into the instrument, the supernatants are filter with 0.22 μm PVDF membranes [8, 45] to remove some of the interferences. The diluent should have adequate properties to not interfere in the instrumental analysis. Dilute-and-shoot procedure is usually applied to the determination of a limited number of analytes, but there are two published multianalyte methods for the determination of synthetic cathinones in urine [42, 45] using this procedure as cleanup. Al-Saffar et al. [42] determined 28 "legal highs," including 16 synthetic cathinones, by the dilution of 50 μL of urine in 1:5 proportion with 0.1% formic acid in water, achieving limits of quantification (LOQ) of 0.5–50 ng/mL depending on the analyte, while Fan et al. [45] determined 73 synthetic cathinones and related metabolites at an LOQs of 0.5–1 ng/mL by the dilution of 50 μL of urine in 1:20 proportion with 50% methanol in water and filtration with 0.22 μm PVDF membranes.

2.1.1 Protein Precipitation

Protein precipitation allows the denaturalization of the blood proteins by the addition of a precipitating agent, usually an organic solvent. The most common organic solvent used for the analysis of synthetic cathinones is acetonitrile [21, 29, 33, 40]; however, the use of a mixture of acetonitrile and methanol in proportion 80:20 (v:v) was also described [39]. The use of cold solvent helps with the denaturalization of the proteins, so it is common practice [21, 33,

39]. After that, centrifugation at high speed (>12000 g) is performed to pellet proteins, and then the supernatant is directly analyzed.

Protein precipitation usually produces a higher number of interferences during the instrumental analysis and more severe matrix effect than other cleanup techniques. Five methods have been published using protein precipitation for determination of synthetic cathinones in plasma [21, 29, 33, 39, 40] but none in whole blood (a more complex matrix). Moreover, three of them have been developed for the analysis of a single analyte [21, 29, 33] although the other two allow the determination of 16 [39] and 19 [40] synthetic cathinones simultaneously. In all these publications, the matrix effect detected was insignificant or low (<34%). Montesano et al. [39] studied the matrix effect caused by different solvents and mixtures of solvents (acetonitrile, methanol, and acetonitrile/methanol). They concluded that similar recoveries are obtained with the three tested solvents, but using acetonitrile/methanol (80:20, v:v), it was possible to obtain a lower matrix effect (−17% to 13%), and by selecting an appropriate mixture of solvent, it was additionally possible to obtain a reduced matrix effect despite the non-specificity of the method [39].

2.1.2 Liquid-Liquid Extraction

The use of liquid-liquid extraction (LLE) has been frequently applied in the analysis of synthetic cathinones in both blood [10, 14, 16, 19, 25, 27, 28, 30–32, 41, 46] and urine [9, 10, 16, 17, 25, 27, 46, 51]. The analytes were isolated from the biological matrix depending on their relative solubility in two different immiscible solvents, an aqueous solution (polar) and an organic solvent (nonpolar), at an adequate pH. The selection of the organic solvent is based in its polarity and should be the less polar solvent that extracts the analytes of interest but avoid extracting the interferences. Some of the preferred characteristics of the organic solvent are low solubility in water, a density lower than water, low flammability and toxicity, moderate volatility, and inertness.

Synthetic cathinones are basic compounds, so firstly the samples are alkalinized to achieve a pH >9. Different buffers have been used for alkalinization, such as ammonia [31], ammonium hydroxide [14, 16], sodium hydroxide [17], potassium hydroxide [51], ammonium acetate [10], carbonate buffer [25, 30, 46], borate buffer [19, 41], sodium tetraborate [28], and ammonium carbonate [32]. Also, different organic solvents have been used, including butyl chloride [14, 19, 25, 30, 51], tert-butylmethyl ether (MTBE) [17, 28, 46], ethyl acetate [31, 32, 41], or mixtures of them (MTBE:butyl acetate, 1:1, and dichloromethane:butyl chloride, 1:4) [10, 16].

After agitation, the binary solution is centrifuged to separate the two phases, and the organic solvent is collected and evaporated to dryness. The base form of the synthetic cathinones is volatile, so to avoid the loss of the compounds during the evaporation step, some authors acidified the extraction solvent with hydrochloric acid solutions prior to evaporation [14, 19, 25, 30, 51].

The use of LLE technique for cleanup proved to be suitable but mainly for the determination of a limited number of analytes simultaneously [9, 10, 14, 16, 19, 25, 27, 28, 30–32, 51] due to the low selectivity of the technique. Only three methods achieved the detection of many analytes in blood [41] or urine [17, 46]. Other disadvantages of this technique are the use of a large amount of solvents, which is time-consuming, less prone to automatization, and non-environmentally friendly.

2.1.3 Solid-Phase Extraction

Solid-phase extraction (SPE) techniques are more robust than other cleanup techniques. Moreover, analyte recovery is generally improved, and they can provide greater selectivity, lower cutoffs, and cleaner extracts [52]. Different SPE protocols have been described for the analysis of synthetic cathinones, all using mixed-mode sorbents [3, 5, 11, 12, 18, 20, 23, 34–38, 43, 52–55, 61]. Mix-mode sorbents act by two mechanisms, reversed-phase for retention of nonpolar or hydrophobic compounds and ion exchange for retention of charged compounds, removing additional endogenous interferences. The basic properties of the synthetic cathinones make mix-mode sorbents with cation exchange properties the most suitable SPE cartridges. These sorbents contain negatively charged groups in their surface which interacts with positively charged analytes by ionic interactions. Samples should be initially conditioned with an acid buffer to achieve the correct pH value (two units below the analyte pK_a is recommended). The most common buffer used to condition the samples is 0.1 M phosphate buffer at pH 6 [3, 5, 11, 12, 18, 23, 34–38, 43, 52, 53, 61] or pH 4.4 [54], but other acid solutions can be used, such as 1 M acetic acid at pH 5.2 [20] or formic acid solutions at pH 3 [55]. It is also recommended to condition the cartridges with an acid solvent to negatively charge the sorbent and improve the retention of the ionized analytes. Cartridges conditioning usually involves three steps: methanol, water, and the acid solvent, being 0.1 M phosphate buffer at pH 6 the most commonly used [3, 5, 12, 18, 23, 34, 38, 43, 53, 61]. Despite this, some authors only conditioned the cartridges with methanol and water [11, 20, 52, 54].

After loading the samples into the cartridges, washing steps to remove interferences are performed. A common washing protocol includes water, 0.1 or 1 M acetic acid, and methanol [3, 5, 11, 23, 35, 37, 38, 53, 61]. Finally, the elution of the analytes is performed using alkalinized organic solvents, being a mixture of dichloromethane/isopropanol/ammonium hydroxide (78:20:2, v:v:v) the most commonly used [3, 5, 11, 12, 18, 23, 34, 38, 52, 54, 61].

2.1.4 Other Cleanup Techniques

The determination of synthetic cathinones in blood and/or urine samples can be performed using other alternative cleanup techniques. Different modifications of the conventional SPE have been described, including solid-phase microextraction, molecularly imprinted polymer, or dispersive solid-phase extraction. Solid-phase microextraction (SPME) is a miniaturized SPE technique which reduces the use of sample and solvents volumes. The analytes are extracted from the sample by adsorption to a sorbent material coating in a fiber [24] or inside a syringe [57]. The analytical protocol follows the same steps performed in SPE, but unlike traditional SPE, the material can be reused multiple times. This technique has been applied to the detection of three synthetic cathinones and other classic drugs of abuse in blood [57] and five synthetic cathinones in urine [24], using 300 μL and 3 μL of sample, respectively.

In molecularly imprinted polymer (MIP) the SPE sorbent is a functional polymer with molecular-level cavities that exhibit selective binding. The copolymerization of functional and cross-linking monomers is made in the presence of a target molecule (template). After the removal of the template, complementary rigid three-dimensional cavities with antibody-like affinity and selectivity allow a highly drug class selective extraction to the target analyte and other closely related compounds [63]. There are commercially available SPE cartridges with MIP technology for the extraction of different target analytes, such as β-agonists, β-blockers, nonsteroidal anti-inflammatory agents, nitrosamines, or polycyclic aromatic hydrocarbons [59]. Murakami et al. [59] used a commercial MIP cartridge for amphetamine derivatives (amphetamine, methamphetamine, and methylenedioxyphenylalkylamines) to the determination of 11 synthetic cathinones in blood and urine. After an adjustment of the pH of the sample solution loaded onto the cartridges at pH 6, due to the different pK_a of the synthetic cathinones and the amphetamine derivatives, they found higher recoveries and less matrix effect for the 11 synthetic cathinones compared to conventional SPE for both biological matrices [59]. On the other hand, the synthesis of MIP sorbents in-house can increase the selectivity for the target synthetic cathinones. Sanchez-Gonzalez et al. [60] synthesized and characterized a MIP sorbent for the detection of ten synthetic cathinones in urine, although the cavities in MIP particles were designed for the recognition of amine and carbonyl groups, and, therefore, they could be effectively recognized other unknown synthetic cathinones and traditional amphetamine derivatives [60].

Other technique used for the analysis of these compounds is dispersive solid-phase extraction (dSPE) [15, 22]. In dSPE the solid sorbent is directly added into the liquid sample, which was previously treated with organic solvent. The contaminants are adsorbed to the sorbent, and the analytes of interest are present in the solvent. The sorbent is separated by a mechanical process such as centrifugation or filtration. This technique has been used for the determination of α-pyrrolidinohexanophenone (α-PHP) [15] in blood and 3,4-dimethylmethcathinone (3,4-DMMC) in blood and urine [22] using QuEChERS, a commercial device for dSPE.

Finally, salting-out assisted liquid-liquid extraction (SALLE) is another cleanup technique used for synthetic cathinones sample preparation [58]. This technique is as simple as protein precipitation but produces cleaner extracts and used less solvents than conventional LLE. It is based on the addition of a water-miscible organic solvent, usually acetonitrile, and a concentrated salt solution to the aqueous sample. The saturation of the aqueous solution with salt causes a decrease in the solubility of the organic molecules present in the sample and their migration to the organic solvent. After mixing and centrifugation, the organic solvent formed a separated phase on top of the aqueous solution. The organic solvent can be analyzed directly, avoiding evaporation and reconstitution steps [63]. Ishii et al. extracted 500 μL of urine for the detection of six synthetic cathinones in 90 s using acetonitrile as the organic solvent and anhydrous sodium carbonate as the salt solution [58].

2.2 Instrumental Analysis

The vast majority of the analytical methods proposed for the target analysis of synthetic cathinones in blood and/or urine are based on gas chromatography (GC) or liquid chromatography (LC) coupled to mass spectrometry (MS). For analytes separation only one exception was found, the use of capillary electrophoresis for the detection of 12 synthetic cathinones in urine [55]. This technique separates ions based on their electrophoretic mobility with the use of an applied voltage, and it was mainly developed for the separation of charged molecules such as peptides and amino acids. In the presence of acid buffers the synthetic cathinones are cations and can migrate; however, the separation of these compounds using this technique is difficult because their structures and pK_a values are very similar and the resolution is not enough good, which makes necessary the addition of different additives to the mobile phase to improve resolution. Although LOQs in Svidrnoch et al. method [55] were comparable or slightly higher than other GC or LC methods, they used three times more sample volume to achieve them.

Regarding the detection, the only exception to MS was the use of ultraviolet-visible (UV-VIS) spectrophotometry, where the analytes are identified by measuring the absorption of light at different wavelengths. Specifically, photodiode array (PDA), which can measure the entire wavelength range simultaneously, was used for the detection of 3 synthetic cathinones and other drugs of abuse in plasma [57], 22 synthetic cathinones in blood and urine [38], and 12 synthetic cathinones in urine [55], using a λ_{max} value of 235 and 260 nm, 270 nm, and 210, 214, and 254 nm, respectively [38, 55, 57].

2.2.1 Chromatographic Separation

GC is widely used technique in the field of forensic and clinical toxicology due to its applicability to a large range of compounds, robustness, and specificity. However, synthetic cathinones may undergo thermal degradation during their analysis by GC, and, for this reason, LC is the main separation technique used for their analysis [35]. Thermal degradation has been observed by several authors [64–69], although, according to Kerrigan et al. [69], it can be minimized using lower inlet temperatures, minimizing residence time in the inlet, and by eliminating active sites during chromatographic analysis. The characteristic species produced by thermal degradation are imine (from the loss of a hydrogen from the nitrogen in the amine group) and enamine (from the loss of a hydrogen from the carbon-carbon bond) [69]. These breakdown products can be difficult to separate from their parent drug and share many of the same characteristic ions [35]. Nevertheless, various GC methods have been described for the determination of synthetic cathinones in blood and/or urine samples [3–5, 9–11, 14, 17–19, 23, 46, 51–54, 61].

In GC, the separation is performed in a column containing a solid stationary phase that has a continuous flow of an inert carrier gas (mobile phase) passing through it and a temperature gradient. The most common column used is a (5%-phenyl)-methylpolysiloxane nonpolar column with helium as carrier gas at a flow rate 0.8–1.8 mL/min [3, 5, 9–11, 14, 17, 52–54, 61]. The injection port temperature ranged from 200 °C to 280 °C [3, 5, 10, 11, 14, 17–19, 23, 51–54, 61], and splitless mode was commonly used in the injection [3, 5, 9–11, 18, 19, 23, 51, 53, 61], although split ratios 1:5–1:16 were also employed [17, 46, 52, 54].

Other main disadvantage of GC is the need to perform derivatization processes, especially due to the low boiling point and the relatively poor mass spectral quality of many of the synthetic cathinones [69]. Derivatization process facilitates chromatographic separation of analytes and increases their volatility, detectability, and stability but is time-consuming and increases sample cost. For the pyrrolidine derivatives, which are tertiary amines without an active hydrogen in the amine group, the derivatization process is prevented using common derivatization approaches [69]. Alexandridou et al. [10] studied the potential utility of derivatization for several NPS including three synthetic cathinones (mephedrone, methylone, and methylenedioxypyrovalerone or MDPV) using two silylation reagents (BSTFA with 1% TMCS and MTBSTFA) and three acetylation reagents (HFBA, TFAA, and PFPA). For these compounds, silylation does not provide detectable derivatives, while acetylation increases the signal of the secondary amines mephedrone and methylone (HFBA > TFAA > PFPA), but not for MDPV [10]. On the contrary, Antunes et al. [52] performed a similar study for four synthetic cathinones (4-chloroethcathinone, α-pyrrolidinovalerophenone, 4-chloro-pyrrolidinovalerophenone, and MDPV) using a silylation reagent (MSTFA with 5% TMCS) and an acetylation reagent (MBTFA). They found better results, lack of interferences, and improved chromatographic peaks, using the silylation reagent even when three of the compounds are tertiary amines. The most common derivatization approaches applied for the analysis of synthetic cathinones are indicated in Table 1.

LC is a more versatile technique and allows separation of non-volatile, thermolabile, and polar analytes, without the need of the derivatization process. In LC, the separation depends on the characteristics of the stationary phase of the analytical column, the composition of the mobile phase, and the column temperature. Reversed-phase columns are the most used stationary phase for synthetic cathinones analysis, with typical C8 or C18 alkyl chains [8, 12, 16, 20, 22, 25, 29–35, 37, 39–43, 56–58, 60, 62]. However, phenyl stationary phases are also employed [21, 28, 45, 59] for its capacity to increase the separation of positional isomers, since phenyl phases interact via pi-pi interactions with some analytes increasing their retention into the column. Moreover, the use of specific chiral stationary phases is described for enantioselective analysis [15, 38].

The typical composition of mobile phases is a mixture of an aqueous solution with a volatile buffer and an organic solution. The most common aqueous solutions used are formic acid in water [8, 12, 21, 25, 30, 34, 35, 37, 39, 40, 42, 43, 56, 57] or ammonium formate with formic acid in water [15, 20, 22, 28, 29, 31–33, 45, 60, 62]. Organic solutions are typically acetonitrile [8, 12, 16, 20, 21, 25, 30–33, 35, 37, 39–43, 57, 58] or methanol [15, 22, 28, 29, 34, 45, 56, 59, 60, 62].

2.2.2 Mass Spectrometry

Mass spectrometry is an excellent tool for the development of sensitive and specific methods for both qualitative and quantitative analysis. First, the molecules present in the sample are ionizing to generated charged molecules or charged molecular fragments of these molecules. Then, these ions are separated by their mass-to-

Table 1
Derivatization approaches for the determination of synthetic cathinones in blood and/or urine

	Derivatization reagents	Incubation	Reconstitution solvent	Compounds	References
Silylation	PFPA:ethyl acetate (2:1, v:v)	Not indicated	Ethyl acetate	Mephedrone, naphyrone, methylenedioxypyrovalerone, α-pyrrolidinovalerophenone	[5]
	PFPA:ethyl acetate (1:1, v:v)	30 min at 65 °C	Ethyl acetate	Ethylone, mephedrone, methylone, 4-bromo-2,5-dimethoxyamphetamine, 4-chloro-2,5-dimethoxyamphetamine, 4-iodo-2,5-dimethoxyamphetamine	[23]
	TFAA	30 min at 65 °C	TBME	3,4-Dimethylmethcathinone, 3,4-methylenedioxypyrovalerone, 4-fluoroamphetamine, 4-fluoromethcathinone, 4-methylethcathinone, mephedrone, buphedrone, butylone, ethcathinone, methedrone, pentylone, 1-naphyrone, 2-naphyrone, β-pentedrone, dimethylcathinone, ethylone, methylone, methcathinone, 3-methylmethcathinone	[17]
	TFAA:ethyl acetate (1:1, v:v)	15 min at 65 °C	Ethyl acetate	Buphedrone, butylone, cathinone, 3,4-dimethylmethcathinone, ethcathinone, N-ethyl-buphedrone, ethylone, eutylone, flephedrone, methcathinone, mephedrone, methedrone, 4-methyl-buphedrone, methylenedioxypyrovalerone, 4-methyl-N-ethyl-cathinone, methylone, naphyrone, pentedrone, pentylone, α-pyrrolidinobutiophenone, α-pyrrolidinopentiophenone	[46]
	HFBA	5 min at room temperature	–	Methcathinone, buphedrone, mephedrone, methamphetamine, methylone, methylenedioxyamphetamine, methylenedioxymethamphetamine, 1,3-methylbenzodioxolylbutanamine, 5,6-methylenedioxy-2-aminoindane, fluoromethcathinone	[51]
	HFBA:ethyl acetate (1:1, v:v)	15 min at 90 °C	Ethyl acetate	Butylone, ethylone, pentylone, mephedrone, methylenedioxypyrovalerone, methylone	[11]
	MBTFA	90 s at 900 W in microwave	–	d-Cathine, ephedrine, methcathinone, 1-(4-methoxyphenyl)-propan-2-amine, mephedrone, methedrone, l-cathinone (l-norephedrone), methylone, 4-methylthioamphetamine, buphedrone, ethylone, flephedrone, methylenedioxypyrovalerone, α-pyrrolidinovalerophenone, pentylone	[6, 54]

Acetylation	MSTFA with 5% TMCS	25 min at 80 °C	–	4-Cloroethcathinone, α-pyrrolidinovalerophenone, 4-chloro-pyrrolidinovalerophenone, methylenedioxypyrovalerone	[52]
Chiral separation	L-TPC	10 min at room temperature	Ethyl acetate	Cathinone, 4-fluoromethcathinone, 4-fluoroethcathinone, nor-mephedrone, buphedrone, 3-methylmethcathinone, 3-methylbuphedrone, 4-methylbuphedrone, 3-ethylmethcathinone, 3-ethylethcathinone, 4-ethylethcathinone, 3,4-dimethylethcathinone, 2,3-methylenedixoymethcathinone, butylone, pentylone	[3]

HFBA heptafluorobutyric anhydride, *L-TPC* (S)-(−)-N-(trifluoroacetyl)pyrrolidine-2-carbonyl chloride, *MSTFA* N-methyl-N-(trimethylsilyl)trifluoroacetamide, *PFPA* pentafluoropropionic anhydride, *TBME* tert-butylmethyl ether, *TFAA* trifluoroacetic anhydride, *TMCS* trimethylchlorosilane

charge ratio (m/z) in high vacuum by the application of either electric or magnetic fields and identified depending on their relative molecular mass and charge. The ionization can be performed using different sources. The most common ionization source coupled to GC is electron impact (EI) [5, 10, 11, 17, 18, 46, 52–54, 61], although negative chemical ionization (NCI) [3] can be also used. EI is a hard ionization technique which provides a characteristic and reproducible mass spectrum for each analyte, while NCI is a soft ionization technique, and its spectrum is simpler and with a low number of fragments, although its use can increase the sensitivity of compounds with high electron affinities [3]. Nevertheless, synthetic cathinones are relatively fragile compounds and can extensively fragment under EI conditions, which results in a mass spectrum with a nonspecific and limited number of diagnostic ions [24, 35]. The use of derivatization can improve the mass spectrum of these compounds, number of characteristic fragments, and sensitivity [10, 24, 52].

On the other hand, the most common ionization source coupled to LC is electrospray ionization (ESI) [8, 12, 15, 16, 20, 22, 25, 28–35, 37, 39–43, 45, 58–60, 62]. ESI employs a softer ionization and can overcome some of the drawbacks of GC-MS methods. Moreover, it is the preferable ionization source for polar analytes and is widely used for synthetic cathinones analysis. The majority LC-MS/MS methods used triple-quadrupole with multiple reaction monitoring for confirmatory and quantitative analysis because of its high sensitivity and specificity [8, 12, 15, 16, 22, 25, 28–34, 37, 40–43, 59, 60]. This is a target analysis, and, therefore, those analytes not included in the methodology are not detected. For that reason, in recent years the use of high-resolution techniques (LC-HRMS/MS) [20, 21, 35, 36, 39, 44, 45, 58, 62] is increasing, since they work in scan mode, collecting high-resolution mass data from all analytes present in the sample. For additional reference about HRMS in NPS analysis, see Chap. 11.

3 Analytical Challenges

3.1 Identification of Enantiomers

Synthetic cathinones have a chiral carbon center which generates two enantiomers, and usually a racemic mixture of both isomers is found in the market, although these enantiomers may have different potency and metabolic pathway [3, 38]. For example, the stimulant effect of (S)-methcathinone is higher than (R)-methcathinone [3], and for this reason not all enantiomers are scheduled [35]. To identify and differentiate the enantiomers by GC or LC-MS(MS), their chromatographic separation is necessary, because enantiomers normally share the same precursor and product ions [35]. Some authors achieved enantiomers separation of some synthetic cathinones using traditional reversed-phase chromatography [23, 35, 37, 43], but generally a more specific approach is necessary for their separation.

Separation of enantiomers can be performed using two different approaches: by direct chromatographic resolution using analytical columns with chiral stationary phases or using chiral additives in the mobile phase or by derivatization with a stereochemically pure reagent that convert the enantiomers to diastereoisomers, which have different physicochemical properties, and can be separated on conventional (achiral) chromatographic columns [38]. The first approach was used by Meetani et al. [38], which compared two commercially available columns, Astec Cellulose DMP column and Amylose-based Chiralpak AS-H column, for enantiomer separation of 22 tertiary amines. The first column coated with dimethylphenyl carbamate-derivatized cellulose was able to separate more enantiomers than the amylose-based column, specifically, 19 versus 7 enantiomers were correctly separated. The second approach was employed by Alremeithi et al. [3] to enantiomerically separate 36 synthetic cathinones using (S)-(−)-N-(trifluoroacetyl)-pyrrolidine-2-carbonyl chloride (L-TPC). L-TPC is a chiral derivatizing agent which reacts with the primary and secondary amines and convert enantiomers into diastereoisomers.

3.2 Stability of Synthetic Cathinones

The possibility of drug instability during transport, storage, handling, and analytical conditions should be taken into consideration during method development. Degradation can produce changes in drug concentrations and in the metabolic profile and, therefore, affect the interpretation of the toxicological results. Cathinone instability in plant material and in plasma was reported [5, 64], but relatively limited data is available for synthetic cathinones. Several publications evaluating different aspects of synthetic cathinones stability in blood and urine samples have been published for some analytes [5–7, 29, 32, 34, 36, 41–44, 54, 70, 71]. Moreover, an interesting revision paper has been recently published summarizing the current knowledge about the stability of synthetic cathinones in different biological samples [72]. The main findings regarding synthetic cathinones stability in blood and urine are summed up as follows:

- The stability of the synthetic cathinones is highly dependent on temperature, being most of them stable at frozen temperatures (>−20 °C) [5, 7, 29, 32, 36, 41–43, 70, 71].
- The pH of the sample has a high impact on synthetic cathinones degradation, and instability of many of these analytes has been reported at neutral to basic pH (7-9) [7, 36, 42, 71].
- Alkalinization of urine samples stored refrigerated or at room temperature increase instability of synthetic cathinones

in this sample [7, 36]. Addition of some preservatives, such as sodium azide, to urine samples can prevent bacterial growth and consequently alkalinization of the samples and degradation of the analytes [7].

- Multiple thawing and freezing cycles have not shown influence on the stability of most of the synthetic cathinones [6, 7, 29, 34, 43, 44, 54].
- Concentration of the analytes in the sample seems to not produce significant differences in synthetic cathinones stability [70, 71].
- Some chemical structures have significant stabilizing effects, such as the methylenedioxy group or the presence of a fluorine atom instead of a chlorine atom [5, 7, 70, 71]. Moreover, substitutions in the methylenedioxy group in both secondary and tertiary amines have a stabilizing effect [70, 71]. In addition, tertiary amines (except for naphyrone) are more stable than secondary amines [7, 36, 43, 70, 71]. On the contrary, other structures such as the presence of fused benzene ring reduce stability [5].

4 Concentrations in Blood Specimens

Quantitative analysis of blood specimens is necessary in clinical and forensic cases to establish the level of impairment of the individual and the contribution of the different xenobiotics to the clinical symptoms or the cause of death. There are none or few data about concentrations ranges for the majority of the synthetic cathinones due to their low prevalence and the lack of specific analytical methods for their determination in many toxicological laboratories. Table 2 shows concentration range in blood for a selection of synthetic cathinones found in authentic clinical and forensic cases. Additional information about concentrations of synthetic cathinones in human performance and postmortem cases can be found in Ellefsen et al. [75].

Table 2
Blood concentrations of synthetic cathinones in authentic clinical and forensic cases

Chemical name	Synonyms	Concentration range (ng/mL)		References
		Living	Postmortem	
1-(1,3-Benzodioxol-5-yl)-2-(1-pyrrolidinyl)-1-butanone	3,4-MD-α-PHP 3,4-MDPHP 3,4-Methylenedioxy-α-pyrrolidinohexanophenone	3.3–140	76 ng/mL (fetus)	[25, 61]
1-Phenyl-2-(1-pyrrolidinyl)-1-hexanone	α-PHP 2-(1-Pyrrolidinyl)-hexanophenone α-Pyrrolidinohexanophenone	<1–175	12 (fetus)–15.3 ng/mL	[15, 25, 34]
4-Methyl-1-phenyl-2-(pyrrolidin-1-yl) pentan-1-one	α-Pyrrolidinoisohexanophenone α-PiHP		69	[26]
4-Chloro-α-pyrrolidinohexanophenone	4-Chloro-PHP 4-Chloro-α-PHP 4-Chloro-pyrrolidinohexanophenone		7.2	[37]
1-(1,3-Benzodioxol-5-yl)-2-(ethylamino)-1-pentanone	N-ethyl pentylone Ephylone		31–4700	[14, 37]
2-(ethylamino)-1-(4-methylphenyl)-1-propanone	4-Methylethcathinone 4-Methyl-N-ethyl cathinone 4-MEC	53–81	14600–43400	[27, 47]
2-(Methylamino)-1-(4-methylphenyl)-1-propanone	Mephedrone 4-MeMC 4-Methylephedrone 4-Methylmethcathinone 4-MMC	86	40–22000	[40, 47, 73]
1-(1,3-Benzodioxol-5-yl)-2-(methylamino)-1-propanone	Methylone M1	7	19–729	[4, 16, 37]

(continued)

Table 2
(continued)

Chemical name	Synonyms	Concentration range (ng/mL)		References
		Living	Postmortem	
	bk-MDMA 3,4-Methylenedioxy-N-methylcathinone			
1-(1,3-Benzodioxol-5-yl)-2-(ethylamino)-1-pentanone	Ephylone N-ethyl pentylone N-ethylnorpentylone	7–149	10–2700	[12, 28]
1-(1,3-Benzodioxol-5-yl)-2-(1-pyrrolidinyl)-1-pentanone	MDVP 3,4-MDPV 3,4-Methylenedioxypyrovalerone	1–530	10–640	[16, 18, 19, 29, 41, 47, 53, 74]
2-(Ethylamino)-1-phenyl-1-hexanone	N-ethylamino-hexanophenone N-ethyl hexedrone		145	[30]
1-(1,3-Benzodioxol-5-yl)-2-(dimethylamino)-1-butanone	bk-DMBDB bk-MMBDB Dibutylone Methylbutylone		6.1–486	[37]
1-(1,3-Benzodioxol-5-yl)-2-(ethylamino)-1-butanone	Eutylone bk-EBDB β-keto-ethylbenzodioxolylbutanamine Ethyl butylone N-ethyl butylone 3,4-methylenedioxy-α-ethylaminobutiophenone		1.6–3.6	[37]
1-(1,3-Benzodioxol-5-yl)-2-(methylamino)-1-butanone	Butylone β-keto MBDB bk-MBDB		1.4–150	[37]

1-(1,3-Benzodioxol-5-yl)-2-(methylamino)-1-pentanone	Pentylone		3–19	[37]
1-(1,3-Benzodioxol-5-yl)-2-(ethylamino)-1-propanon	Ethylone		38–2572	[23]
1-Phenyl-2-(propylamino)-1-pentanone	α-Propylaminopentiophenone 2-Propylamino-1-phenylpentanone 2-(Propylamino)-valerophenone N-propylpentedrone		3200	[31]
1-(4-Chlorophenyl)-2-(methylamino)-1-propanone	4-Chloromethcathinone Clephedrone 4-CMC		66.2	[32]
1-Phenyl-2-(1-pyrrolidinyl)-1-heptanone	α-PHPP PV8		70–260	[33]
1-(4-Fluorophenyl)-2-(methylamino)propan-1-one	4-Fluoromethcathinone Flephedrone 4-FMC	346		[74]
1-(3,4-Dimethylphenyl)-2-(methylamino)-1-propanone	3,4-Dimethylmethcathinone 3,4-DMMC		27000	[22]

References

1. European Monitoring Centre for Drugs and Drug Addiction and Europol (2019) EU Drug Markets Report 2019, Publications Office of the European Union, Luxembourg. https://www.emcdda.europa.eu/system/files/publications/12078/20192630_TD0319332ENN_PDF.pdf. Accessed date: 10 July 2021
2. United Nations Office on Drugs and Crime (2021) World Drug Report 2021, Drug market trends: cocaine, amphetamine type stimulants, United Nations Publication, Vienna, 2021. https://www.unodc.org/res/wdr2021/field/WDR21_Booklet_4.pdf Accessed date: 10 July 2021
3. Alremeithi R, Meetani MA, Alaidaros AA, Lanjawi A, Alsumaiti K (2018) Simultaneous quantitative determination of synthetic cathinone enantiomers in urine and plasma using GC-NCI-MS. J Anal Methods Chem 2018:4396043
4. Carbone PN, Carbone DL, Carstairs SD, Luzi SA (2013) Sudden cardiac death associated with methylone use. Am J Forensic Med Pathol 34(1):26–28
5. Ciallella HL, Rutter LR, Nisbet LA, Scott KS (2020) Extended stability evaluation of selected cathinones. Front Chem 8:597726
6. Margalho C, Almeida P, Rosado T, Corte Real F, Gallardo E (2020) Determination of new psychoactive substances in whole blood using microwave fast derivatization and gas chromatography/mass spectrometry. J Anal Toxicol 44(1):92–102
7. Adamowicz P, Malczyk A (2019) Stability of synthetic cathinones in blood and urine. Forensic Sci Int 295:36–45
8. Carrola J, Duarte N, Florindo P, Henriques S, da Silva G, Bijlsma L, Moreira R, Correia C, Perry MJ, Lopes A, de Mello-Sampayo C, do Rosário Bronze M (2020) Metabolism of N-ethylhexedrone and buphedrone: an in vivo study in mice using HPLC-MS/MS. J Chromatogr B Analyt Technol Biomed Life Sci 1159:122340
9. Klavž J, Gorenjak M, Marinšek M (2016) Suicide attempt with a mix of synthetic cannabinoids and synthetic cathinones: Case report of non-fatal intoxication with AB-CHMINACA, AB-FUBINACA, alpha-PHP, alpha-PVP and 4-CMC. Forensic Sci Int 265:121–124
10. Alexandridou A, Mouskeftara T, Raikos N, Gika HG (2020) GC-MS analysis of underivatised new psychoactive substances in whole blood and urine. J Chromatogr B Analyt Technol Biomed Life Sci 1156:122308
11. Hong WY, Ko YC, Lin MC, Wang PY, Chen YP, Chiueh LC, Shih DYC, Chou HH, Chen HF (2016) Determination of synthetic cathinones in urine using gas chromatography–mass spectrometry techniques. J Anal Toxicol 40(1):12–16
12. Giachetti AD, Kahl JH, Zaney ME, Hime GW, Boland DM (2020) Method validation of seven synthetic cathinones by LC-MS/MS analysis and the prevalence of N-ethylpentylone in postmortem casework. J Anal Toxicol Bkaa194. Online ahead of print.
13. Adamowicz P (2021) Blood concentrations of synthetic cathinones. Clin Toxicol (Phila) 59(7):648–654
14. Atherton D, Dye D, Robinson CA, Beck R (2019) N-ethyl pentylone-related deaths in Alabama. J Forensic Sci 64(1):304–308
15. Fujita Y, Mita T, Usui K, Kamijo Y, Kikuchi S, Onodera M, Fujino Y, Inoue Y (2018) Toxicokinetics of the synthetic cathinone α-pyrrolidinohexanophenone. J Anal Toxicol 42(1):e1–e5
16. Marinetti LJ, Antonides HM (2013) Analysis of synthetic cathinones commonly found in bath salts in human performance and postmortem toxicology: method development, drug distribution and interpretation of results. J Anal Toxicol 37(3):135–146
17. Gerace E, Caneparo D, Borio F, Salomone A, Vincenti M (2019) Determination of several synthetic cathinones and an amphetamine-like compound in urine by gas chromatography with mass spectrometry. Method validation and application to real cases. J Sep Sci 42(8):1577–1584
18. Wyman JF, Lavins ES, Engelhart D, Armstrong EJ, Snell KD, Boggs PD, Taylor SM, Norris RN, Miller FP (2013) Postmortem tissue distribution of MDPV following lethal intoxication by "bath salts". J Anal Toxicol 37(3):182–185
19. Spiller HA, Ryan ML, Weston RG, Jansen J (2011) Clinical experience with and analytical confirmation of "bath salts" and "legal highs" (synthetic cathinones) in the United States. Clin Toxicol (Phila) 49(6):499–505
20. Bertol E, Mari F, Berto RB, Mannaioni G, Vaiano F, Favretto D (2014) A mixed MDPV and benzodiazepine intoxication in a chronic drug abuser: determination of MDPV metabolites by LC–HRMS and discussion of the case. Forensic Sci Int 243:149–155

21. Anizan S, Ellefsen K, Concheiro M, Suzuki M, Rice KC, Baumann MH, Huestis MA (2014) 3,4-Methylenedioxypyrovalerone (MDPV) and metabolites quantification in human and rat plasma by liquid chromatography–high resolution mass spectrometry. Anal Chim Acta 827:54–63
22. Usui K, Aramaki T, Hashiyada M, Hayashizaki Y, Funayama M (2014) Quantitative analysis of 3,4-dimethylmethcathinone in blood and urine by liquid chromatography–tandem mass spectrometry in a fatal case. Leg Med (Tokyo) 16(4):222–226
23. Lee D, Chronister CW, Hoyer J, Goldberger BA (2015) Ethylone-Related Deaths: Toxicological Findings. J Anal Toxicol 39(7): 567–571
24. LaPointe J, Musselman B, O'Neill T, Shepard JRE (2015) Detection of "bath salt" synthetic cathinones and metabolites in urine via DART-MS and solid phase microextraction. J Am Soc Mass Spectrom 26(1):159–165
25. Adamowicz P, Hydzik P (2019) Fetal death associated with the use of 3,4-MDPHP and α-PHP. Clin Toxicol (Phila) 57(2):112–116
26. Adamowicz P, Jurczyk A, Gil D, Szustowski S (2020) A case of intoxication with a new cathinone derivative α-PiHP – a presentation of concentrations in biological specimens. Leg Med (Tokyo) 42:101626
27. Braham MY, Franchi A, Cartiser N, Bévalot F, Bottinelli C, Fabrizi H, Fanton L (2021) Fatal 4-MEC intoxication: case report and review of literature. Am J Forensic Med Pathol 42(1): 57–61
28. Costa JL, Francisco Cunha K, Lanaro R, Leal Cunha R, Walther D, Baumann MH (2019) Analytical quantification, intoxication case series, and pharmacological mechanism of action for N-ethylnorpentylone (N-ethylpentylone or ephylone). Drug Test Anal 11(3): 461–471
29. Diestelmann M, Zangl A, Herrle I, Koch E, Graw M, Paul LD (2018) MDPV in forensic routine cases: psychotic and aggressive behavior in relation to plasma concentrations. Forensic Sci Int 283:72–84
30. Domagalska E, Banaszkiewicz L, Woźniak MK, Kata M, Szpiech B, Kaliszan M (2021) Fatal N-ethylhexedrone intoxication. J Anal Toxicol 45(6):e1–e6
31. Majchrzak M, Celiński R, Kowalska T, Sajewicz M (2018) Fatal case of poisoning with a new cathinone derivative: α-propylaminopentiophenone (N-PP). Forensic Toxicol 36(2):525–533
32. Nowak C, Szpot P, Zawadzki M (2020) The stability of 4-chloromethcathinone in blood and vitreous humor. J Forensic Sci 65(5): 1784–1790
33. Pieprzyca E, Skowroneka R, Korczyńska M, Kulikowska J, Chowaniec M (2018) A two fatal cases of poisoning involving new cathinone derivative PV8. Leg Med (Tokyo) 33: 42–47
34. Vignali C, Moretti M, Groppi A, Osculati AMM, Tajana L, Morini L (2019) Distribution of the synthetic cathinone α-pyrrolidinohexiophenone in biological specimens. J Anal Toxicol 43(1):e1–e6
35. Glicksberg L, Bryand K, Kerrigan S (2016) Identification and quantification of synthetic cathinones in blood andurine using liquid chromatography-quadrupole/time of flight (LC-Q/TOF) mass spectrometry. J Chromatogr B Analyt Technol Biomed Life Sci 1035: 91–103
36. Glicksberg L, Rana S, Kerrigan S (2018) Cathinone stability in authentic urine specimens. Forensic Sci Int 286:54–60
37. Lau T, Concheiro M, Cooper G (2020) Determination of 30 synthetic cathinones in postmortem blood using LC–MS-MS. J Anal Toxicol 44(7):679–687
38. Meetani MA, Alremeithi RH, Mousa MK (2019) Enantioseparation of synthetic cathinones enantiomers with tertiary amine structure in urine and plasma. J Chromatogr Sci 57(4):361–368
39. Montesano C, Vannutelli G, Gregori A, Ripani L, Compagnone D, Curini R, Sergi M (2016) Broad screening and identification of novel psychoactive substances in plasma by high-performance liquid chromatography–high-resolution mass spectrometry and postrun library matching. J Anal Toxicol 40(7): 519–528
40. Vaiano F, Busardò FP, Palumbo D, Kyriakou C, Fioravanti A, Catalani V, Maria F, Bertol E (2016) A novel screening method for 64 new psychoactive substances and 5 amphetamines in blood by LC–MS/MS and application to real cases. J Pharm Biomed Anal 129:441–449
41. Wang Y, Shi Y, Yu Y, Chen L, Jiang J, Long J, Xiang P, Duan G (2021) Screening of synthetic cathinones and metabolites in dried blood spots by UPLC-MS-MS. J Anal Toxicol 45(7):633–643
42. Al-Saffar Y, Stephanson NN, Beck O (2013) Multicomponent LC–MS/MS screening method for detection of new psychoactive drugs, legal highs, in urine—Experience from

the Swedish population. J Chromatogr B Analyt Technol Biomed Life Sci 930:112–120

43. Concheiro M, Anizan S, Ellefsen K, Huestis MA (2013) Simultaneous quantification of 28 synthetic cathinones and metabolites in urine by liquid chromatography-high resolution mass spectrometry. Anal Bioanal Chem 405(29):9437–9448

44. Concheiro M, Castaneto M, Kronstrand R, Huestis MA (2015) Simultaneous determination of 40 novel psychoactive stimulants inurine by liquid chromatography–high resolution mass spectrometryand library matching. J Chromatogr A 1397:32–42

45. Fan SY, Zang CZ, Shih PH, Ko YC, Hsu YH, Lin MC, Tseng SH, Wang DY (2020) A LC-MS/MS method for determination of 73 synthetic cathinones and related metabolites in urine. Forensic Sci Int 315:110429

46. Uralets V, Rana S, Morgan S, Ross W (2014) Testing for designer stimulants: metabolic profiles of 16 synthetic cathinones excreted free in human urine. J Anal Toxicol 38(5):233–241

47. Mercolini L, Protti M, Catapano MC, Rudge J, Sberna AE (2016) LC-MS/MS and volumetric absorptive microsampling for quantitative bioanalysis of cathinone analogues in dried urine, plasma and oral fluid samples. J Pharm Biomed Anal 123:186–194

48. Thornton SL, Gerona RR, Tomaszewski CA (2012) Psychosis from a bath salt product containing flephedrone and MDPV with serum, urine, and product quantification. J Med Toxicol 8(3):310–313

49. Zaami S, Giorgetti R, Pichini S, Pantano F, Marinelli E, Busardò FP (2018) Synthetic cathinones related fatalities: an update. Eur Rev Med Pharmacol Sci 22(1):268–274

50. Jones AW (2016) Introduction to forensic toxicology and the value of a nationwide database. In: Davies S, Johnston A, Holt D (eds) Forensic toxicology. drug use and misuse, 1st edn. The Royal Society of Chemistry, pp 1–29

51. Hagan KS, Reidy L (2015) Detection of synthetic cathinones in victims of sexual assault. Forensic Sci Int 257:71–75

52. Antunes M, Sequeira M, de Caires PM, João Caldeira M, Santos S, Franco J, Barroso M, Gaspar H (2021) Determination of selected cathinones in blood by solid-phase extraction and GC–MS. J Anal Toxicol 45(3):233–242

53. Grapp M, Kaufmann C, Ebbecke M (2017) Toxicological investigation of forensic cases related to the designer drug 3,4-methylenedioxypyrovalerone (MDPV): Detection, quantification and studies on human metabolism by GC–MS. Forensic Sci Int 273:1–9

54. Margalho C, Castanheira A, Corte Real F, Gallardo E, López-Rivadulla M (2016) Determination of "new psychoactive substances" in postmortem matrices using microwave derivatization and gas chromatography-mass spectrometry. J Chromatogr B Analyt Technol Biomed Life Sci 1020:14–23

55. Švidrnoch M, Lněníčková L, Válka I, Ondra P, Maier V (2014) Utilization of micellar electrokinetic chromatography-tandem mass spectrometry employed volatile micellar phase in the analysis of cathinone designer drugs. J Chromatogr A 1356:258–265

56. Strickland EC, Cummings OT, Mellinger AL, McIntire JL (2019) Development and validation of a novel all-inclusive LC–MS-MS designer drug method. J Anal Toxicol 43(3): 161–169

57. Fernández P, González M, Regenjo M, Aresa AM, Fernández AM, Lorenzo RA, Carro AM (2017) Analysis of drugs of abuse in human plasma using microextraction bypacked sorbents and ultra-high-performance liquid chromatography. J Chromatogr A 1485:8–19

58. Ishii A, Kusakabe K, Na K, Sasaki S, Tsujikawa K, Wada T (2020) Studies on the phase I metabolites of the new designer drug 1-(2,3-dihydro-1H-inden-5-yl)-2-(pyrrolidine-1-yl)butan-1-one (5-PPDI) in human urine. Forensic Sci Int 310:110214

59. Murakami Y, Iwamuro Y, Ishimaru R, Chinaka S, Hasegawa H (2018) Molecularly imprinted polymer solid-phase extraction of synthetic cathinones from urine and whole blood samples. J Sep Sci 41(24):4506–4514

60. Sánchez-González J, Odoardi S, Bermejo AM, Bermejo-Barrera P, Romolo FS, Moreda-Piñeiro A, Strano-Rossi S (2019) HPLC-MS/MS combined with membrane-protected molecularly imprinted polymer micro-solid-phase extraction for synthetic cathinones monitoring in urine. Drug Test Anal 11(1):33–44

61. Grapp M, Kaufmann C, Schwelm HM, Neukamm MA, Blaschke S, Eidizadeh A (2020) Intoxication cases associated with the novel designer drug 3′,4-′-methylenedioxy–αpyrrolidinohexanophenone and studies on its human metabolism using high-resolution mass spectrometry. Drug Test Anal 12(9):1320–1335

62. Kavanagh P, Gofenberg M, Shevyrin V, Dvorskaya O, Dowling G, Grigoryev A (2020) Tentative identification of the phase I and II metabolites of two synthetic cathinones, MDPHP and α-PBP, in human urine. Drug Test Anal 12(10):1442–1451

63. Lendoiro E, de Castro A, Fernández-Vega H, Cela-Pérez MC, López-Vilariño JM, González-Rodríguez MV, Cruz A, López-Rivadulla M (2014) Molecularly imprinted polymer for selective determination of Δ9-tetrahydrocannabinol and 11-nor-Δ9-tetrahydrocannabinol carboxylic acid using LC-MS/MS in urine and oral fluid. Anal Bioanal Chem 406(15): 3589–3597
64. Morad A, Al-Meshal I, Nasir M, El-Feraly F (1989) High performance liquid chromatographic determination of (−)-cathinone in plasma. Chromatographia 27:201–204
65. Paul BD, Cole KA (2001) Cathinone (Khat) and methcathinone (CAT) in urine specimens: a gas chromatographic-mass spectrometric detection procedure. J Anal Toxicol 25:525–530
66. Sørensen LK (2011) Determination of cathinones and related ephedrines in forensic whole-blood samples by liquid-chromatography-electrospray tandem mass spectrometry. J Chromatogr B 879:727–736
67. Johnson RD, Botch-Jones SR (2013) The stability of four designer drugs: MDPV, mephedrone, BZP and TFMPP in three biological matrices under various storage conditions. J Anal ToxicoL 37:51–55
68. Soh YN, Elliott S (2013) An investigation of the stability of emerging new psychoactive substances. Drug Test Anal 6:696–704
69. Kerrigan S, Savage M, Cavazos C, Bella P (2016) Thermal degradation of synthetic cathinones: implications for forensic toxicology. J Anal Toxicol 40:1–11
70. Glicksberg L, Kerrigan S (2017) Stability of synthetic cathinones in blood. J Anal toxicol 41(9):711–719
71. Glicksberg L, Kerrigan S (2018) Stability of synthetic cathinones in urine. J Anal Toxicol 42(2):77–87
72. Aldubayyan AA, Castrignanò E, Elliott S, Abbate V (2021) Stability of synthetic cathinones in clinical and forensic toxicological analysis—where are we now? Drug Test Anal 13(1): 44–68
73. Busardò FP, Kyriakou C, Napoletano S, Marinelli E, Zaami S (2015) Mephedrone related fatalities: a review. Eur Rev Med Pharmacol Sci 19(19):3777–3790
74. Thornton SL, Gerona RR, Tomaszewski CA (2012) Psychosis from a bath salt product containing flephedrone and MDPV with serum, urine, and product quantification. J Med Toxicol 8:310–313
75. Ellefsen KN, Concheiro M, Huestis MA (2016) Synthetic cathinones pharmacokinetics, analytical methods, and toxicological findings from human performance and postmortem cases. Drug Metab Rev 48(2):237–265

Chapter 3

Target Analysis of Phenethylamines, Tryptamines, and Piperazines in Blood and Urine

Steven Towler and Marta Concheiro

Abstract

Phenethylamines, tryptamines, and piperazines constitute a complex group of hallucinogenic and stimulant drugs. Their analysis in toxicological laboratories may be not routinely performed, and it constitutes a challenge due to the high number of substances involved, the constant appearance of novel derivatives in the market, and the low concentrations found in biological samples due to their high potency. In this chapter we summarized the current methodologies available for the target analysis of this type of substances in plasma, serum, blood, and urine samples. This critical review focuses mainly on recently published confirmation methods, and it highlights practical aspects of these methods, such as what are the recommended target analytes, expected concentrations, stability of these compounds in blood and urine samples, and analytical aspects, including sample hydrolysis, extraction procedures, and instrumental analysis.

Key words Phenethylamines, Tryptamines, Piperazines, Serum, Plasma, Blood, Urine, GC-MS, LC-MS

1 Introduction

Phenethylamines and tryptamines are potent psychedelic substances that produce their hallucinogenic effects by binding to serotonin receptors, 5-HT_{2A}, and others. The 5-HT_{2A} serotonin receptor has been linked to complex behaviors including working memory, cognitive processes, and affective disorders such as schizophrenia [1]. Specifically, phenethylamines have a high affinity for 5-HT_{2A} serotonin receptors, as well as partial agonism at the 5-HT_{2C} receptor [2]. On the other hand, tryptamines act as agonists of 5-HT_{2A} and 5-HT_{1A} serotonin receptors.

Phenethylamines are chemically related to the natural hallucinogenic alkaloid mescaline. Several hallucinogenic phenethylamines were synthesized by Shulgin during the 1970s and the 1980s [3]. These substances, as well as novel derivatives, have reappeared and appeared in the market since the 2000s

Marta Concheiro and Karl B. Scheidweiler (eds.), *Methods for Novel Psychoactive Substance Analysis*,
Methods in Pharmacology and Toxicology, https://doi.org/10.1007/978-1-0716-2605-4_3,

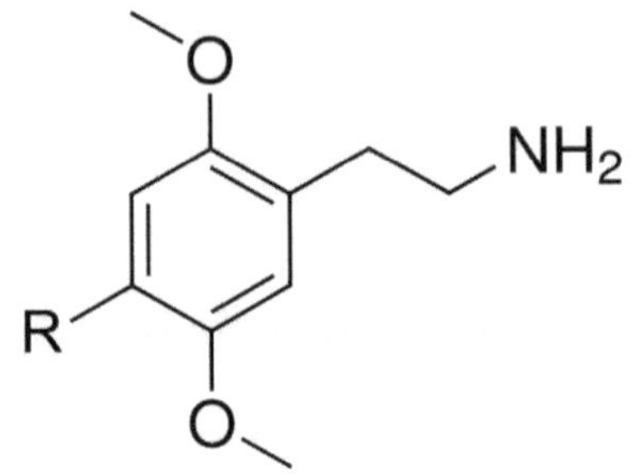

2C Compound	R
2C-B	Br-
2C-D	H_3C-
2C-E	H_3C
2C-H	H-
2C-I	I-
2C-N	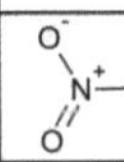
2C-P	H_3C

Fig. 1 Chemical structures of selected 2C compounds (phenethylamines)

[2, 4]. Members of this hallucinogen class can be further chemically subdivided into three subclasses: the 2C, the 2C-T, and the DO-series (*see* Figs. 1, 2, 3 and 4). The 2C series are dimethoxyphenethylethanamines containing two methoxy groups at the 2 and 5 positions of the benzene ring and two carbons (2C) between the amine and benzene ring. In the 2C family, novel derivatives belong to the NBOMe group (N-2-methoxybenzyl phenethylamines). The NBOMe class of drugs includes 25B-, 25C-, 25D-, 25H-, 25I-, and 25 T2-NBOMe. In this group, the addition of an N-benzyl substituent to phenethylamine hallucinogens dramatically increases 5-HT_{2A} receptor affinity [5]. The 2C-T series are dimethoxyphenethylethanamines that typically contain an alkylated thio-group on the 4 position of the ring. The DO-series are dimethoxyphenylpropanamines containing a methyl on the aminoethyl chain and a halogen or alkyl group on the 4 position of the benzene ring. The DO-series produce more powerful hallucinogenic effects compared with either the 2C or 2C-T series [6].

Whereas some phenethylamines are listed in Schedules I and II of the 1971 United Nation Convention on Psychotropic Substances, most of the new substances are not under international control. For example, in the USA, most of the 2C-x chemicals are Schedule I under the American Food and Drug Administration Safety and Innovation Act of 2012 and the Federal Analogue Act [7], and the NBOMes, 25I-NBOMe, 25B-NBOMe, and

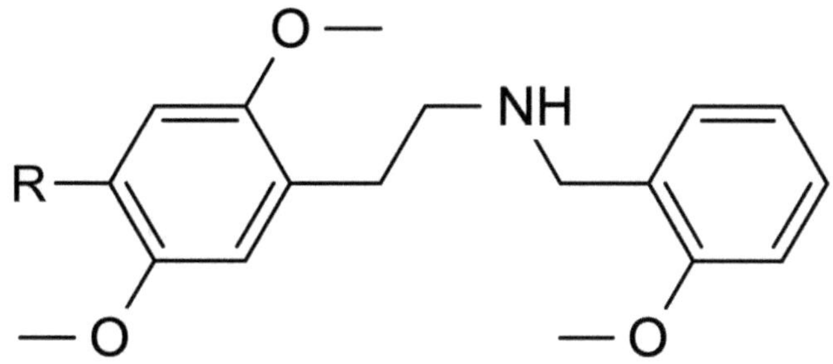

NBOMe Compound	R
25B-NBOMe	Br-
25C-NBOMe	Cl-
25D-NBOMe	H_3C-
25E-NBOMe	H_3C
25H-NBOMe	H-
25I-NBOMe	I-

Fig. 2 Chemical structures of selected NBOMe compounds (phenethylamines)

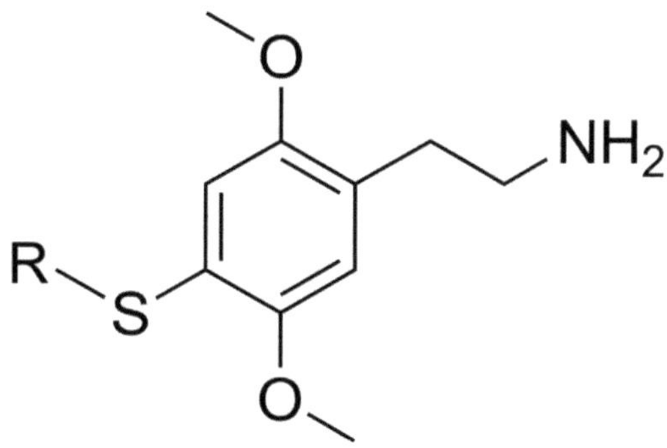

2C-T Compound	R
2C-T	H_3C-
2C-T-2	H_3CH_2C-
2C-T-5	
2C-T-7	$H_3CH_2CH_2C$-

Fig. 3 Chemical structures of selected 2C-T compounds (phenethylamines)

25C-NBOMe, are currently Schedule I Substances according to 21 CFR 1308.11(d) [8].

Well-known tryptamines are psilocin, psilocybin, N, N-dimethyltryptamine (DMT), and lysergic acid diethylamide (LSD). Chemically, they are related to the neurotransmitter

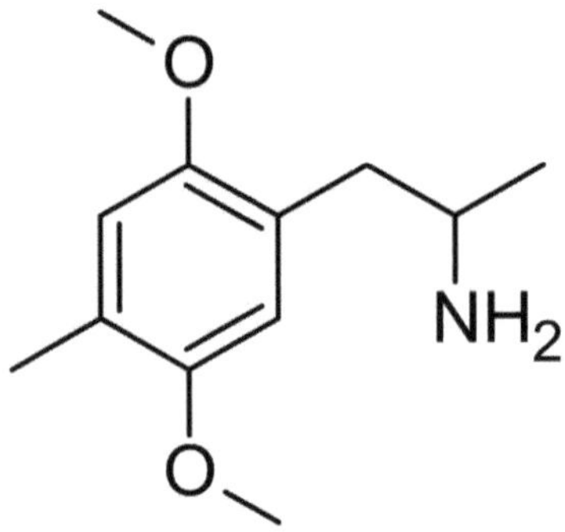

DO- Compound	R
DOB	Br-
DOC	Cl-
DOET	H_3CH_2C-
DOI	I-
DOM	H_3C-

Fig. 4 Chemical structures of selected DO compounds (phenethylamines)

serotonin. The use of the naturally occurring psilocybin became widespread in the late 1950s in the USA, whereas synthetic tryptamines appeared on illicit drug markets throughout the 1990s. More recently, emerging tryptamine hallucinogens are alpha-methyltryptamine (AMT), 5-methoxy-N,N-dimethyltryptamine (5-MeO-DMT), and 5-methoxy-N,N-diisopropyltryptamine (5-MeO-DIPT). Among the novel LSD derivatives that recently appeared in the market include 1-propionyl-LSD (1P-LSD), lysergic acid 2,4-dimethylazetidide (LSZ), N-allyl-nor-LSD (AL-LAD), N- ethyl-nor-LSD (ETH-LAD), lysergic acid morpholide (LSM-775), N-ethyl-N-cyclopropyllysergamide (ECPLA), and 1-propionyl-ETH- LAD (1P-ETH-LAD) [9]. See Figs. 5 and 6. Psilocin and psilocybin are the only tryptamines under international control (listed in Schedule I of the 1971 Convention), while some others are controlled at the national level in several countries.

The third group of substances included in this chapter, piperazines, is psychostimulant drugs which contain a piperazine ring and produce effects like 3,4-methylenedioxymethamphetamine (MDMA), amphetamine, and methamphetamine (*see* Figs. 7 and 8). Piperazines are chemically divided into two categories, namely, benzylpiperazines (*see* Fig. 7), such as 1-benzylpiperazine (BZP), 1-(3,4-methylenedioxybenzyl) piperazine (MDPP), and 1,4-dibenzylpiperazine (DBZP), and phenylpiperazines (*see* Fig. 8), like 1-(3-tri-fluoromethylphenyl) piperazine (TFMPP), 1-(4-fluorophenyl) piperazine (pFPP), and 1-(3-chlorophenyl) piperazine (mCPP) [10]. According to the United Nations Office on Drugs and Crimes (UNODC), BZP itself was initially developed

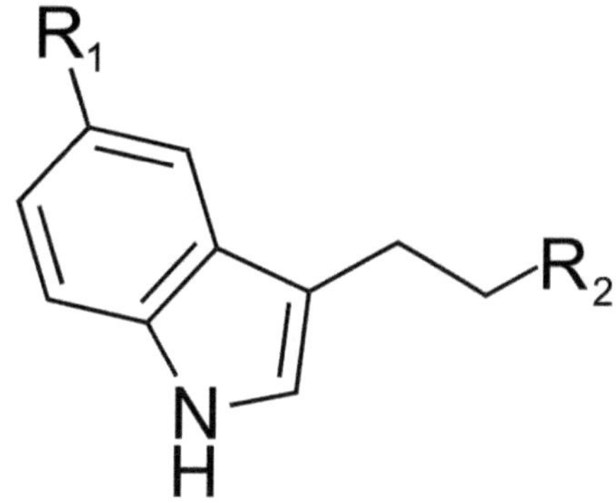

DMT Compound	R_1	R_2
DMT	H-	
5-MeO-DMT		
5-MeO-DIPT		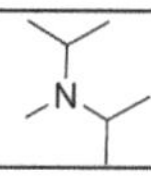

Fig. 5 Chemical structure of N,N-dimethyltryptamine (DMT) and novel derivatives (tryptamines)

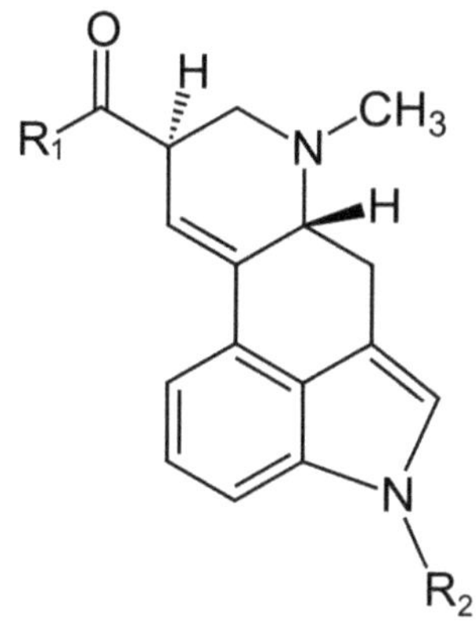

LSD Compound	R_1	R_2
LSD	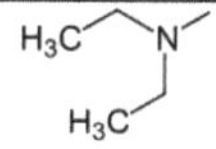	H-
1P-LSD	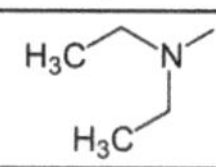	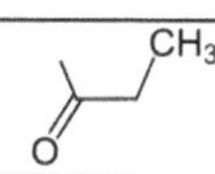
LSZ	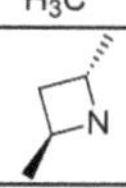	H-

Fig. 6 Chemical structure of lysergic acid diethylamide (LSD) and novel derivatives (tryptamines)

Benzylpiperazine Compound	Structure
BZP	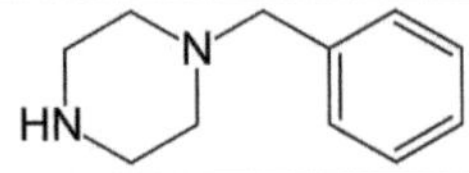
MDPP	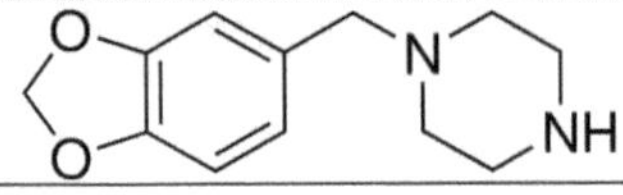
DBZP	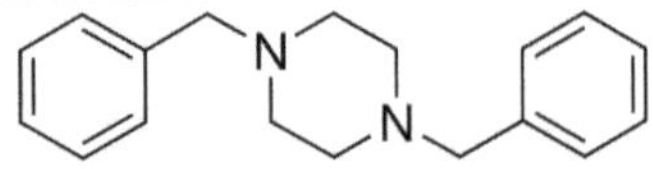

Fig. 7 Chemical structures of selected benzylpiperazine compounds (piperazines)

Phenylpiperazine Compound	Structure
TFMPP	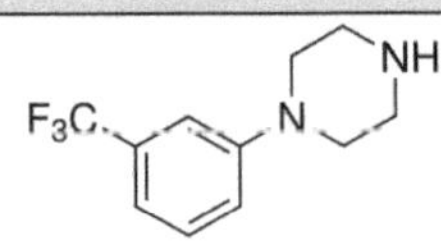
pFPP	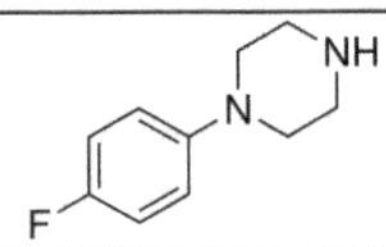
mCPP	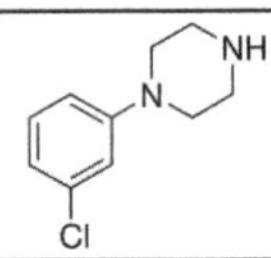

Fig. 8 Chemical structures of selected phenylpiperazine compounds (piperazines)

as a potential antidepressant drug but was found to have similar properties to amphetamine and therefore liable to abuse. In the late 1990s, BZP emerged in New Zealand as a "legal alternative" for MDMA and methamphetamine. In Europe, its use was first reported in Sweden in 1999, but it only became widespread as a NPS from 2004 onwards until controls over the substance were introduced in 2008. Neither BZP nor any other piperazines are under international control, although several countries have introduced national control measures over piperazines.

Phenethylamines, piperazines, and tryptamines, although they belong to different drug groups, all share similar physicochemical properties. Phenethylamines, piperazines, and tryptamines are small organic molecules (molecular weight less than 500 Da), basic with pKas between 8 and 11, and with low lipophilicity, log P less than 5 (Table 1). These physicochemical properties are key for the design of the extraction procedures from biological matrices and for their detection methods by different analytical techniques.

Table 1
Monoisotopic molecular weight (MW, g/mol), pKa, and Log P of selected phenethylamines, tryptamines, and piperazines

Group	Analyte	MW (g/mol)	pKa	Log P
Phenethylamines	2C-B	259.0208	9.68	1.84
	25I-NBOMe	427.0644	8.91	4.3
	2C-T-7	255.1293	9.74	3.08
	DOB	273.0364	9.4	2.58
Tryptamines	LSD	323.1997	7.98	2.28
	DMT	188.1313	9.55	2.3
	5-MeO-DMT	218.1419	9.58	2.14
	5-MeO-DiPT	274.2045	10.68	3.69
	AMT	174.1157	9.96	1.9
Piperazines	BZP	176.1313	9.02	1.37
	mCPP	196.0767	8.87	2.15
	TFMPP	230.1030	8.88	2.63

Data obtained from Drugbank (https://go.drugbank.com/), Pubchem (https://pubchem.ncbi.nlm.nih.gov/), Monteiro et al. [15], and Dias et al. [16]

25I-NBOMe 4-iodo-2,5-dimethoxy-N-(2-methoxybenzyl)phenethylamine, *2C-B* 2,5-dimethoxy-4-bromophenethylamine, *2C-T-7* 2,5-dimethoxy-4-(propylthio)-benzeneethanamine, *5-MeO-DiPT* 5-methoxy-N,N-diisopropyltryptamine, *5-MeO-DMT* 5-methoxy-N,N-dimethyltryptamine, *AMT* alpha-methyltryptamine, *BZP* 1-benzylpiperazine, *DMT* N,N-dimethyltryptamine, DOB 4-bromo-2,5-dimethoxyamphetamine, *LSD* lysergic acid diethylamide, *mCPP* 1-(3-chlorophenyl) piperazine, *TFMPP* 1-(3-tri-fluoromethylphenyl) piperazine

In the literature, review papers have been published focusing on different pharmacological and analytical aspects of specific phenethylamines (2C-series, NBOMe-series) [11, 12] and tryptamines [13, 14]. In this chapter, we provide practical guidelines for the quantitative analysis of phenethylamines, tryptamines, and piperazines in blood, serum, plasma, and urine, including an overview of published methods. We highlight the recommended target analytes, what are the expected concentrations, and the most common sample extraction techniques, analyte chromatographic separation and instrumental detection.

2 Determination of Phenethylamines in Blood and Urine

Table 2 summarizes the main characteristics of selected analytical methods for the determination of phenethylamines, including the 2C-, 2C-T-, and DO-series, in whole blood, plasma, serum, and urine. All these methodologies showed high sensitivity and specificity and were fully validated as quantification methods, and most of them were applied to authentic samples.

Table 2
Analytical methods for the determination of phenethylamines in whole blood, plasma, serum, and urine

Analytes	Matrix	Sample Volume (mL)	Linearity (ng/mL)	Extraction	Instrument	Separation	MS mode	References
2C-D, 2C-E, 2C-P, 2C-B, 2C-I, 2C-T-2, 2C-T-7, mescaline	Plasma	1	5–500	SPE Mixed mode	GC-MS (Q)	HP-5 MS capillary (30 m × 0.25 mm, 250 nm)	SIM	[4]
2C-T, 2C-T-2, 2C-T-5, 2C-T-7	Plasma	1	10–1000	PP	CE-ESI-MS (Q)	Uncoated fused-silica capillary (120 cm × 50 m)	Scan (100–400 *m/z*)	[17]
2CC-NBOMe, 25I-NBOMe	Serum	1	0.03–2	SPE Mixed mode	LC-MSMS (QTrap)	Luna C8 (3 μm, 100 × 2.0 mm)	MRM	[1]
2C-H, 2C-I, 2C-B, 2C-T-2, 2C-T-7, 2C-D, 2C-E, 2C-P, 2C-N, 2C-B-fly, 25H-NBOMe, 25C-NBOMe, 25B-NBOMe, 25I-NBOMe, 25E-NBOMe, LSD, LSZ, AL-LAD, 1P-LSD, fentanyl, butyryl fentanyl	Plasma	1	0.1–40	2 × LLE diethyl ether-ethyl acetate (1:1)	LC-HRMS (Q-Orbitrap)	Accucore Phenyl Hexyl LC column (2.6 μm, 100 × 2.1 mm)	Scan (100–600 *m/z*) and all ions fragmentation (AIF) (50–600 *m/z*)	[18]
25H-NBOMe, 2CC-NBOMe, 25I-NBF, 25D-NBOMe, 25B-NBOMe, 2CT-NBOMe, 25I-NBMD, 25G-NBOMe, 25I-NBOMe	Urine	0.5	1–100	Fast SPE	LC-MSMS (QTrap)	Allure Biphenyl (5 μm, 100 × 3.2 mm)	MRM	[19]

2C-B, 2C-C, 2C-D, 2C-E, 2C-H, 2C-I, 2C-T-2, 2C-T-4, 2C-T-7, DOB, DOC, DOET, DOI, DOM, 4-MTA	Urine	2	1–2000	SPE Mixed mode	LC-MSMS (QqQ)	Luna C18 (5 μm, 100 × 2 mm)	MRM	[6]
25B-, 25C-, 25D-, 25H-, 25I-, 25 T2-NBOMe	Whole blood, plasma, urine	0.5	0.1–20	SPE Mixed mode	LC-MSMS (QTrap)	Kinetex PFP (1.7 μm, 50 × 2.1 mm)	MRM	[20]

1P-LSD 1-propionyl-lysergic acid diethylamide, *2C-B* 2,5-dimethoxy-4-bromophenethylamine, *2C-B-FLY* 8-bromo-2,3,6,7-benzo-dihydro-difuran-ethylamine, *2C-C* 2-(4-chloro-2,5-dimethoxyphenyl)ethan-1-amine, *2CC-NBOMe* {4-chloro-2,5-dimethoxyphenethyl-N-[(2-methoxyphenyl) methyl]ethanamine}, *2C-D* 2,5-dimethoxy-4-methylphenethylamine, *2C-E* 2,5-dimethoxy-4-ethylphenethylamine, *2C-H* 2,5-dimethoxyphenethylamine, *2C-I* 2-(4-iodo-2,5-dimethoxyphenyl)ethan-1-amine, *2C-N* 2,5-dimethoxy-4-nitrophenethylamine, *2C-P* 2-(2,5-dimethoxy-4-propylphenyl)ethanamine, *2C-T* 2,5-dimethoxy-4-methylthiophenethylamine, *2C-T-NBOMe* 2-(2,5-dimethoxy-4-(methylthio)phenyl)-N-(2-methoxybenzyl)ethanamine, *2C-T-2* 2,5-dimethoxy-4-ethylthiophenethylamine, *2C-T-4* 2,5-dimethoxy-4-isopropylthiophenethylamine, *2C-T-5* 5-dimethoxy-4-cyclohexylthiophenethylamine, *2C-T-7* 2,5-dimethoxy-4-(n)-propylthiophenethylamine, *25I-NBOMe* {2-(4-iodo-2,5-dimethoxyphenyl)-N-[(2-methoxyphenyl) methyl]ethanamine, *4-MTA* 4-methylthioamphetamine, *25B-NBOMe* 4-bromo-2,5-dimethoxy-N-(2-methoxybenzyl)phenethylamine, *25C-NBOMe* N-(2-methoxybenzyl)-2-(4-chloro-2,5-dimethoxyphenyl)ethanamine, *25D-NBOMe* 2-(2,5-dimethoxy-4-methylphenyl)-N-(2-methoxybenzyl)ethanamine, *25E-NBOMe* 2-(4-ethyl-2,5-dimethoxyphenyl)-N-(2-methoxybenzyl)ethanamine, *25G-NBOMe* 2-(2,5-dimethoxy-3,4-dimethylphenyl)-N-[(2-methoxyphenyl)methyl]ethanamine, *25H-NBOMe* 2,5-dimethoxy-N-((2-methoxyphenyl)methyl)benzeneethanamine, *25I-NBF* 2-(4-iodo-2,5-dimethoxyphenyl)-N-[(2-fluorophenyl)methyl]ethan-1-amine, *25I-NBMD* N-[(2H-1,3-benzodioxol-4-yl)methyl]-2-(4-iodo-2,5-dimethoxyphenyl)ethan-1-amine, *25 T2-NBOMe* 4-(ethylthio)-2,5-dimethoxy-N-[(2-methoxyphenyl)methyl]-benzeneethanamine, *AL-LAD* 6-allyl-6-nor-LSD, *CE-ESI-MS* capillary electrophoresis coupled with electrospray ionization-mass spectrometry, *DOB* 4-bromo-2,5-dimethoxyamphetamine, *DOC* 2,5-dimethoxy-4-chloroamphetamine, *DOET* 2,5-dimethoxy-4-ethylamphetamine, *DOI* 2,5-dimethoxy-4-iodoamphetamine, *DOM* 2,5-dimethoxy-4-methylamphetamine, *LSD* lysergic acid diethylamide, *LSZ* lysergic acid 2,4-dimethylazetidide

LLE liquid-liquid extraction, *GC-MS* gas chromatography-mass spectrometry, *LC-HRMS* liquid chromatography-high-resolution mass spectrometry, *LC-MSMS* liquid chromatography-tandem mass spectrometry, *MRM* multiple reaction monitoring, *PP* protein precipitation, *Q* quadrupole, *QqQ* triple quadrupole, *QTrap* quadrupole-ion trap, *SIM* single ion monitoring, *SPE* solid-phase extraction

2.1 Target Analytes

All these methodologies had as target analyte the parent drugs, regardless of the biological sample analyzed. This is a common approach with this type of substances because limited data are available about their metabolism in humans, and/or the metabolites are not commercially available [6]. Pichini et al. [21], which procedure is summarized in Table 5, analyzed hydrolyzed vs. not hydrolyzed urine samples and reported higher concentrations for 2C compounds in hydrolyzed samples. This indicates that these compounds can be found in urine as glucuronide and/or sulfate conjugates. In that study, urine samples were hydrolyzed with β-glucuronidase from *Helix pomatia* (114,400 β-glucuronidase units/mL; 3290 sulfatase units/mL), 16 h at 37 °C.

2.2 Doses and Concentrations

Due to their high potency, the typical reported doses of phenethylamines are low. For the 2C drugs the doses range from 6 to 60 mg, for the 2C-T-compounds 8–30 mg, and DO compounds 1–10 mg, and for the NBOMe derivative the doses are in the μg range (500–800 μg) [3, 6, 22–24]. Because of this, the expected blood and urine concentrations are low as well (pg/mL and low ng/mL). For example, an intake of 2C hallucinogens leads to plasma concentrations of 27.3–43 ng/mL [17, 25], and consumption of NBOMes resulted in blood concentrations of 0.034–2.78 ng/mL [1, 19, 23]. The analytical methods summarized in Table 2 achieved limits of quantification (LOQ) from 0.03 to 10 ng/mL in blood, plasma, serum, and urine.

2.3 Stability

A critical preanalytical aspect that must be considered in the analysis of phenethylamines is their stability in biological samples. Several studies from Table 2 evaluated the stability in fortified but not in authentic samples. Caspar et al. [18] found reduced stability for 2C-N, 2C-B-fly, and 2C-E after freeze-thaw cycles, as well as 2C-E and NBOMes at room temperature in plasma. Habrdova et al. [4] also observed stability problems with 2C-I and 2C-D in plasma after freeze-thaw cycles. Cunha et al. [26] observed stability issues for NBOMes analytes in whole blood. At room temperature, 25B-, 25C-, 25D-, 25E-, 25G-, 25H-, and 25I-NBOMe showed poor stability over 15 days and decreased in most cases by more than 20% from the initial concentration at 4 °C after 180 days. At −20 °C, all analytes were stable for 180 days. On the other hand, Curtis et al. [27] reported good stability of 2C-T-7 up to 70 days in sodium fluoride-preserved blood at 4 °C. Poklis et al. [19] reported NBOMe derivatives were stable in frozen (1 month) or room temperature (72 h) in urine. The sole exception was 2CT-NBOMe which had a 232–243% deviation from the target concentrations of the quality control specimens stored at room temperature for 72 h. Due to these stability issues and the low concentrations present in the biological samples, the samples for the analysis of phenethylamines should be stored frozen until analysis.

2.4 Internal Standards

As internal standards, most published methodologies utilized deuterated analogs of chemically related compounds, except Poklis et al. that employed the deuterated analog of one of the target analytes (25I-NBOMe-d_3) [19]. These internal standards employed were trimipramine-d_3 [18], MDPV-d_8 [20], amphetamine-d_5, methamphetamin-d_5, MDA-d_5, MDMA-d_5, MDEA-d_5 [4], and mescaline-d_9 [4, 6].

2.5 Extraction Procedures

Due to the basic nature of the phenethylamines, the most common extraction procedure was solid-phase extraction (SPE) with mixed mode-cation exchange cartridges. In these procedures the sample was loaded at pH 6 or 7, and after washing the cartridges with water, acidic buffer, and organic solvents, the elution was performed in organic solvent containing 2% ammonium hydroxide. Habrdova et al. [4] diluted 1 mL plasma in 2 mL water, and after washing with water, 0.01 M HCl, and methanol, the sample was eluted with methanol-ammonium hydroxide (98:2, v/v). Johnson et al. [20] and Poklis et al. [1] conditioned the biological sample in phosphate buffer (pH 6), and after washing with water, 0.1 M acetic acid, and methanol, the elution was performed with 2% ammonium hydroxide in methylene chloride/isopropanol (80:20, v/v). Kerrigan et al. [6] also conditioned the samples with a phosphate buffer, but a more exhaustive washing was performed with water, 1 M acetic acid, hexane, ethyl acetate, and methanol. The elution was carried out by 2% ammonium hydroxide in methylene chloride/isopropanol (95:5, v/v). Poklis et al. [19] performed a different type of SPE, employing a special type of cartridges that allowed a faster procedure; 0.5 mL sample and 0.5 mL methanol were mixed, and 300 μL were transferred to Clean Screen Fast® cartridges and directly aspirated to vials for injection.

Caspar et al. [18] employed two consecutive liquid-liquid extraction (LLE) procedures to clean up the plasma. One mL of plasma was mixed with 2 mL saturated aqueous sodium sulfate solution and 5 mL diethyl ether-ethyl acetate mixture (1:1, v/v). The mixture was shaken manually for at least 20 s and centrifuged until phase separation, and the upper organic extract was transferred into a flask. A volume of 0.5 mL sodium hydroxide (1 M, pH 8–9) and 5 mL diethyl ether-ethyl acetate mixture (1:1, v/v) were added to the remaining liquid and again manually mixed for at least 20 s, followed by 2 min centrifugation. The upper solvent phase was again transferred into a flask. Solvents were then evaporated until dryness. Boatto et al. [17] utilized the simpler cleanup procedure, which consisted in a protein precipitation (PP). Pretreated plasma (1 mL) with 20 μL of phosphoric acid was precipitated with acetonitrile.

After the described extraction procedures (SPE, LLE, and PP), the extracts were evaporated to dryness, reconstituted in mobile phase, and injected in the instrument. Before the evaporation step,

Poklis et al. [1] and Kerrigan et al. [6] recommended the addition of acidic methanol to reduce the loss of phenethylamines during evaporation. During the development of the assay, Poklis et al. [1] observed a consistent loss of both 2CC-NBOMe and 25I-NBOMe if the basic SPE elution solvent was evaporated to dryness. The addition of methanolic HCl and water to the elution solvent combined with limited evaporation of the solvent resulted in acceptable recoveries. Kerrigan et al. [6] investigated this effect of "salting out" prior to evaporation to determine whether sample losses could occur due to the volatility of some target drugs in the base (uncharged) form. The abundance of 2C, 2C-T, and DO compounds evaporated in the base form (uncharged) was significantly lower (19–75%), relative to those treated with acidic methanol prior to evaporation. However, Caspar et al. [18] and Habrdova et al. [4] did not perform that step and did not acidify the extracts before evaporation.

2.6 Derivatization

Among the different publications in Table 2, only Habrdova et al. [4] performed the analysis by GC-MS, and, therefore, a derivatiza tion step was performed. The selected derivatization reagent was heptafluorobutyric anhydride (HFBA), using 20 μL and a microwave at 450 W, for 5 min. After derivatization, the extract was mixed with hexane and 0.5 M Na_3PO_4, and 1 μL of the organic layer was injected into the GC-MS. It should be indicated that the authors observed reproducibility problems for the derivatization of 2C-T-2 and 2C-T-7. They suggested that the sulfur moiety in these two drugs may be the reason.

2.7 Chromatographic Separation

In the GC method, Habrdova et al. [4] used a nonpolar capillary column, a HP-5MS capillary column (30 m × 0.25 mm i.d.). The run time was 10 min. In the LC methods, the chromatographic separations were performed in reversed-phase columns, and the mobile phases were employed in gradient mode with a mixture of water with modifiers such as formic acid and ammonium acetate or formate and organic solvents, mainly acetonitrile. Caspar et al. [18] employed an Accucore Phenyl Hexyl LC column (100 mm × 2.1 mm, 2.6 μm). The mobile phase consisted of 2 mM aqueous ammonium formate at pH 3.4 containing 0.1% formic acid (eluent A) and methanol/acetonitrile (50:50, v/v) containing 0.1% formic acid (eluent B). The total run time was 15 min, and the 2C compounds retention times (RT) were 1.7–4.7 min, and NBOMe compounds RT were 7.1–10.2 min. Johnson et al. [20] used a Phenomenex Kinetex PFP (50 × 2.1 mm, 1.7 um), with mobile phase 2 mM ammonium formate containing 0.2% formic acid and acetonitrile containing 0.1% formic acid. The total run time was 4.51 min, and the NBOMe compounds eluted between 1.5 and 3 min. Poklis et al. employed a Phenomenex Luna C8(2) (100 × 2 mm, 3 μm) [1] or a

Restek Allure Biphenyl (100 × 3.2 mm, 5 μm) column [19]. In both studies, the mobile phase was water with 10 mM ammonium acetate and 0.1% formic acid (eluent A) and methanol (Eluent B). The total run time was 10 min (RT 3.5–5.5 min) or 13 min (RT 7–12 min), respectively. Kerrigan et al. [6] used a Phenomenex Luna C18 column (100 × 2 mm, 5 μm), with mobile phase eluent A 50 mM ammonium acetate in water/methanol (95:5, v/v), and eluent B 50 mM ammonium acetate in acetonitrile/water (90:10, v/v). The run time was 6 min plus equilibration time.

2.8 Mass Spectrometry

Most of the publications employed LC-MS and as detection method low-resolution tandem mass spectrometry in multiple reaction monitoring (MRM) mode. Two MRM transitions were monitored per analyte, and all of them were ionized by electrospray (ESI) in positive mode [1, 6, 19, 20]. Boato et al. [17] employed a single MS, and the protonated molecule was selected as the quantifier ion. Caspar et al. [18] employed high-resolution mass spectrometry (HRMS). HRMS methods are explained in more detail in Chap. 11. Briefly, the authors monitored the accurate mass of the protonated molecule acquired in scan mode and at least two characteristic fragment ions with mass accuracies ≤5 ppm acquired in the all-ion-fragmentation (AIF) mode, and all had to be present with signal to noise ratios of 3:1. The authors identified group-indicating fragments, such as the *m/z* 121.0653 indicating NBOMes, *m/z* 188.1439 indicating fentanyl derivatives, or *m/z* 223.1235 indicating LSD derivatives.

Harbdova et al. [4] employed gas chromatography-electron impact-mass spectrometry (GC-EI-MS). The MS method worked in SIM mode, monitoring three *m/z* per compound. The authors also employed this methodology as a screening method for other phenethylamine derivatives by extraction of characteristic fragment ions. The following ions were used with that purpose: *m/z* 181, 190, 244, 290, 391, 405, 419, 437, and 451.

3 Tryptamines in Blood and Urine

Table 3 summarizes the main characteristics of selected analytical methods for the determination of tryptamines in plasma, serum, and urine. All these methodologies showed high sensitivity and specificity and were fully validated, and some were applied to authentic samples. The most comprehensive methodology was published by Meyer et al. [28], which included 37 tryptamines. Caspar et al. [18], which method was described in the previous section (Table 2), also analyzed LSD derivatives, LSZ, AL-LAD, and 1P-LSD.

Table 3
Analytical methods for the determination of tryptamines in plasma, serum, and urine

Analytes	Matrix	Sample Volume (mL)	Linearity (ng/mL)	Extraction	Instrument	Separation	MS Mode	References
5-MeO-DMT, bufotenine	Rat serum	0.02	LOQ (0.90–2.52)-ULOQ (5510-5890)	PP	LC-MSMS (QqQ)	Phenyl-hexyl column, (3 μm, 50 × 4.6 mm)	MRM	[29]
5-MeO-DiPT, 5-OH-DiPT, 5-MeO-iPT	Rat urine	0.3	10–1000	PP	LC-MSMS (QqQ)	BDS HYPERSIL C18 column (5 μm, 50 × 2.0 mm)	MRM	[30]
5-MeO-DiPT	Urine	1	2–300	LLE, ethyl acetate	GC-MS (Orbitrap)	TG-5SILMS column (30 m × 0.25 mm, 0.25 μm)	Scan (50–500 *m/z*)	[31]
37 tryptamines, 5 β-carbolines, ibogaine, yohimbine[a]	Plasma/urine	0.5 (plasma) 0.1 (urine)	1–100	LLE, butyl acetate/ethyl acetate, 50:50 (plasma); PP (urine)	LC-MS^n (linear ion trap)	TF Hypersil GOLD C18 column (1.9 5 μm, 100 × 2.1 mm)	MS^1 scan (100–800 *m/z*), MS^2 and MS^3 IDA; full MS^2 product ion scan	[28]
1P LSD, LSD	Serum/urine	0.5	0.015–0.4	LLE, 1-chlorobutane	LC-MSMS (QTrap)	Biphenyl column (2.6 5 μm, 100 × 2.1 mm)	MRM	[9]

5-MeO-DiPT 5-methoxy-N,N-diisopropyltryptamine, *5-MeO-DMT* 5-methoxy-N,N-dimethyltryptamine, *5-OH-DiPT* 5-hydroxy-N,N-diisopropyltryptamine, *5-MeO-iPT* 5-methoxy-N,N-isopropyltryptamine, *LSD* lysergic acid diethylamide, *1P-LSD* 1-propionyl-lysergic acid diethylamide

LLE liquid-liquid extraction, *GC-MS* gas chromatography-mass spectrometry, *IDA* information-dependent acquisition, *LC-MSMS* liquid chromatography-tandem mass spectrometry, *MRM* multiple reaction monitoring, *PP* protein precipitation, *Q* quadrupole, *QqQ* triple quadrupole, *QTrap* quadrupole-ion trap.

[a]Analytes included in Meyer et al. method: *MiPT* N-methyl-N-isopropyltryptamine, *NiPT* N-isopropyltryptamine, *DALT* N,N-diallyltryptamine, *DPT* N,N-dipropyltryptamine, *2-Ph-DMT* 2-phenyl-N,N-dimethyltryptamine, *2-Ph-DALT* 2-phenyl-N,N-diallyltryptamine, *2-Ph-DiPT* 2-phenyl-N,N-diisopropyltryptamine, *5-BnO-DMT* 5-benzyloxy-N,N-dimethyltryptamine, *5-BnO-DALT* 5-benzyloxy-N,N-diallyltryptamine, *5-BnO-DPT* dipropyltryptamine, *5-BnO-DiPT* diisopropyltryptamine, *5,6-MD-DALT* 5,6-methylenedioxy-N,N-diallyltryptamine, *5-Me-DALT* 5-methyl-N,N-diallyltryptamine, *7-Me-DALT* 7-methyl-N,N-diallyltryptamine, *7-Et-DALT* 7-ethyl-N,N-diallyltryptamine, *5-MeO-DALT* 5-methoxy-N,N-diallyltryptamine, *5-MeO-DPT* 5-methoxy-N,N-dipropyltryptamine, *5-EtO-DALT* 5-ethoxy-N,N-diallyltryptamine, *5-MeO-2-Me-DMT* 5-methoxy-2-methyl-N,N-dimethyltryptamine, *5-MeO-2-Me-DPT* 5-methoxy-2-methyl-N,N-dipropyltryptamine, *5-MeO-2-Me-DiPT* 5- methoxy-2-methyl-N,N-diisopropyltryptamine, *5-MeO-2-Me-DALT* 5-methoxy-2-methyl-N,N-diallyltryptamine, *5-MeO-2-Me-EPT* 5-methoxy-2-methyl-N-ethyl-N-propyltryptamine, *5-MeO-2-Me-EiPT* 5-methoxy-2-methyl-N-ethyl-N-isopropyltryptamine, *5-MeO-2-Me-ALCHT* 5-methoxy-2-methyl-N-allyl-N-cyclohexyltryptamine, *5-MeO-2-Me-MiPT* 5-methoxy-2-methyl-N-methyl-N-isopropyltryptamine, *5-MeO-2-Me-2-MALET* 5-methoxy-2-methyl-N-(2-methylallyl)-N-ethyltryptamine, *5-MeO-2-Me-Pip-T* 5-methoxy-2-methyl-N,N-pentamethylenetryptamine, *5-MeO-2-Me-Pyr-T* 5-methoxy-2-methyl-N,N-tetramethylenetryptamine, *4-HO-MiPT* 4-hydroxy-N-methyl-N-isopropyltryptamine, *4-HO-DiPT* 4-hydroxy-N,N-diisopropyltryptamine, *DMT* N,N-dimethyltryptamine, *NMT* N-methyltryptamine, *4-HO-MET* 4-hydroxy-N-methyl-N-ethyltryptamine, *psilocin* 4-hydroxy-N,N-dimethyltryptamine, *bufotenin* 5-hydroxy-N,N-dimethyltryptamine, and *4-AcO-DiPT* 4-acetoxy-N,N-diisopropyltryptamine. B-carolines: harmalol, harmaline, harmol, harmine, and tetrahydroharmine. Yohimbine and ibogaine

3.1 Target Analytes

The recommended target analyte for tryptamines in most of the publications in Table 3 is the parent compound; however, metabolites have been identified in urine samples. 5-MeO-DiPT is mainly metabolized to 5-hydroxy-N,N-diisopropyltryptamine (5-OH-DiPT) and 5-methoxy-N-diisopropyltryptamine (5-MeO-iPT) in humans [31, 32]. Although hydroxy-tryptamines and hydroxy-metabolites have been reported and, therefore, conjugated metabolites are expected, none of the cited publications in Table 3 performed a hydrolysis procedure in the sample preparation. Pichini et al. [21] observed an increase in the synthetic tryptamine 4-hydroxy-N,N-diisopropyltryptamine (4-HO-DiPT) concentrations after hydrolysis (*Helix pomatia*, 16 h, 37 °C). In that study, no changes were observed for 4-acetoxy-N,N-diisopropyltryptamine (4-acetoxy-DiPT) after hydrolysis of authentic samples. In the case of the novel derivative of LSD, 1P-LSD, Grumann et al. [9] could not detect 1P-LSD neither in urine nor in serum samples of a drug user, which support earlier suggestion that 1P-LSD might serve as a precursor of LSD in vivo [33].

3.2 Doses and Concentrations

Due to the low dosage of this type of substances, the concentrations in biological samples are expected to be in the pg/mL-low ng/mL range. Common doses of LSD and some of its derivatives are 50–200 μg [34], while 5-MeO-DiPT appears to be orally active at a 6–12 mg level [35]. Plasma levels reported after ingestion of tryptamine compounds were between 6.2 and 115 ng/mL [36]. Yan et al. [31] analyzed urine samples from 5-MeO-DiPT users, and the concentrations ranged from 1 to 2.8 ng/mL.

3.3 Stability

Stability in fortified samples was investigated for 5-MeO-DiPT and metabolites. Jin et al. [30] reported that 5-MeO-DiTP, 5-OH-DiPT, and 5-MeO-iPT were stable in rat urine after 6 h at room temperature, 14 days at −70 °C, and after three freeze-thaw cycles. However, Yan et al. [31] observed that the concentration of 5-MeO-DiPT in urine decreased by 22.8–38.2% after samples were stored for 24 h at 25 °C and 5 and 7 days at 4 °C. As a general recommendation, samples should be stored at −20 °C until analysis. Regarding 1P-LSD, Grumann et al. [9] demonstrated that serum and urine samples with 1P-LSD could be stored at room temperature, 5 °C, or −20 °C for 3–5 days either with or without light protection and could undergo at least two freeze-thaw cycles without major degradation of the analyte. However, hydrolysis of 1P-LSD to LSD was observed to some extent (up to 21% and 4% in serum and urine, respectively) after 3–5 days when the samples were stored at room temperature or 5 °C, regardless of protection from light. The authors recommended urine and serum samples should therefore be stored at −20 °C, but it appeared that no precautionary measures need to be undertaken to protect samples from light. Also, they observed that the addition of NaF to serum samples prevented enzymatic hydrolysis to LSD and was therefore recommended.

3.4 Internal Standards

In most of the publications in Table 3, the internal standards employed were the deuterated analogs of some of the compounds included in the method. Grumann et al. [9] used LSD-d_3, Jin et al. [30] used 5-MeO-DiPT-d_4, 5-OH- DiPT-d_4, and 5-MeO-iPT-d_4, and Meyer et al. [28] used three deuterated tryptamines 5-ethoxy-N-allyl-N-cyclohexyltryptamine (5-EtO-ALCHT-d_4), N, N-diisopropyltryptamine (DiPT-d_4), and 5-ethoxy-N,N-diallyl-tryptamine (5-EtO-DALT-d_4) and citalopram-d_6, trimipramin-d_3, and zolpidem-d_6. Shen et al. [29] employed 5-Methyl-N,N-dimethyltrypamine and Yan et al. [31] skf-525.

3.5 Extraction Procedures

Regarding sample extraction or cleanup procedures, three authors employed protein precipitation (PP), and three authors employed liquid-liquid extraction (LLE). PP was employed in plasma and in urine samples. Shen et al. [29] analyzed 5-MeO-DMT and bufotenine in 20 μL of plasma by adding 60 μL of acetonitrile. After centrifugation, the sample was directly injected into the LC-MSMS, avoiding the evaporation step. Jin et al. [30] and Meyer et al. (2014) [28] performed protein precipitation in urine samples (300 μL urine and 600 μL acetonitrile or 100 μL urine and 500 μL acetonitrile, respectively). In both methodologies, the supernatant was evaporated to dryness and reconstituted before injection in the instrument.

In the LLE procedures, the water phase was conditioned at basic pH, from pH 7.4 to >12, and the organic solvents employed were 1-chlorobutane, butyl acetate/ethyl acetate (50:50, v/v), or ethyl acetate. Grumman et al. [9] analyzed 1P-LSD in 0.5 mL of sample. They added 1 mL borate buffer pH 9 and 2 mL 1-chlorobutane. Meyer et al. [28] extracted 37 tryptamines from 0.5 mL plasma samples after adjusting the pH to 7.4 with a 10 mM phosphate buffer. The organic solvent employed was 600 μL of butyl acetate/ethyl acetate (50:50, v/v). Yan et al. [31] employed 1 mL urine, and the pH was adjusted to greater than 12 with 10% NaOH solution, and the LLE was performed with 3 mL ethyl acetate.

3.6 Chromatographic Separation

Most of the procedures were developed in LC and only one in GC. The GC procedure [31] did not perform any derivatization, and the column employed was a Thermo Fisher Scientific TG-5SILMS 30 m × 0.25 mm I.D., 0.25 μm column. The chromatographic separations in the LC procedures employed reversed-phase columns. Grumann et al. [9] used a Phenomenex biphenyl column (100 × 2.1 mm, 2.6 μm). Mobile phase A consisted of deionized water with 1% acetonitrile, 0.1% formic acid, and 2 mM ammonium formate, and mobile phase B consisted of 0.1% formic acid and 2 mM ammonium formate in acetonitrile. Total run time was 22 min. Shen et al. [29] used a Phenomenex phenyl-hexyl column (50 × 4.6 mm, 3 μm). The mobile phase included 0.1%

formic acid in water (A) and 0.1% formic acid in methanol (B). The run time was 9 min. Jin et al. [30] performed the chromatographic separation on a BDS Hypersil C18 column (50 × 2.1 mm, 5 μm; Thermo Fisher Scientific Inc.). The mobile phases consisted of 0.1% formic acid (A) and methanol (B). Total run time was 6 min. For the most comprehensive method developed by Meyer et al. [28], the column employed was a TF Hypersil GOLD C18 column (100 × 2.1 mm, 1.9 μm). The mobile phases were 10 mM aqueous ammonium formate plus 0.1% formic acid pH 3.4 (A) and acetonitrile plus 0.1% formic acid (B). The run time was 25 min.

3.7 Mass Spectrometry

In the methods developed in LC-MSMS (QTrap or QqQ), the data was acquired in ESI in positive mode and in MRM mode. Grumann et al. [9] monitored two transitions per compound, which is the recommended approach; however, Shen et al. [29] and Jin et al. [30] monitored just one transition per compound. Meyer et al. [28] performed a more complex MS analysis using a linear ion trap (LIT). The data was also acquired in ESI in positive mode. In the urine method, data were acquired in full scan (MS^1 stage), and collision-induced dissociation-MS^n experiments (MS^2 and MS^3) were performed on precursor ions selected from MS^1 using information-dependent acquisition (IDA). The analytes were identified based on full mass spectra recorded in the stages of MS^2 and MS^3 and matching with the reference spectra of the library. In plasma, instead of an IDA, full MS^2 product ion spectra (PIS) of predefined protonated molecules of all target analytes and internal standards were recorded, and a specific fragment ion was used as quantifier. Detection in plasma was based on the recorded full PIS spectra of previously defined target compounds (protonated molecules) and subsequent library match based on the same settings used for urine. The urine method was a screening method, and the plasma method was a quantification method.

Yan et al. [31] employed a GC instrument with electron impact (EI) ion source and a high-resolution mass spectrometer (Orbitrap). The data was acquired in full scan in positive polarity, and three *m/z* were monitored per compound.

4 Piperazines in Blood and Urine

Table 4 summarizes the main characteristics of selected analytical methods for the determination of piperazines in whole blood, plasma, and urine. All these methodologies showed high sensitivity and specificity and were fully validated, and some were applied to authentic samples.

Table 4
Analytical methods for the determination of piperazines in whole blood, plasma, and urine

Analytes	Matrix	Sample Volume (mL)	Linearity (ng/mL)	Extraction	Instrument	Separation	MS Mode	References
BZP, FBZP, MBZP, MeOPP, pFPP, mCPP, TFMPP, DCPP	Whole blood	0.1	20–2000	SPE Mixed mode	LC-MSMS (QTrap)	Kinetex F5 (2.6 μm, 3 × 150 mm)	MRM	[37]
BZP, TFMPP, *p*-OH-BZP, *m*-OH-BZP, *p*-OH TFMPP	Urine	1	5–40 (scan mode), 0.2–1 (SIM)	SPE Reversed phase	LC-MS (Q)	SCX column (2.0 mm × 150 mm)	Scan (80–300 *m/z*), SIM	[38]
MeOPP, 4-HO-PP, 2-MeOPP, 2-HO-PP	Urine	2.5	LOD (50–100)	LLE, DCM-IPA-ethyl acetate(1: 1:3)	GC-MS (Q)	HP-1 capillary (12 m × 0.2 mm i.d.)	Scan 50–550 m/z	[39]
BZP, TFMPP	Urine	0.02–0.1	100-ULOQ (5000–10,000)	LLE, n-butyl chloride	LC-MS (Q)	Synergi TM polar-RP C18 (4 μm, 150 × 2.0 mm)	SIM	[40]
BZP, TFMPP, mCPP	Urine	2	25–2000	LLE, chlorobutane	GC-MS (Q)	J&W DB-5MS (20 m × 0.18 mm, 0.18 μm)	SIM	[41]
BZP, TFMPP, mCPP, MeOPP	Urine	0.1	100–5000	MEPS, cation exchange sorbent	HPLC-DAD	Zorbax 300 SB-C18 (5 μm, 4.6 × 150 mm)	Selected wavelength monitored	[42]
BZP, MDPP, DBZP, TFMPP, mCPP, pFPP, mMPP, oCPP, pCPP, DFPP	Urine	1	10–1500	DLLME, n-hexane	LC-MSMS (QqQ)	DB-5MS capillary (30 m × 0.32 mm, 0.25 mm)	MRM	[10]
BZP, TFMPP, 3-OH BZP, 4-OH BZP, 4-OH TFMPP	Plasma/urine	0.1 (plasma) 0.2 (urine)	1–50	PP (plasma); filtration (urine)	LC-MS (Q)	Zorbax C18 HPLC (5 μm, 4.6 × 150 mm)	SIM	[43]

BZP 1-benzylpiperazine, *DBZP* dibenzylpiperazine, *DCPP* 2,3-dichlorophenylpiperazine, *DFPP* 4,4′-difluorobenzhydrylpiperazine, *FBZP* 1-(4-fluorobenzyl)-piperazine, *MBZP* 1-(4-methylbenzyl)-piperazine, *mCPP* 1-(3-chlorophenyl)-piperazine, *MDPP,* 3′,4′-methylenedioxy-α-pyrrolidinopropiophenone, *MeOPP* 1-(4-methoxyphenyl)piperazine, *mMPP* 1-(3-methylphenyl)piperazine, *oCPP* 1-(2-chlorophenyl)piperazine, *pCPP* 1-(4-chlorophenyl)piperazine, *pFPP* para-fluorophenylpiperazine, *TFMPP* trifluoromethylphenylpiperazine

DLLME dispersive liquid-liquid microextraction, *LLE* liquid-liquid extraction, *GC-MS* gas chromatography-mass spectrometry, *HPLC-DAD* high performance liquid chromatography-diode array detector, *LC-MS* liquid chromatography-mass spectrometry, *LC-MSMS* liquid chromatography-tandem mass spectrometry, *MEPS* microextraction by packed sorbent, *MRM* multiple reaction monitoring, *PP* protein precipitation, *Q* quadrupole, *QqQ* triple quadrupole, *QTrap* quadrupole-ion trap, *SIM* single ion monitoring, *SPE* solid phase extraction

4.1 Target Analytes

As reflected in Table 4, most of the methodologies were developed in urine, and the preferred target analytes were the precursor drugs. However, the metabolites of BZP, TFMPP, and MeOPP were also analyzed in several publications [38, 39, 43]. The hydrolysis of the urine samples to increase the detectability of the hydroxylated metabolites of BZP and TFMPP was performed enzymatically for 12 h at 37 °C [38, 43]. Staack et al. [39] performed a cost-effective acid hydrolysis for MeOPP hydroxylated metabolites.

Although the addition of these metabolites in the analysis is recommended to extend the window of detection and add valuable information in the interpretation of a case, it is recommended to always include the precursor drug as well to avoid misinterpretations [39]. Certain piperazine-related medicines, such as the cough suppressant dropropizine and the antipsychotic oxypertine, show the same metabolite as MeOPP, the 1-(4-hydroxyphenyl)piperazine (4-OH-PP). The determination of the precursor drug clarifies the origin of the common metabolite. It should also be considered that the piperazine mCPP is a piperazine drug as well as a metabolite of the antidepressants, trazodone and nefazodone [44]. The inclusion of these antidepressants in the methodology could avoid mCPP misinterpretations.

4.2 Doses and Concentrations

Piperazines are consumed in higher doses than phenethylamines and tryptamines, and, therefore, higher concentrations are expected in the biological samples. The average doses can range from 75 to 150 mg for BZP and 25 to 100 mg for TFMPP, which are like MDMA doses [40]. BZP and TFMPP are the most encountered active constituents of what is called "party pills" or "herbal highs." BZP and TFMPP could be present at different ratios in those pills and receive different names, like "Charge" (50 mg BZP and 200 mg TFMPP) or "Bliss" (100 mg BZP and 50 mg TFMPP) [43]. The reason why BZP and TFMPP are often consumed together is because they show a synergistic effect causing a substantial increase in dopamine and serotonin levels [45].

4.3 Stability

Short-term stability studies of piperazines in fortified urine samples showed no significant loss of this type of compound. Antia et al. [43] investigated the stability of BZP, TFMPP, and hydroxy-metabolites in fortified samples at room temperature for 24 h, freeze-thaw, and −20 °C for 3 months. Similarly, Moreno et al. [42] found no losses for MeOPP, BZP, mCPP, and TFMPP in fortified urine 24 h at room temperature. Lau et al. [37] investigated the stability of eight piperazines in whole blood. In general, benzylpiperazines were more stable than phenylpiperazines under all storage conditions, in which MBZP had more than 70% remaining after 12 months at −20 °C. MeOPP was not detected under room and refrigerated temperatures after 6 months and was the least stable. Considering this information, biological samples for the analysis of piperazines should be stored at −20 °C until analysis.

4.4 Internal Standards

As internal standards, most of the publications employed a piperazine derivative, such as 1-(2-chlorophenyl)-piperazine [42], 1-(4-chlorophenyl)piperazine [38], 1-(3-chlorophenyl)-piperazine [40], or 1-(2-methoxyphenyl)piperazine [10]. Dickson et al. [41] and Lau et al. [37] used the deuterated analogs of several piperazines included in the method (BZP-d_7, mCPP-d_8, TFMPP-d_4). Although Antia et al. [43] did not use any internal standard and performed external calibration, it should be kept in mind that the external calibration approach is not recommended in clinical and forensic toxicology.

4.5 Extraction Procedures

Regarding the extraction procedure in plasma, Antia et al. [43] performed a simple protein precipitation, where 100 μL of plasma was deproteinized by adding $ZnSO_4$ (20 μL, 35%) and methanol (100 μL). After vortexing and centrifugation, the supernatant was directly analyzed into the instrument.

In whole blood, Lau et al. [37] performed a mixed mode cation exchange SPE. Approximately 100 μL of whole blood was conditioned at pH 6 and loaded in the cartridges. After washing with water, 0.1 N HCl, and methanol, the elution was performed by adding 20% 2-propanol, 3% of concentrated ammonium hydroxide, and 77% of methylene chloride. After evaporation and reconstitution, the samples were injected into the LC-MSMS.

In urine samples, the authors employed a simple filtration [43], LLE [39–41], SPE [38], and microextraction procedures [10, 42]. In all LLE, the urine sample pH was adjusted to basic conditions, with KOH [40, 41], or ammonium sulfate and NaOH [39] and extracted with the organic solvent, such as chlorobutane [41], dichloromethane-isopropanol-ethyl acetate (1:1:3, v/v/v) [39], or n-butyl chloride [40].

Tsutsumi et al. [38] employed reversed-phase SPE. The hydrolyzed urine sample (pH 5) was loaded in the SPE cartridges. After washing with 0.1% HCl, water, and methanol, the elution was performed with methanol. In this study, they compared that procedure to cation exchange SPE and LLE with chloroform-2-propanol (3:1, v/v). SPE procedures showed better recoveries than LLE, especially for the hydroxy-metabolites. The recoveries ranged from 81% to 109% for reversed phase SPE, from 20% to 103% for mixed mode, and from 6% to 77% for LLE.

Piperazines have also been extracted from urine by microextraction procedures, MEPS (microextraction by packed sorbent), and UA-LDS-DLLME (ultrasound-assisted low-density solvent dispersive liquid-liquid microextraction). In the MEPS procedure [42], the packed sorbent employed was cation exchange material; 0.1 mL urine was diluted with 0.1 mL deionized water and vortexed. The sample was manually drawn through the sorbent and ejected eight times. The sorbent was washed with 250 μL 1% acetic acid and 100 μL 10% methanol in water. The analytes were eluted with 50 μL 5% ammonia in methanol and injected into the instrument (no evaporation). The sorbent was reused after washing with methanol and water (five and four times 250 μL, respectively). The sorbent could be reused for 100 extractions. In the UA-LDS-DLLME procedure [10], 1 mL urine sample pH was adjusted to 12 by 0.2 M NaOH, and then 100 μL n-hexane was added as extraction solvent. After sonication in an ultrasound bath for 3 min and centrifugation, the supernatant was directly analyzed by GC-MSMS.

To avoid any loss of the analytes in the evaporation step before reconstitution, several authors acidified the extracts with methanolic HCl [40, 41]; however, other authors did not do this step [37, 38].

4.6 Derivatization

In the GC methods, different derivatization procedures were used. Dickson et al. [41] performed the derivatization of BZP and TFMPP with pentafluoropropionic anhydride (PFPA). Staack et al. [39] derivatized MeOPP and metabolites by acetylation with acetic anhydride-pyridine mixture (3:2, v/v), and Zhu et al. [10] analyzed ten piperazine derivatives by GC, but they did not perform any derivatization.

4.7 Chromatographic Separation

All GC procedures employed similar nonpolar chromatographic columns for the analysis of piperazines, although with different dimensions (Table 5). In the LC procedures, most of the chromatographic separations were performed in reversed-phase mode with the columns indicated in Table 5. As mobile phases, Antia et al. [43] employed ammonium formate buffer (pH 4.5, 0.01 M, eluent A) and acetonitrile (eluent B), Moreno et al. [42] 5 mM ammonium formate (pH 6.4) and methanol (55:45, v/v), Vorce et al. [40] 0.1% formic acid in water (A) and acetonitrile (B), and Lau et al. [37] 2 mM ammonium formate buffer with 0.2% formic acid (A) and methanol with 0.1% formic acid (B). All the methods were in gradient mode, except Moreno et al. [42] in isocratic mode and Tsutsumi et al. [38], who employed a different type of chromatographic separation, using a cation exchange LC column, a semi-micro SCX column (2.0 mm I.D. × 150 mm; Shiseido, Tokyo, Japan) with a mobile phase consisting of 40 mM ammonium acetate buffer (pH 4) and acetonitrile (25:75, v/v).

4.8 Mass Spectrometry and Diode Array Detector

As seen in Table 4, piperazine analysis is commonly performed by MS, but Moreno et al. [42] employed HPLC-DAD. In their method, each piperazine was monitored at a specific wavelength. Among the methods highlighted in Table 4, three employed LC-MS single Q [38, 40, 43]. Antia et al. [43] and Tsutsumi et al. [38] monitored just one *m/z* per compound (the protonated molecule), but Vorce et al. [40] were able to perform in source collision-induced fragmentation, and three *m/z* per compound

Table 5
Analytical methods for the determination of multiple NPS, including phenethylamines, piperazines, and tryptamines in blood, serum, and urine

Analytes	Matrix	Sample Volume (mL)	Linearity (ng/mL)	Extraction	Instrument	Separation	MS Mode	References
Synthetic cathinones, phenethylamines, tryptamines, piperazines (31 NPS)	Serum	1	1–10	SPE Mixed mode	LC-MSMS (QqQ)	Synergi Polar RP column (4 μm, 150 × 2 mm)	MRM	[46]
Phenethylamines, piperazines, tryptamines (32 NPS)	Serum	1	10–250	SPE Mixed mode	LC-MSMS (QqQ)	Zorbax Rapid Resolution HD Eclipse Plus C18 LC column (1.8 μm, 50 × 2.1 mm)	MRM	[47]
Phenethylamines, synthetic cathinones, aminoindanes, PCP-type substances, tryptamines, piperazines (74 NPS)	Serum	0.15	5–250	SPE Mixed mode	LC-MSMS (QqQ)	Kinetex Biphenyl Core-Shell (2.6 μm, 100 × 2.1 mm)	MRM	[48]
Synthetic cathinones, phenylethylamines, piperazines (37 NPS)	Blood	0.1	50–2000	PP (acetonitrile)	LC-QTOF	Waters ACQUITY UPLC HSS C18 (1.8 μm, 150 × 2.1 mm)	Full scan (50–1000 *m/z*)	[49]
Synthetic cathinones, synthetic cannabinoids, phenethylamines, piperazines, ketamine and analogues, benzofurans, tryptamines (78 NPS)	Blood	0.5	5–100 (cannabinoids), 5–1000 (stimulants)	DLLME	LC-MSMS (QqQ)	Kinetex C18 column (2.6 μm, 100 × 2.1 mm)	MRM	[50]
Cathinones, phenethylamines, tryptamines, piperazines, piperidines, synthetic cannabinoids, arylalkylamines, arylcyclohexylamines, aminoindanes (143 NPS)	Blood	0.2	LOD 0.01–3.09	PP (acetonitrile)	LC-MSMS (QqQ)	Perlan Zorbax SB-C18 (1.8 μm, 2.1 × 50 mm)	MRM	[51]

28 cannabinoids, 19 cathinones, 5 phenethylamines, 3 indanes, 2 piperazines, 2 tryptamines, 2 phencyclidine, methoxetamine, ketamine (64 NPS)	Blood	0.2	1–100	PP (acetonitrile)	LC-MSMS (QqQ)	Zorbax Eclipse Plus C18 (1.8 μm, 2.1 × 50 mm)	MRM	[52]
MDMA, 2C-D, 2C-B, 2C-B-Fly, 2CT-2, 2C-I, 2C-E, m-CPP, 4-OH-DIPT, and 4-acetoxy-DIPT (10 NPS)	Urine	1	LOQ (20–60)-2000	SPE Mixed mode	LC-MS (Q)	Electron-Hypersil Gold ultrapure silica column (5 μm, 150 × 4.6 mm)	SIM	[21]
Cathinones, piperazines (8 NPS)	Urine	1	LOQ (6.5–96.0)-1000	Dilution	LC-MSMS (QqQ)	Acquity UPLC BEH C18 (1.7 μm, 50 × 2.1 mm)	MRM	[53]
Methylenedioxymethamphetamines, piperazines (16 NPS)	Urine	3	LOQ (0.5–9)-1000	SPE Reversed phase	LC-MSMS (QqQ)	Gemini C18 column (3 μm, 100 × 2 mm)	MRM	[54]
Cathinones, phenethylamines, piperazines, tryptamines (26 NPS)	Urine	0.05	100–1000	Dilution	LC-MSMS (QqQ)	Ethylene Bridged Hybrid (BEH) C18 (1.7 μm, 100 × 2.1 mm)	SIM	[55]
8 piperazines, 4 designer Amphetamines and 28 synthetic cathinones and 4 metabolites (40 NPS)	Urine	0.1	LOQ (2.5–5)-500	SPE Mixed mode	LC-Q Exactive MS	Accucore C18 (2.6 μm, 100 × 2.1 mm)	Full scan (100–400 *m/z*)	[56]
Amphetamine derivatives, 2C compounds, aminoindanes, cathinones, piperazines, tryptamines, dissociatives (56 NPS)	Blood/urine	0.25	LOQ (1–10)-1000	SPE Mixed mode (blood)/LLE with ethyl acetate (urine)	LC-MSMS (QqQ)	Synergi Polar-RP column (2.5 μm, 100 × mm)	sMRM	[57]
Synthetic cathinones, phenethylamines, tryptamines (50 NPS)	Blood/urine	1	LOD (0.3–425)	LLE, 70:30 n-butyl chloride/ethyl acetate	LC-MSMS (QqQ)	Poroshell 120 EC-C18 columns (2.7 μm, 3 × 50 mm)	MRM	[58]

(continued)

Table 5 (continued)

Analytes	Matrix	Sample Volume (mL)	Linearity (ng/mL)	Extraction	Instrument	Separation	MS Mode	References
Synthetic cathinones, phenethylamines (23 NPS)	Blood/urine	1	0.5–50 (NBOMe compounds), 50–2000 (all others)	SPE Mixed mode	GC-MS (Q)	J&W DB-5 ms Low-bleed column (30 m × 0.32 mm, 0.25 μm)	SIM	[59]

DLLME dispersive liquid-liquid microextraction, *LLE* liquid-liquid extraction, *LOD* limit of detection, *LOQ* limit of quantification, *GC-MS* gas chromatography-mass spectrometry, *LC-MS* liquid chromatography-mass spectrometry, *LC-MSMS* liquid chromatography-tandem mass spectrometry, *MRM* multiple reaction monitoring, *PP* protein precipitation, *Q* quadrupole, *SIM* single ion monitoring, *SPE* solid-phase extraction

were monitored. Lau et al. [37] employed LC-MSMS monitoring two MRM transitions per compound. Dickson et al. [41] and Staack et al. [39] employed GC-EI-MS and monitored three *m/z* per compound. Zhu et al. [10] employed GC-EI-MSMS, and three MRM transitions were monitored per analyte.

5 Multianalyte Methods in Blood and Urine

With the goal of analyzing several groups of NPS in just one analysis, several laboratories have developed multianalyte methods in blood and urine. These methods include several phenethylamines, tryptamines, and piperazines, besides other NPS such as synthetic cathinones. In order to cover in this chapter all the analytical procedures available for the analysis of the phenethylamines, tryptamines, and piperazines, Table 5 summarizes multianalyte NPS screening and confirmation procedures in blood, serum, and urine that include part of these analytes. Dickson et al. [41], besides the confirmation of piperazines described in Table 4, also screened for tryptamines and phenethylamines and other drugs (32 analytes). The screening method is similar to the one described in the previous section.

References

1. Poklis JL, Charles J, Wolf CE, Poklis AL (2013) High-performance liquid chromatography tandem mass spectrometry method for the determination of 2CC-NBOMe and 25I-NBOMe in human serum. Biomed Chromatogr 27:1794–1800. https://doi.org/10.1002/bmc.2999
2. Acuña-Castillo C, Villalobos C, Moya PR, Sáez P, Cassels BK, Huidobro-Toro JP (2002) Differences in potency and efficacy of a series of phenylisopropylamine/phenylethylamine pairs at 5-HT(2A) and 5-HT(2C) receptors. Br J Pharmacol 136:510–519. https://doi.org/10.1038/sj.bjp.0704747
3. Shulgin A (1992) PIHKAL: a chemical love story. Nation Publishing Group Co. Ltd
4. Habrdova V, Peters FT, Theobald DS, Maurer HH (2005) Screening for and validated quantification of phenethylamine-type designer drugs and mescaline in human blood plasma by gas chromatography/mass spectrometry. J Mass Spectrom 40:785–795. https://doi.org/10.1002/jms.853
5. Braden MR, Parrish JC, Naylor JC, Nichols DE (2006) Molecular interaction of serotonin 5-HT2A receptor residues Phe339(6.51) and Phe340(6.52) with superpotent n-benzyl phenethylamine agonists. Mol Pharmacol 70: 1956–1964. https://doi.org/10.1124/mol.106.028720
6. Kerrigan S, Mott A, Jatzlau B, Ortiz F, Perrella L, Martin S, Bryand K (2014) Designer psychostimulants in urine by liquid chromatography-tandem mass spectrometry. J Forensic Sci 59:175–183. https://doi.org/10.1111/1556-4029.12306
7. Synthetic Drug Control Act of 2011. http://www.justice.gov/ola/views-letters/112/093011-ltr-re-hr1254-synthetic-drug-control-act-2011.pdf
8. 21 CFR 1308.11(d) https://www.ecfr.gov/cgi-bin/text-idx?SID=b632b274cf6322a0450af69d7c7a4f46&node=pt21.9.1308&rgn=div5#se21.9.1308_111
9. Grumann C, Henkel K, Stratford A, Hermanns-Clausen M, Passie T, Brandt SD, Auwärter V (2019) Validation of an LC-MS/MS method for the quantitative analysis of 1P-LSD and its tentative metabolite LSD in fortified urine and serum samples including stability tests for 1P-LSD under different storage conditions. J Pharm Biomed Anal 174: 270–276. https://doi.org/10.1016/j.jpba.2019.05.062
10. Zhu B, Meng L, Cao J, Yang W, Conlan XA (2021) Simultaneous determination of 10 new

psychoactive piperazine derivatives in urine using ultrasound-assisted low-density solvent dispersive liquid-liquid microextraction combined with gas chromatography-tandem mass spectrometry. J Forensic Sci 66:748–757. https://doi.org/10.1111/1556-4029.14624

11. Dean B, Stellpflug S, Burnett A, Engebretsen K (2013) 2C or not 2C: phenethylamine designer drug review. J Med Toxicol 9: 172–178. https://doi.org/10.1007/s13181-013-0295-x

12. Moreira AM dos S, Oliveira HL de, Filho JFA, Florez DHÂ, Borges MMC, Lacerda V, Romão W, Borges KB (2019) NBOMe compounds: an overview about analytical methodologies aiming their determination in biological matrices. TrAC Trends Anal Chem Regul Ed 114:260–277. https://doi.org/10.1016/j.trac.2019.02.034

13. Araújo AM, Carvalho F, Bastos MDL, Pinho PGD, Carvalho M (2015) The hallucinogenic world of tryptamines: an updated review. Arch Toxicol 89:1151–1173. https://doi.org/10.1007/s00204-015-1513-x

14. Malaca S, Faro AFL, Tamborra A, Pichini S, Busardò FP, Huestis MA (2020) Toxicology and analysis of psychoactive tryptamines. Int J Mol Sci 21:9279. https://doi.org/10.3390/ijms21239279

15. Monteiro M, de Bastos M, de Pinho PG, Carvalho M (2013) Update on 1-benzylpiperazine (BZP) party pills. Arch Toxicol 87:929–947. https://doi.org/10.1007/s00204-013-1057-x

16. da Silva DD, Silva M, Moreira P, Martins M, Valente M, Carvalho F, de Bastos M, Carmo H (2017) In vitro hepatotoxicity of 'Legal X': the combination of 1-benzylpiperazine (BZP) and 1-(m-trifluoromethylphenyl)piperazine (TFMPP) triggers oxidative stress, mitochondrial impairment and apoptosis. Arch Toxicol 91:1413–1430. https://doi.org/10.1007/s00204-016-1777-9

17. Boatto G, Nieddu M, Dessì G, Manconi P, Cerri R (2007) Determination of four thiophenethylamine designer drugs (2C-T-series) in human plasma by capillary electrophoresis with mass spectrometry detection. J Chromatogr A 1159:198–202. https://doi.org/10.1016/j.chroma.2006.10.051

18. Caspar AT, Kollas AB, Maurer HH, Meyer MR (2018) Development of a quantitative approach in blood plasma for low-dosed hallucinogens and opioids using LC-high resolution mass spectrometry. Talanta 176:635–645. https://doi.org/10.1016/j.talanta.2017.08.063

19. Poklis JL, Clay DJ, Poklis A (2014) High-performance liquid chromatography with tandem mass spectrometry for the determination of nine hallucinogenic 25-NBOMe designer drugs in urine specimens. J Anal Toxicol 38: 113–121. https://doi.org/10.1093/jat/bku005

20. Johnson RD, Botch-Jones SR, Flowers T, Lewis CA (2014) An evaluation of 25B-, 25C-, 25D-, 25H-, 25I- and 25T2-NBOMe via LC–MS-MS: method validation and analyte stability. J Anal Toxicol 38:479–484. https://doi.org/10.1093/jat/bku085

21. Pichini S, Pujadas M, Marchei E, Pellegrini M, Fiz J, Pacifici R, Zuccaro P, Farré M, Torre R de la (2008) Liquid chromatography–atmospheric pressure ionization electrospray mass spectrometry determination of "hallucinogenic designer drugs" in urine of consumers. J Pharm Biomed Anal 47:335–342. https://doi.org/10.1016/j.jpba.2007.12.039

22. P 3rd Jacob, Shulgin AT (1994) Structure-activity relationships of the classic hallucinogens and their analogs. NIDA Res Monogr 146:74–91

23. Poklis JL, Nanco CR, Troendle MM, Wolf CE, Poklis A (2014) Determination of 4-bromo-2,5-dimethoxy-N-[(2-methoxyphenyl)methyl]-benzeneethanamine (25B-NBOMe) in serum and urine by high performance liquid chromatography with tandem mass spectrometry in a case of severe intoxication. Drug Test Anal 6:764–769. https://doi.org/10.1002/dta.1522

24. Sargent T, Shulgin AT, Naranjo C (1969) Structure-activity relationships of one-ring psychotomimetics. Nat Lond 221:537–541. https://doi.org/10.1038/221537a0

25. Vorce SP, Sklerov JH (2004) A general screening and confirmation approach to the analysis of designer tryptamines and phenethylamines in blood and urine using GC-EI-MS and HPLC-Electrospray-MS. J Anal Toxicol 28: 407–410. https://doi.org/10.1093/jat/28.6.407

26. da Cunha K, Eberlin M, Huestis M, Costa J (2019) NBOMe instability in whole blood. Forensic Toxicol 37:82–89. https://doi.org/10.1007/s11419-018-0438-5

27. Curtis B, Kemp P, Harty L, Choi C, Christensen D (2003) Postmortem identification and quantitation of 2,5-dimethoxy-4-n-propylthiophenethylamine using GC-MSD and GC-NPD. J Anal Toxicol 27:493–498. https://doi.org/10.1093/jat/27.7.493

28. Meyer M, Meyer M, Caspar A, Caspar A, Brandt S, Brandt S, Maurer H, Maurer H (2014) A qualitative/quantitative approach

for the detection of 37 tryptamine-derived designer drugs, 5 β-carbolines, ibogaine, and yohimbine in human urine and plasma using standard urine screening and multi-analyte approaches. Anal Bioanal Chem 406: 225–237. https://doi.org/10.1007/s00216-013-7425-9

29. Shen H-W, Jiang X-L, Yu A-M (2009) Development of a LC-MS/MS method to analyze 5-methoxy-N,N-dimethyltryptamine and bufotenine, and application to pharmacokinetic study. Bioanalysis 1:87–95. https://doi.org/10.4155/bio.09.7
30. Jin MJ, Jin C, Kim JY, In MK, Kwon OS, Yoo HH (2011) A quantitative method for simultaneous determination of 5-methoxy-N,N-diisopropyltryptamine and its metabolites in urine using liquid chromatography–electrospray ionization–tandem mass spectrometry. J Forensic Sci 56:1044–1048. https://doi.org/10.1111/j.1556-4029.2011.01753.x
31. Yan X, Xiang P, Zhao Y, Yu Z, Yan H (2020) Determination of 5-MeO-DIPT in human urine using gas chromatography coupled with high-resolution Orbitrap mass spectrometry. J Anal Toxicol 44:461–469. https://doi.org/10.1093/jat/bkaa005
32. Kamata T, Katagi M, Kamata HT, Miki A, Shima N, Zaitsu K, Nishikawa M, Tanaka E, Honda K, Tsuchihashi H (2006) Metabolism of the psychotomimetic tryptamine derivative 5-methoxy-n,n-diisopropyltryptamine in humans: identification and quantification of its urinary metabolites. Drug Metab Dispos 34:281–287. https://doi.org/10.1124/dmd.105.005835
33. Brandt SD, Kavanagh PV, Westphal F, Stratford A, Elliott SP, Hoang K, Wallach J, Halberstadt AL (2016) Return of the lysergamides. Part I: analytical and behavioural characterization of 1-propionyl-d-lysergic acid diethylamide (1P-LSD). Drug Test Anal 8: 891–902. https://doi.org/10.1002/dta.1884
34. Nichols DE (2004) Hallucinogens. Pharmacol Ther 101:131–181. https://doi.org/10.1016/j.pharmthera.2003.11.002
35. Brandt SD, Freeman S, Fleet IA, Mcgagh P, Alder JF (2004) Analytical chemistry of synthetic routes to psychoactive tryptamines. Part I. Characterisation of the Speeter and Anthony synthetic route to 5-methoxy-N, N-diisopropyltryptamine using ESI-MS-MS and ESI-TOF-MS. Analyst 129:1047–1057. https://doi.org/10.1039/b407239c
36. Callaway JC, Raymon LP, Hearn WL, McKenna DJ, Grob CS, Brito GS, Mash DC (1996) Quantitation of N, N-Dimethyltryptamine and Harmala Alkaloids in Human Plasma after Oral Dosing with Ayahuasca. J Anal Toxicol 20:492–497. https://doi.org/10.1093/jat/20.6.492
37. Lau T, LeBlanc R, Botch-Jones S (2018) Stability of synthetic piperazines in human whole blood. J Anal Toxicol 42:88–98. https://doi.org/10.1093/jat/bkx090
38. Tsutsumi H, Katagi M, Miki A, Shima N, Kamata T, Nishikawa M, Nakajima K, Tsuchihashi H (2005) Development of simultaneous gas chromatography–mass spectrometric and liquid chromatography–electrospray ionization mass spectrometric determination method for the new designer drugs, N-benzylpiperazine (BZP), 1-(3-trifluoromethylphenyl)piperazine (TFMPP) and their main metabolites in urine. J Chromatogr B Analyt Technol Biomed Life Sci 819:315–322. https://doi.org/10.1016/j.jchromb.2005.02.016
39. Staack RF, Maurer HH (2003) Toxicological detection of the new designer drug 1-(4-methoxyphenyl)piperazine and its metabolites in urine and differentiation from an intake of structurally related medicaments using gas chromatography–mass spectrometry. J Chromatogr B Analyt Technol Biomed Life Sci 798:333–342. https://doi.org/10.1016/j.jchromb.2003.10.004
40. Vorce SP, Holler JM, Levine B, Past MR (2008) Detection of 1-benzylpiperazine and 1-(3-trifluoromethylphenyl)-piperazine in urine analysis specimens using GC-MS and LC-ESI-MS. J Anal Toxicol 32:444–450. https://doi.org/10.1093/jat/32.6.444
41. Dickson AJ, Vorce SP, Holler JM, Lyons TP (2010) Detection of 1-benzylpiperazine, 1-(3-trifluoromethylphenyl)-piperazine, and 1-(3-chlorophenyl)-piperazine in 3,4-methylenedioxymethamphetamine-positive urine samples. J Anal Toxicol 34:464–469. https://doi.org/10.1093/jat/34.8.464
42. Moreno IED, da Fonseca BM, Barroso M, Costa S, Queiroz JA, Gallardo E (2012) Determination of piperazine-type stimulants in human urine by means of microextraction in packed sorbent and high performance liquid chromatography-diode array detection. J Pharm Biomed Anal 61:93–99. https://doi.org/10.1016/j.jpba.2011.12.004
43. Antia U, Tingle MD, Russell BR (2010) Validation of an LC-MS method for the detection and quantification of BZP and TFMPP and their hydroxylated metabolites in human plasma and its application to the pharmacokinetic study of TFMPP in humans. J Forensic Sci 55:1311–1318. https://doi.org/10.1111/j.1556-4029.2010.01457.x

44. Baselt RC (2004) Disposition of toxic drugs and chemicals in man, Seventh edition. Biomedical Publications, Foster City
45. Baumann MH, Clark RD, Budzynski AG, Partilla JS, Blough BE, Rothman RB (2004) Effects of "Legal X" piperazine analogs on dopamine and serotonin release in rat brain. Ann N Y Acad Sci 1025:189–197. https://doi.org/10.1196/annals.1316.024
46. Wohlfarth A, Weinmann W, Dresen S (2010) LC-MS/MS screening method for designer amphetamines, tryptamines, and piperazines in serum. Anal Bioanal Chem 396: 2403–2414. https://doi.org/10.1007/s00216-009-3394-4
47. Swortwood M, Swortwood M, Boland D, Boland D, DeCaprio A, DeCaprio A (2013) Determination of 32 cathinone derivatives and other designer drugs in serum by comprehensive LC-QQQ-MS/MS analysis. Anal Bioanal Chem 405:1383–1397. https://doi.org/10.1007/s00216-012-6548-8
48. Lehmann S, Kieliba T, Beike J, Thevis M, Mercer-Chalmers-Bender K (2017) Determination of 74 new psychoactive substances in serum using automated in-line solid-phase extraction-liquid chromatography-tandem mass spectrometry. J Chromatogr B Analyt Technol Biomed Life Sci 1064:124–138. https://doi.org/10.1016/j.jchromb.2017.09.003
49. Pasin D, Bidny S, Fu S (2015) Analysis of new designer drugs in post-mortem blood using high-resolution mass spectrometry. J Anal Toxicol 39:163–171. https://doi.org/10.1093/jat/bku144
50. Odoardi S, Fisichella M, Romolo FS, Strano-Rossi S (2015) High-throughput screening for new psychoactive substances (NPS) in whole blood by DLLME extraction and UHPLC–MS/MS analysis. J Chromatogr B Analyt Technol Biomed Life Sci 1000:57–68. https://doi.org/10.1016/j.jchromb.2015.07.007
51. Adamowicz P, Tokarczyk B (2016) Simple and rapid screening procedure for 143 new psychoactive substances by liquid chromatography-tandem mass spectrometry. Drug Test Anal 8: 652–667. https://doi.org/10.1002/dta.1815
52. Vaiano F, Busardò FP, Palumbo D, Kyriakou C, Fioravanti A, Catalani V, Mari F, Bertol E (2016) A novel screening method for 64 new psychoactive substances and 5 amphetamines in blood by LC–MS/MS and application to real cases. J Pharm Biomed Anal 129: 441–449. https://doi.org/10.1016/j.jpba.2016.07.009
53. Bell C, George C, Kicman AT, Traynor A (2011) Development of a rapid LC-MS/MS method for direct urinalysis of designer drugs. Drug Test Anal 3:496–504. https://doi.org/10.1002/dta.306
54. Montesano C, Sergi M, Moro M, Napoletano S, Romolo FS, Carlo MD, Compagnone D, Curini R (2013) Screening of methylenedioxyamphetamine- and piperazine-derived designer drugs in urine by LC-MS/MS using neutral loss and precursor ion scan. J Mass Spectrom 48:49–59. https://doi.org/10.1002/jms.3115
55. Al-Saffar Y, Stephanson NN, Beck O (2013) Multicomponent LC–MS/MS screening method for detection of new psychoactive drugs, legal highs, in urine—experience from the Swedish population. J Chromatogr B Analyt Technol Biomed Life Sci 930:112–120. https://doi.org/10.1016/j.jchromb.2013.04.043
56. Concheiro M, Castaneto M, Kronstrand R, Huestis MA (2015) Simultaneous determination of 40 novel psychoactive stimulants in urine by liquid chromatography–high resolution mass spectrometry and library matching. J Chromatogr A 1397:32–42. https://doi.org/10.1016/j.chroma.2015.04.002
57. Ambach L, Redondo AH, König S, Angerer V, Schürch S, Weinmann W (2015) Detection and quantification of 56 new psychoactive substances in whole blood and urine by LC-MS/MS. Bioanalysis 7:1119–1136. https://doi.org/10.4155/bio.15.48
58. Fagiola M, Hahn T, Avella J (2018) Screening of novel psychoactive substances in postmortem matrices by liquid chromatography–tandem mass spectrometry (LC–MS-MS). J Anal Toxicol 42:562–569. https://doi.org/10.1093/jat/bky050
59. Nisbet LA, Wylie FM, Logan BK, Scott KS (2019) Gas chromatography-mass spectrometry method for the quantitative identification of 23 new psychoactive substances in blood and urine. J Anal Toxicol 43:346–352. https://doi.org/10.1093/jat/bky109

Chapter 4

Target Analysis of Synthetic Opioids in Blood and Urine

Luke N. Rodda

Abstract

The most toxic class of NPS are those derivatives that act on the opioid receptors known as novel synthetic opioids (NSO). The pharmacology of different types of NSO classes is typically either equipotent or up to hundreds and even thousands of times more potent than morphine. Following the rapid increase in NSO usage, clinical and forensic postmortem toxicology laboratories have seen a significant increase in their detection in survived and fatal overdoses occurring predominantly from 2015 onwards. Although opioids were traditionally morphinans, NSO with morphine-like properties consist of a structurally diverse chemical class. In this chapter the testing regimes required to detect this NPS class are discussed including the sensitivities that must be achieved, selectivity aspects of structurally similar NSO compounds, typical targets for their detection, and protocols used. Sample preparation techniques and developed instrumental analyses are also reviewed, with LC-MS/MS more recently being favored.

Key words Fentalog, Potency, Mu-opioid receptor, Utopioids, Blood, Urine, LC-MS/MS

1 Introduction

Novel psychoactive opioids (NSO) consist of highly potent fentanyl analogs and other non-fentanyl synthetic opioids to produce opioid-like effects by acting on opioid receptors [1, 2]. The pharmacology of different types of NSO classes is typically either equipotent or up to hundreds and even thousands of times more potent than morphine [3, 4]. NSO are typically detected in counterfeit pills, powder, liquids and nasal sprays, and illicit drug products such as white powder and black-tar heroin [5, 6].

One of the first reported cases of NSO use was in 1979 following several opioid overdoses in California from the use of "China White", thought to be heroin. However, toxicological analysis determined no heroin or other common opioids in biological specimens, but alpha-methylfentanyl (AMF) [7–9]. Five years later 3-methylfentanyl was a causative agent for 16 fatal overdose cases in Pennsylvania [10]. Internationally, Sweden reported in the later 6 months of 2015 nine acetylfentanyl, three

Marta Concheiro and Karl B. Scheidweiler (eds.), *Methods for Novel Psychoactive Substance Analysis*,
Methods in Pharmacology and Toxicology, https://doi.org/10.1007/978-1-0716-2605-4_4,

4-methoxybutyrfentanyl, one furanylfentanyl, and one combined 4-methoxybutyrfentanyl and furanylfentanyl, overdoses [6]. Although NSO-related cases are predominantly postmortem accidental overdose in nature, NSO has been shown to be prevalent in human performance casework also. From 2016 to 2017 in Palm Beach County, Florida, a third of all driving under the influence of drug (DUID) casework detected carfentanil, more than any other drug including fentanyl and other impairing drugs such as alprazolam [11]. For these reasons, alongside postmortem casework, laboratories should perform screening of NSO (and other NPS) in human performance casework.

Following a rapid increase of new NPS reports, the United Nations Office on Drugs and Crime (UNODC) established an Early Warning Advisory in 2013 to monitor NPS trends [12]. An NPS portal database was subsequently developed to enable reporting from laboratories globally into a centralized system. Although this database revealed that the growth rate of NPS in general has flattened or reduced in the recent years of 2016–2020, new NPS opioids (i.e., NSO) substances continued to be increasingly detected with an almost fourfold increase from 2016 to 2020 in forensic casework including fatal and nonfatal drug overdoses, and driving under the influence of drugs [13]. This is consistent with known offenders of fentanyl analog trafficking commencing at 2016 and continuing to increase sharply, even after fentanyl-related substances scheduling measures were enacted [14]. The UNODC and the 2021 European Monitoring Centre for Drugs and Drug Addiction (EMCDDA) report on new drugs listed over 80 and 67 NSO, respectively, between 2009 and 2020 [15, 16].

2 Types of Novel Psychoactive Opioids (Target and Derivatives)

Although opioids were traditionally morphinans, NSO with morphine-like properties consist of classes such as phenethylpiperidines (e.g., isobutyrfentanyl), phenylpiperidines (e.g., desmethylprodine/MPPP), cyclohexylbenzamides (e.g., U-47700, AH-7921), diphenethylpiperzines (e.g., MT-45), and, more recently, cinnamylpiperazines (e.g., bucinnazine/AP-237, 2-methyl AP-237), thiambutenes (e.g., piperidylthiambutene), and nitrobenzimidazoles (e.g., isotonitazene). The generic structures of fentanyl analogs aka "fentalogs" and U-opioid drugs aka "utopioids" are provided in Fig. 1.

Briefly described below are the pharmacological properties, prevalence, and analytical detection details for some more common NSO substances, including fentalogs (acetylfentanyl, carfentanil, ocfentanil, furanylfentanyl, butyrylfentanyl) and non-fentanyl analogs (U-47700 and MT-45).

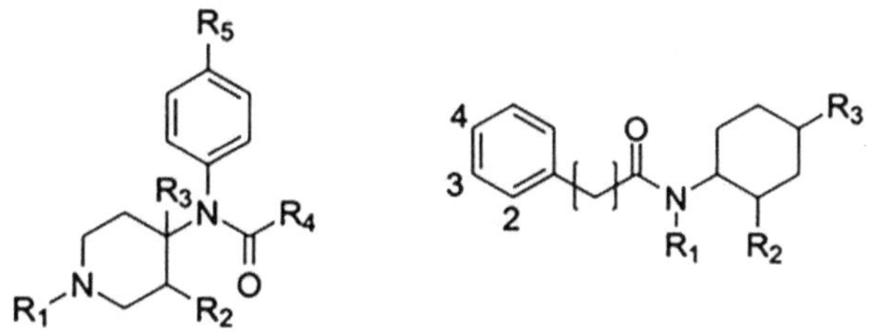

Fig. 1 Examples of some generic structures of NSO: fentanyl analogs aka "fentalogs" (left); U-opioid drugs aka "utopioids" originally manufactured by the Upjohn Company (right) with a methylene (CH2) spacer

Acetylfentanyl is structurally very similar to fentanyl and is also known as N-(1-phenylethylpiperidin-4-yl)-N-phenylacetamide, with a 5–15 times more potency than heroin [17]. In 2013 this NSO caused 10 fatalities and 41 deaths between 2015 and 2016 in North Eastern USA, and has since continually been involved in overdoses around the country and throughout the world including 9 hospitalized cases in Sweden where one subsequently died during 7 months in 2015 and in the same year an insufflation case in Japan [6, 18–24]. Thirteen acetylfentanyl-positive cases were recorded in Miami-Dade County, Florida, between 2014 and 2017 [25]. Acetylfentanyl was detected at concentrations 235 ng/mL, 2400 ng/g, 131 ng/mL, and 234 ng/mL in subclavian blood, liver, vitreous fluid, and urine, respectively [26]. The relatively high concentrations were likely due to intravenous administration. In these documented case series, acetylfentanyl was predominantly detected with fentanyl after its illicit production, or with heroin, and its prevalence enhanced due to its ease of production. Due to the limited screening targeting acetylfentanyl, it is likely that the number of actual fatalities contributed by this NSO might be underappreciated [5]. However, due to its structural similarity, its detection via immunoassay techniques that target fentanyl may be improved compared to other NSO. Chromatographical-based techniques include the extraction of blood via liquid-liquid acid/base, and GC-MS analysis has been described [20]. In addition, a solid-phase extraction (SPE) extraction with LC-MS/MS for detecting acetylfentanyl and its metabolite acetyl norfentanyl was described in urine [27]. In LC-MS techniques, poor fragmentation may cause product ion transitions that are considerably low in abundance [28]. Other analytical approaches involved 24 synthetized fentanyl-related compounds including acetylfentanyl, p-fluorofentanyl, p-chlorofentanyl, p-tolylfentanyl, alpha-methylacetylfentanyl, alpha-methylfentanyl, N-methylfentanyl, N-benzylfentanyl, and 3-methylfentanyl, and other unnamed NSO synthetized compounds were analyzed by thin-layer chromatography (TLC), direct-inlet electron-impact mass spectrometry (DIEIMS), gas chromatography (GC), infrared spectrometry, and GC-MS in a 1999 study [29].

Carfentanil (methyl 1-(2-phenylethyl)-4-(N-propanylanilino) piperidine-4-carboxylate) is a highly potent μ opioid agonist with a said potency of 10,000 times greater than morphine and 100 times greater than fentanyl [4]. Case reports of carfentanil deaths have been commonly reported in the USA and internationally [30, 31]. In Miami-Dade County, Florida, between 2014 and 2017, approximately 200 carfentanil-positive cases were recorded [25]. Due to the high potency, particularly sensitive limits of detections are required for effective detection of carfentanil toxicity and impact in casework. In a terrorist attack in Moscow, carfentanil and remifentanil were detected in two British survivors [32].

Ocfentanil (N-(2-fluorophenyl)-2-methoxy-N-[1-(2-phenylethyl)-piperidin-4-yl]acetamide) is also highly potent, approximately twice to that of fentanyl [33]. A comprehensive review of both NSO is detailed in a 2017 study by Misailidi et al. [34]. In 2017, ocfentanil has been described in a case of the death of young man from Switzerland [35].

A critical review on furanylfentanyl was prepared by the WHO in 2017 in response to a growing number of cases in previous years [36]. Over a 3-year period, approximately 40 postmortem cases detected furanylfentanyl [25]. In a case study involving a furanylfentanyl overdose where the subject believed he was consuming oxycodone, 1.9 and 4.3 ng/mL of furanylfentanyl and 4-ANPP were detected in blood, respectively, with oxycodone or metabolites not detected [37]. Another case report involves furanylfentanyl that also had combined carfentanil toxicity [31].

In vitro studies suggest butyrylfentanyl (or butyrfentanyl) is 31-fold more potent than morphine [3]. Two postmortem cases involving butyrylfentanyl occurred in Virginia, with death solely due to the NSO, and the other was a mixed toxicity with acetylfentanyl, alprazolam, and ethanol [38]. In the fatal overdose caused only by butyrylfentanyl, concentrations of 99, 220, 32, 590, 260, and 64 ng/mL in peripheral blood, heart blood, vitreous humor, gastric contents, bile, and urine, respectively, were observed. Combined toxicity due to butyrylfentanyl, acetylfentanyl, and cocaine was observed in a man from San Diego, USA, who had a history of intravenous drug abuse and was found unresponsive on the bathroom floor of his home with drug paraphernalia in the near vicinity [39]. Analytically, ELISA screened positive for fentanyl; however, GC-MS confirmed the two NSO. Three butyrfentanyl-positive cases were recorded in Miami-Dade County, Florida, between 2014 and 2017 [25]. In four nonfatal intoxications involving butyrfentanyl and 4-fluorobutyrfentanyl (*para*-fluorobutyrfentanyl) from Sweden between May 2014 to January 2015, typical opioid symptoms such as unconsciousness, respiratory depression, and apnea were observed [40].

Non-fentanyl NSO opioid-receptor agonists, such as the N-substituted benzamides and acetamides, were developed by the Upjohn Company starting in the 1970s and are typically known as "U-drugs" [1]. In vitro studies suggest U-47700 is 12-fold more potent than morphine [3]. The non-fentanyl NSO U-47700 and its analogs' impact in the recreational drug community were described by Baumann et al. [41]. In a case report from Central California, USA, three perceived Norco tablets contained U-47700 and were consumed by a 41 female who subsequently experienced opioid-like toxicity however, survived after naloxone administration and discharged after 4 h [42]. Detection of this NSO has been observed via screening of urine via a LC-QTOF/MS where a positive detection of U-47700 and presumptive detections of N-desmethyl-U-47700 and N,N-didesmethyl-U-47700 were obtained, with subsequent quantitation of U-47700 using LC-MS/MS in MRM mode [43]. In a series of 19 death investigation cases assessed originally for opioid-like overdoses were positive for either U-47700 (n = 11), furanylfentanyl (n = 3), or both (n = identified in blood by LC-MS/MS MRM and SPE) [44].

The NSO MT-45 was detected via GC-MS in biological material (Papsun et al. 2016). In vitro studies suggest MT-45 is equipotent with morphine [3]. In 2020, the novel benzimidazole opioid etazene (N,N-diethyl-2-ethan-1-amine (dihydrochloride)) was identified in seized material with its crystal structure and spectroscopic characterization studies performed [45]. It was reported to be 70 times more potent than morphine [46]. In recent years, isotonitazene and brorphine have been reported across the USA, and it is suggested they mimic the potency of fentanyl [47]. Flunitazene, another recently observed NSO, is reported to be less potent than fentanyl [48].

3 Testing Regime

3.1 Sensitivity

An effective testing regime must first detect that an NSO has been consumed. Due to their high potency, many NSO substances are present in relatively low concentrations in biological specimens compared to other forensically relevant drugs (pg/mL or low ng/mL). In 355 blood specimens positive for carfentanil in forensic casework, 83.6% casework detected the highly potent fentanyl analog below 1 ng/mL [49]. When cases below the 0.10 ng/mL limit of quantitation were excluded, the mean and median concentrations were ~0.6 and ~0.4 ng/mL, respectively, with a range of <0.1–14 ng/mL. Another study of 160 carfentanil-positive decedents resulted in mean and median blood concentrations of ~0.5 and ~0.2 ng/mL, respectively, with a range of <0.1–9.2 ng/mL [50]. Subsequently, employing some of the most sensitive instrumentation and techniques is required when investigating NSO.

Further, detection of NSO use, either via the active drug or their specific and common metabolites as targets, must then also be achievable by methods used to screen casework if routine screening is desired. Although immunoassay and GC-MS may detect some NSO, challenges can be practically overcome by LC-MS/MS and LC-HRMS techniques.

3.2 Selectivity

Chromatographic separation is a major challenge for specific NSO detection. This is owed to the structural similarity of NSO isomers and analogs which often co-elute on commonly used and generic LC programs. Co-elution is very common with screening methods such as those employed as general unknown screening or multi-class techniques that also focus on other drugs of abuse. However, better separation to specifically identify isomers with positional non-hydrogen substitutes on aromatic hydrocarbons can be achieved with tailored LC-MS/MS programs. Further, specific identification with MS may also not be achieved, particularly nominal MS, with precursor and product ions identical masses and the same fragmentation patterns. High Resolution Mass Spectrometry (HRMS) may further elicit the specific NSO entity(s).

3.3 Scope and Targets for Analysis

Standards or recommendations for targeting NSO in forensic casework are minimal, namely, due to the rapid shifting market of NPS compounds in general. There is a set of recommended NPS drugs that is reestablished quarterly, of which approximately 25% of those listed are NSO in a 2021 report [51]. The scope is broken into three tiers that imply to either strongly recommend, recommend, or consider the listed NSO for your casework screening inclusion. Due to the changing environment, it is suggested that laboratories consider additional NPS into their scope based on local, national, and international trends. For example, although the 2021 Q2 program strongly recommends *N*-pyrrolidino etonitazene, *o/m/p*-fluorofentanyl, metonitazene, 2-methyl AP-237, and carfentanil as tier 1 NSO, promotion of protonitazene currently in tier 2 may occur in Q3 as its use increased. These challenges clearly demonstrate the requirement for laboratories to communicate findings of both new and existing NPS trends.

Additionally, it is similarly crucial to establish a collaboration between drug seizure and healthcare organizations [52]. An increase in drug seizure data leads one jurisdiction to investigate the carfentanil routinely with subsequently high detectability in forensic casework, including impaired drivers [11].

Both blood and urine analysis may readily detect active drug metabolites, with urine enabling better ability to identify metabolites, including common metabolites such 4-ANPP. Targets within the NSO group challenge classic analytical workflows owing to their swift rise and growth, chemical diversity, and lack of rapid available reference standards typically commercially available [12, 16].

3.4 Protocol

Detection of NSO necessitates specific testing with either forensic and clinical practitioners suspecting opioid toxicity via respiratory depression, pupil constriction, and other classic opioid symptoms or comprehensive analytical testing on all casework [53]. Relevant information from the scene investigation or observations at autopsy indicating an opioid-like toxicity event should trigger such testing, although, to avoid bias, this regime should be consistently applied with minimal variations based on case type where possible [54].

The best-established testing regime protocol will likely consist of a systematic combination of LC-HRMS and LC-MS/MS methods. The use of immunoassay has been shown to have limitations with a third of NSO-positive cases in one study returning a negative fentanyl targeted immunoassay and concluding that the generic screening technique has been suggested as an ineffective tool for the detection of NSO [52, 55].

If resources are limited, then NSO testing should at least be performed on likely drug-involved casework. Care must be taken to ensure that, in addition to NSO detection, a sufficiently wide screening scope including drugs of abuse, with an adequate quality assurance program, is the basis of effective forensic toxicology testing.

Quantitation presents challenges due to the required resources to perform complete validation for often rarely detected substances. This may be further exacerbated due to the lack of deuterated NSO reference materials for internal standard use to enable accurate quantitation. A solution for laboratories may exist in the form of standard addition workflows that utilize the case sample for calibration samples with the advantage mitigating matrix effects differences between blank sample typically used for calibrators and the case sample [56, 57]. Mochizuki et al. performed standard addition for acetylfentanyl in postmortem samples [58]. However, this technique does require considerably more sample use and thus low sample volume methodologies, or a large starting volume, is required.

3.5 Combined Substance Use

Polysubstance use of NSO and other recreational drugs has been readily reported. A large-scale testing of urine specimens from patients presenting to a sizeable healthcare system the study described the prevalence of NSO through preliminary screening of fentanyl and opiates, followed by identification and confirmation of NSO by HPLC-MS/MS [55]. In 65% of samples, at least one NSO was detected, and 26% of samples were positive for two or more NSO substances. In one study where carfentanil-positive forensic casework was also screened for other drugs, almost half of the cases were positive for another opioid(s) with morphine and/or fentanyl making up the majority but also the heroin-marker 6-monacetylmorpine being present in 10% of casework [49]. In the same study, cocaine was co-detected in approximately a third of the cases, demonstrating stimulant and opioid combined use. In another study of 160 carfentanil-positive deaths, over half showed co-cocaine use and just under half with fentanyl co-detected along with a range of other drugs of abuse detected [50]. In a case of furanylfentanyl overdose, polysubstance use was demonstrated with codeine, amphetamine, methamphetamine, delta 9-THC, and delta 9-carboxy-THC co-detection [37].

4 Sample Preparation Techniques

The extraction of NSO from biological matrices, namely, blood and urine, should consider recovery as a main focus due to the relatively low concentrations seen in casework. However, as is the nature of attempting to detect NPS compounds in general, the extraction should not be too specific to ensure general unknown screening methods can readily detect new NSO as they enter the recreational drug-using community, typically before a drug reference material is commercially available. In general, regular sample preparation techniques employed in forensic toxicology and analytical laboratories can be applied for the extraction of NSO. A summary of the employed extraction techniques in blood and urine is shown in Table 1.

4.1 Liquid-Liquid Extraction (LLE)

The use of liquid-liquid extraction (LLE) for NSO sample preparation has typically been the technique of choice and used with screening, confirmation, and quantitation purposes, across both GC-MS [21, 22, 26, 31, 38, 39, 59–65] and LC-MS/MS analyses [11, 23]. For GC-MS analysis, derivatization of samples may be required [62]. An alkaline LLE technique was employed for the extraction of carfentanil ahead of screening analysis via GC-MS [31]. Remifentanil was extracted from blood by n-butyl chloride following methanolic PP, back extracted into acid, and re-extracted into a methanolic n-butyl chloride solvent [59].

4.2 Solid-Phase Extraction (SPE)

Solid-phase extraction (SPE) offers a typically cleaner extract for introduction onto instrumentation; however, it is also more selective and thus risks exclusions of unknown and new NSO substances that vary in chemical composition. For targeted analyses such a confirmation testing, such extraction technique is useful, although for screening purposes such risks should be mitigated by other protocol means. U-47700, U-50488, and furanylfentanyl were identified in blood by DAU mixed-mode SPE columns (UCT, Bristol, PA) characterized by both hydrophobic and cation exchange [44]. Acetylfentanyl and its metabolite acetyl norfentanyl were assessed in urine using Strata-X-drug B polymeric strong cation exchange SPE cartridges (Phenomenex, Torrance, CA) with subsequent LC-MS/MS analysis in urine [27]. A mixed-

Table 1
Analytical techniques including sample preparation and instrumental analyses for the detection of novel psychoactive opioids (NSO) in biological specimens

Technique	Sample preparation	Column	Detection	#NSO	NSO	Matrix (mL)	LOD (ng/mL)	LOQ/Analytical range (ng/mL)	Described validation	References
GC-NPD	LLE	BPX-5 (25 m 0.22 mm i.d., 0.25 μm)	ND	1	Remifentanil	Blood (1)	ND	0.2	Linearity, precision, accuracy, recovery	[59]
GC-MS	SPE	DB-1MS (30 m × 0.25 mm i.d., 0.25 μm)	EI; full scan (40–550)	1	Acrylfentanyl	Blood	5	ND	ND	[68]
LC-MS/MS	PP	BEH C18 column (2.1 mm × 100 mm, 1.7 μm)	ESI (pos); MRM	1	Acrylfentanyl	Blood (0.25)	0.05	0.1	Linearity, LOD, LOQ, precision, accuracy, interferences, selectivity, carryover, matrix effects	[68]
GC-MS	LLE	HP-5 (30 m × 0.25 mm i.d., 0.25 μm)	EI; FS	1	Acetylfentanyl	Blood; liver; vitreous fluid; urine	ND	ND	ND	[26]
GC-MS	LLE	ND	EI; SIM	1	Acetylfentanyl	Blood	ND	ND	ND	[21]
LC-MS/MS	LLE	Poroshell 120 EC-C18 column (3 mm × 50 mm, 2.7 μm)	ESI (pos); MRM	1	Acetylfentanyl	Blood	2.5	2.5	LOD/LOQ, matrix effects, exogenous drug interferences, carryover, and ion suppression/ enhancement	[23]
GC-MS	LLE	DB5 (15 m × 0.25 mm i.d., 1 μm)	EI	1	Sufentanil	Blood; urine; liver; vitreous; kidney	ND	ND	ND	[60]
LC-MS/MS	SPE	Poroshell 120 EC-C18 (2.1 mm × 100 mm, 2.7 μm)	ESI (pos); MRM	1	U-47700	Urine	1	Quant (1–2500)	Specificity/interference, LOD, LOQ, linearity, extraction efficiency, accuracy, precision (inter- and intra-day), carryover, and stability	[43]
LC–QTOF/ MS	Dilution	Kinetex C18 (100 mm × 3.0 mm, 2.6 μm)	ESI (pos); IDA (100–1000 m/ z)	1	U-47700	Urine	ND	1	ND	[43]

(continued)

Table 1
(continued)

Technique	Sample preparation	Column	Detection	#NSO	NSO	Matrix (mL)	LOD (ng/mL)	LOQ/Analytical range (ng/mL)	Described validation	References
GC-MS	LLE	RTX-1-ms (15 m 0.25 mm i. d., 0.25 μm)	EI; SIM	1	Acetylfentanyl	Blood; urine; brain; liver; vitreous	0.0625	0.125	LOD, LOQ, matrix effects, interferences, linearity	[61]
Paper spray-MS	None	NA	MRM; FS; ddMS2	8	Acetylfentanyl; acrylfentanyl; carfentanil; norcarfentanil; fentanyl; norfentanyl; sufentanil; desmethyl U-47700	Urine	0.5 (2 pg)	Qual; quant (0.5–50)	Linearity, repeatability, bias, matrix effects, applicability	[70]
GC-MS	LLE	Ultra-inert Zebron (ZB-Drug-1, 15 m × 250 μm i.d., film thickness 0.25 μm)	EI; full-scan mode (40–550 m/z)	19	4-ANPP; acetylfentanyl; AH-7921; alfentanil; alpha-methylfentanyl; beta-hydroxyfentanyl; carfentanil; despropionyl Para-fluorofentanyl; fentanyl; fluorofentanyl; isotonitazene; MT-45; N-methyl norfentanyl; norfentanyl; ocfentanil; remifentanil; sufentanil; thienyl fentanyl; U-47700	Urine (0.5)	ND	ND	Applicability	[62]
LC-Orbitrap	LLE	Accucore phenyl hexyl (100 × 2.1 mm, 2.6 μm)	ESI; full-scan in ddMS2	See above	See above	See above	ND	ND	Applicability	[62]
GC-MS	LLE	Zebron capillary ZB-5MS (15 m, 0.25 mm i.d., 0.25 mm)	EI; SIM	1	Acetylfentanyl; acetyl norfentanyl	Blood; liver; vitreous humor; urine	50 (observed 1)	100	LOD, LOQ, matrix effects, interferences	[22]
GC-MS	LLE	ZB-5MS (15 m 0.25 mm i.d., 0.25 μm)	EI; SIM	2	Acetylfentanyl; butyrylfentanyl	Blood; urine; liver; vitreous; gastric	2	10	LOD, LOQ, matrix effects, interferences	[39]

GC-MS	SPE	DB-5MS (30 m × 0.25 mm i. d., 0.25)	EI; SIM	4	Acetylfentanyl; butyrylfentanyl; furanylfentanyl; ocfentanil	Blood	0.15–1	ND	Selectivity, specificity, LOD, LOQ, linearity, recovery, accuracy, precision, stability	[75]
LC-MS/MS	SPE	Zorbax eclipse plus C18 column (4.6 × 100 mm, 3.5 μm)	ESI (pos); MRM	3	U-47700; U-50488; furanylfentanyl	Blood (0.5)	0.5	Quant (1–100/ 500)	LOD, linearity, bias, accuracy, precision, stability	[44]
LC-MS/MS	PP; SPE	ND	ND	1[a]	Carfentanil	Blood	0.03	0.1–10	ND	[49]
LC-MS/MS	β-Glucuronidase prior to SPE	Phenyl-hexyl (50 mm × 4.6 mm, 2.6 μm, 100 Å)	ESI (pos); MRM EPI triggered by IDA	2	Acetylfentanyl; acetyl norfentanyl	Urine	0.5–1	~1–1.5	ND	[27]
LC-MS/MS	β-Glucuronidase prior to SPE	Raptor biphenyl (50 mm × 3.0 mm, 2.7 μm)	ESI (pos); MRM EPI triggered by IDA	2	Acetylfentanyl; acetyl norfentanyl	Urine	0.5–1	~1–1.5	ND	[27]
GC-MS	LLE	Rtx-5 (30 m × 0.25 mm i.d., 0.25 μm)	EI; full scan	2	Acetylfentanyl; butyrylfentanyl	Blood; urine; vitreous (2)	ND	ND	ND	[38]
LC-MS/MS	SPE	Biphenyl (100 × 2.5 mm, 5 μm)	ESI; MRM	3	Acetylfentanyl; acetyl norfentanyl; butyrylfentanyl	Blood	ND	1	Selectivity, specificity, linearity, LOD, LOQ, accuracy, precision, ion suppression/ enhancement, matrix effects; recovery, stability, dilution integrity	[38]
LC-ion trap	SPE	Acclaim RSLC 120 C18 column (100 mm × 2.1 mm × 2.2 μm)	ESI (pos); full-scan MS^n	27	Isobutyryl fentanyl; 4-methylphenethyl acetylfentanyl; (±)-trans-3-methyl fentanyl; Para-fluorofentanyl; despropionyl Para-fluorofentanyl; cyclopentyl fentanyl; ortho-fluorofentanyl; benzyl fentanyl; despropionyl ortho-fluorofentanyl; meta-fluorofentanyl; ocfentanil;	PM samples	> ~ 0.1–0.5	ND	Stability, selectivity/ specificity, matrix effect, carryover, applicability (for selected NSO analytes)	[25]

(continued)

Table 1
(continued)

Technique	Sample preparation	Column	Detection	#NSO	NSO	Matrix (mL)	LOD (ng/mL)	LOQ/Analytical range (ng/mL)	Described validation	References
					4-methoxy-butyryl fentanyl; alpha-methyl acetylfentanyl; valeryl fentanyl; 3-methylfentanyl; despropionyl fentanyl; acetylfentanyl; furanylfentanyl; Para-fluorobutyryl fentanyl; alfentanil; Para-fluoroisobutyryl fentanyl; beta-hydroxythio fentanyl; sufentanil; butyrylfentanyl; norfentanyl; carfentanil; U-47700; W-18					
LC-MS/MS	PP; SPE	ND	ESI (pos); MRM	1	Carfentanil	Blood	0.05	0.1	LOD, linearity, bias, precision	[50]
GC-MS	LLE	HP-Si (12.5 m × 0.25 mm i. d.)	EI	1	Sufentanil	Blood; CSF	ND	ND	ND	[63]
GC-MS	LLE	J&W DB5 fused Si (10 m × 0.18 mm, 0.18 mm)	EI	2	Alfentanil; sufentanil	Urine	2–5	5–10	LOD, LOQ, specificity, linearity, precision, accuracy, recovery, stability	[65]
GC-MS	LLE	Rtx-5 (30 m × 0.25 mm i.d., 0.25 μm)	EI; full and SIM	2	Carfentanil; furanylfentanyl	Blood; urine; vitreous	ND	ND	ND	[31]
GC-MS	SPE	DB-5MS (30 m 0.25 mm i.d., 0.25 μm)	EI	1	Acetylfentanyl	Blood; urine; gastric	ND	ND	ND	[24]

LC-MS/MS	LLE	Biphenyl (50 × 2.1 mm, 2.7 μm)	ESI (pos); MRM	7	Acetylfentanyl; beta-hydroxythiofentanyl; butyrylfentanyl; carfentanil; furanylfentanyl; MT-45; U-47700	Blood	0.02–0.2	ND	Selectivity/specificity matrix effect, sensitivity, precision, linearity, carryover, processed sample stability, ruggedness/robustness	[11]
GC-MS	SPE	DB- 5MS (J&W) (30 m × 0.25 mm i.d., 0.1 μm)	EI	3	Alfentanil; sufentanil; norsufentanil	Urine	0.0025–0.0075		LOD, specificity, accuracy, precision, linearity, stability, applicability	[74]
GC-MS	LLE	Bonded methyl/silicone (12.5 m × 0.25 mm)	EI	1	Sufentanil	Blood; urine; liver; vitreous; kidney	0.03–0.05	0.03–0.05	ND	[64]

CSF cerebrospinal fluid, *ddMS2* data-dependent MS2, *EI* electron impact ionization, *EPI* enhanced product ion, *ESI* electrospray ionization, *FS* full scan, *GC-MS* gas chromatography-mass spectrometry, *IDA* independent data acquisition, *LC-MS/MS* liquid chromatography tandem mass spectrometry, *LC-QTOF/MS* liquid chromatography-quadrupole-time of flight mass spectrometry, *LLE* liquid-liquid extraction, *LOD* limit of detection, *LOQ* limit of quantification, *MRM* multiple reaction monitoring, *ND* not described, *NPD* nitrogen-phosphorous detector, *PP* protein precipitation, *SIM* single ion monitoring, *SPE* solid-phase extraction

mode SPE technique was developed for the analysis of 27 fentanyl analog NSO targets [25]. Fourteen NSO (fentanyl analogs) were extracted from bovine blood via the more specific SPE technique; however, they were analyzed using the more nonselective data-independent HRMS acquisition [66].

4.3 Protein Precipitation (PP)

A protein precipitation (PP) method for the extraction of NSO was performed on 0.1 g of blood via acetonitrile agitation and centrifugation with particularly sensitive limits of detection (LOD) [67]. Acrylfentanyl was targeted in blood in a similar PP technique and quantified using LC-MS/MS [68].

5 Chromatographic Separation and Detection Techniques

Detection of NSO can be performed by instrumental techniques typically employed by analytical laboratories in forensic toxicological services. This includes immunoassay, GC-MS, LC-MS, and HRMS. The detection of NSO using HRMS techniques is described in more detail in Chap. 9. Boronet Ena et al. provide a comprehensive review on the use of ambient ionization mass spectrometry for NPS detection, including NSO [69]. Of note was a paper spray mass spectrometry (PS-MS) method developed to detect eight NSO in urine [70]. In this chapter, developed methods employed to detect NSO substances in blood and urine using techniques typically employed in forensic and clinical laboratories are described below and in Table 1.

5.1 Immunoassay

The most simplistic of all analytical techniques, immunoassay, has shown to be able to detect NSO in some cases. A comprehensive analysis of 19 commercially available immunoassay kits was performed using 30 fentanyl analogs chosen due to their relevance during 2015–2018 [71]. Specifically, nine lateral flow assays, seven heterogeneous immunoassays, and three homogenous immunoassays were assessed. The majority of the kits were designed to target fentanyl, which achieved detection of fentanyl in 100% of casework. Their ability to detect fentanyl analogs varied considerably. In order from poorest performers, the following averaged less than 20% cross-reactivity with fentanyl-targeted immunoassays: N-phenethyl-4-piperidinone, norcarfentanil, despropionyl-p-fluorofentanyl, 4-ANPP, acetyl norfentanyl, carfentanil, norfentanyl, 4-methoxy-butyryl fentanyl, 3-methoxyfentanyl, and 2-thiofuranylfentanyl. It can be suggested that the more structurally diverse the analog, the less chance of a successful detection. Carfentanil essentially only had cross-reactivity with specifically designed carfentanil kits. It was summarized that analogs with considerable group modifications to the piperidine, or bulky rings, or long alkyl chain modifications in the N-aryl or alkyl amide regions, were poorly detected compared to other types of modifications.

Conversely, a cross-reactivity study for a fentanyl-specific immunoassay via spiking studies showed acetylfentanyl had a cross-reactivity starting at 2.5 ng/mL, at an 80% cross-reactivity [72]. Another study showed a lack of NSO to sufficiently cross-react with a fentanyl ELISA screening assay. Subsequently, the postmortem toxicology laboratory commenced using an LC-MS/MS NSO method on all cases with a history of illicit drug use with 42 cases not detecting fentanyl, but an NSO, 17 of which were carfentanil [73].

Considering the variety of NSO available, their detection using immunoassay is vastly limited, requiring the need for better techniques to screen this novel drug class. An example of this was reported where ELISA testing in apparent accidental overdose cases revealed fentanyl positivity in blood; however, subsequent gas chromatography-mass spectrometry (GC-MS) confirmatory testing only detected acetylfentanyl [19].

5.2 Gas Chromatography-Mass Spectrometry (GC-MS)

Numerous NSO utilize GC-MS techniques [21, 22, 24, 26, 31, 32, 38, 39, 59–65, 68, 74, 75]. Acrylfentanyl was screened with GC-MS at a relatively high LOD in blood following SLE sample preparation [68]. A comprehensive review for the detection of NSO analytes by GC-MS techniques was assessed by Valdez [76]. In general, the relatively low LODs required for NSO detection may be a challenge for GC-MS use.

5.3 Liquid Chromatography with Tandem Mass Spectrometry (LC-MS/MS)

LC-MS/MS methodologies for NSO detection have been utilized in more recent years compared to GC-MS [11, 23, 25, 27, 38, 43, 44, 49, 50, 62, 68, 70, 77]. The importance of confirmation testing was observed where one study showed that over a third of hospital patient samples that screened positive for opiates but negative for fentanyl detected NSO following confirmatory LC-MS/MS testing [55]. A run time of 11 min was successfully in separating 27 fentanyl analog NSO targets using a C18 column and regular ammonium formation and acetonitrile mobile phases Schemes [25].

Acetylfentanyl and acetyl norfentanyl were quantitatively assessed in human urine by LC-MS/MS using two complimentary columns (phenyl-hexyl and biphenyl) for increased selectivity and conformational purposes [27]. Quantitation of acrylfentanyl was achieved using LC-MS//MS in blood following PP sample preparation [68]. A targeted LC-MS/MS method for the detection and quantitation of U-47700, U-50488, and furanylfentanyl in blood was developed on an LC-MS/MS operating in positive MRM with separation via a C18 column and a commonly used mobile phase program [44]. Recently, Garneau and authors described a validated comprehensive analytical workflow to systematic screen for 31 NSO, 22 NSO metabolites, and other NPS substances [78]. Of just over 5000 postmortem cases, 5% were positive for an NSO.

References

1. Sharma KK, Hales TG, Rao VJ, NicDaeid N, McKenzie C (2019) The search for the "next" euphoric non-fentanil novel synthetic opioids on the illicit drugs market: current status and horizon scanning. Forensic Toxicol 37(1): 1–16. https://doi.org/10.1007/s11419-018-0454-5
2. Varshneya NB, Hassanien SH, Holt MC, Stevens DL, Layle NK, Bassman JR, et al (2021) Respiratory depressant effects of fentanyl analogs. American Society for Pharmacology and Experimental Therapeutics Virtual Meeting
3. Baumann MH, Majumdar S, Le Rouzic V, Hunkele A, Uprety R, Huang XP et al (2018) Pharmacological characterization of novel synthetic opioids (NSO) found in the recreational drug marketplace. Neuropharmacology 134: 101–107. https://doi.org/10.1016/j.neuropharm.2017.08.016
4. Van Bever WF, Niemegeers CJ, Schellekens KH, Janssen PA (1976) N-4-Substituted 1-(2-arylethyl)-4-piperidinyl-N-phenylpropanamides, a novel series of extremely potent analgesics with unusually high safety margin. Arzneimittelforschung 26(8):1548–1551
5. Armenian P, Vo KT, Barr-Walker J, Lynch KL (2018) Fentanyl, fentanyl analogs and novel synthetic opioids: a comprehensive review. Neuropharmacology 134:121–132. https://doi.org/10.1016/j.neuropharm.2017.10.016
6. Helander A, Bäckberg M, Beck O (2016) Intoxications involving the fentanyl analogs acetylfentanyl, 4-methoxybutyrfentanyl and furanylfentanyl: results from the Swedish STRIDA project. Clin Toxicol 54(4): 324–332. https://doi.org/10.3109/15563650.2016.1139715
7. Kram TC, Cooper DA, Allen AC (1981) Behind the identification of China White. Anal Chem 53(12):1379a–1386a. https://doi.org/10.1021/ac00235a003
8. Henderson GL (1988) Designer drugs: past history and future prospects. J Forensic Sci 33(2):569–575
9. Henderson GL (1991) Fentanyl-related deaths: demographics, circumstances, and toxicology of 112 cases. J Forensic Sci 36(2): 422–433
10. Hibbs J, Perper J, Winek CL (1991) An outbreak of designer drug—related deaths in Pennsylvania. JAMA 265(8):1011–1013. https://doi.org/10.1001/jama.1991.03460080081037
11. Tiscione NB, Alford I (2018) Carfentanil in impaired driving cases and the importance of drug seizure data. J Anal Toxicol 42(7): 476–484. https://doi.org/10.1093/jat/bky026
12. Tettey J, Crean C (2015) New psychoactive substances: catalysing a shift in forensic science practice? Philos Trans R Soc B Biol Sci 370(1674):20140265. https://doi.org/10.1098/rstb.2014.0265
13. United Nations Office on Drugs and Crime (UNODC) (2020) Current NPS Threats, Volume III. Austria
14. Weedn VW, Elizabeth Zaney M, McCord B, Lurie I, Baker A (2021) Fentanyl-related substance scheduling as an effective drug control strategy. J Forensic Sci. https://doi.org/10.1111/1556-4029.14712
15. European Monitoring Centre for Drugs and Drug Addiction (EMCDDA) (2021) European drug report 2021: trends and developments. Luxembourg: Publications Office of the European Union
16. United Nations Office on Drugs and Crime (UNODC) (2020) The growing complexity of the opioid crisis. Global SMART Update, Vienna
17. Higashikawa Y, Suzuki S (2008) Studies on 1-(2-phenethyl)-4-(N-propionylanilino)piperidine (fentanyl) and its related compounds. VI. Structure-analgesic activity relationship for fentanyl, methyl-substituted fentanyls and other analogues. Forensic Toxicol 26(1):1–5. https://doi.org/10.1007/s11419-007-0039-1
18. Stogner JM (2014) The potential threat of acetyl fentanyl: legal issues, contaminated heroin, and acetyl fentanyl "disguised" as other opioids. Ann Emerg Med 64(6):637–639. https://doi.org/10.1016/j.annemergmed.2014.07.017
19. Ogilvie L, Stanley C, Lewis L, Boyd M, Lozier M (2013) Acetyl fentanyl overdose fatalities — Rhode Island, march–may 2013. Morbidity and Mortality Weekly Report (MMWR) 62(34):703–704
20. Lozier MJ, Boyd M, Stanley C, Ogilvie L, King E, Martin C et al (2015) Acetyl fentanyl, a novel fentanyl analog, causes 14 overdose deaths in Rhode Island, March–May 2013. J Med Toxicol 11(2):208–217. https://doi.org/10.1007/s13181-015-0477-9
21. Dwyer JB, Janssen J, Luckasevic TM, Williams KE (2018) Report of increasing overdose deaths that include acetyl fentanyl in multiple counties of the Southwestern region of the commonwealth of Pennsylvania in

2015–2016. J Forensic Sci 63(1):195–200. https://doi.org/10.1111/1556-4029.13517

22. McIntyre IM, Trochta A, Gary RD, Malamatos M, Lucas JR (2015) An acute acetyl fentanyl fatality: a case report with postmortem concentrations. J Anal Toxicol 39(6):490–494. https://doi.org/10.1093/jat/bkv043
23. Fagiola M, Hahn T, Avella J (2019) Evaluation of acetylfentanyl following suspected heroin overdose when complicated by the presence of toxic fentanyl and alprazolam concentrations. Academic Forensic Pathol 9(3–4):191–199. https://doi.org/10.1177/1925362119892005
24. Takase I, Koizumi T, Fujimoto I, Yanai A, Fujimiya T (2016) An autopsy case of acetyl fentanyl intoxication caused by insufflation of 'designer drugs'. Legal Med 21:38–44. https://doi.org/10.1016/j.legalmed.2016.05.006
25. Shoff E (2017) Qualitative identification of fentanyl analogs and other opioids in postmortem cases by UHPLC-Ion Trap MSn. SOFT-TIAFT Annual Joint Meeting 2017. Boca Raton, Florida, U.S.
26. Cunningham SM, Haikal NA, Kraner JC (2016) Fatal intoxication with acetyl fentanyl. J Forensic Sci 61(S1):S276–S280. https://doi.org/10.1111/1556-4029.12953
27. Patton AL, Seely KA, Pulla S, Rusch NJ, Moran CL, Fantegrossi WE et al (2014) Quantitative measurement of acetyl fentanyl and acetyl norfentanyl in human urine by LC-MS/MS. Anal Chem 86(3):1760–1766. https://doi.org/10.1021/ac4036197
28. Farley M, Tran H, Towler S, Gevorkyan J, Pearring S, Rodda LN (2021) A single method for 127 recommended and additional DUID drugs in blood and urine by LC-MS/MS. J Anal Toxicol. https://doi.org/10.1093/jat/bkab075
29. Ohta H, Suzuki S, Ogasawara K (1999) Studies on fentanyl and related compounds IV. Chromatographic and spectrometric discrimination of fentanyl and its derivatives. J Anal Toxicol 23(4):280–285. https://doi.org/10.1093/jat/23.4.280
30. Müller S, Nussbaumer S, Plitzko G, Ludwig R, Weinmann W, Krähenbühl S et al (2018) Recreational use of carfentanil – a case report with laboratory confirmation. Clin Toxicol 56(2):151–152. https://doi.org/10.1080/15563650.2017.1355464
31. Swanson DM, Hair LS, Strauch Rivers SR, Smyth BC, Brogan SC, Ventoso AD et al (2017) Fatalities involving carfentanil and furanyl fentanyl: two case reports. J Anal Toxicol 1–5. https://doi.org/10.1093/jat/bkx037
32. Riches JR, Read RW, Black RM, Cooper NJ, Timperley CM (2012) Analysis of clothing and urine from Moscow theatre siege casualties reveals carfentanil and remifentanil use. J Anal Toxicol 36(9):647–656. https://doi.org/10.1093/jat/bks078
33. Fletcher JE, Sebel PS, Murphy MR, Mick SA, Fein S (1991) Comparison of ocfentanil and fentanyl as supplements to general anesthesia. Anesth Analg 73(5):622–626
34. Misailidi N, Papoutsis I, Nikolaou P, Dona A, Spiliopoulou C, Athanaselis S (2017) Fentanyls continue to replace heroin in the drug arena: the cases of ocfentanil and carfentanil. Forensic Toxicol. https://doi.org/10.1007/s11419-017-0379-4
35. Dussy FE, Hangartner S, Hamberg C, Berchtold C, Scherer U, Schlotterbeck G et al (2016) An acute ocfentanil fatality: a case report with postmortem concentrations. J Anal Toxicol 40(9):761–766. https://doi.org/10.1093/jat/bkw096
36. World Health Organization (WHO) (2017) Furanyl fentanyl: critical review report. Expert Committee on Drug Dependence Thirty-ninth Meeting, Geneva
37. Martucci HFH, Ingle EA, Hunter MD, Rodda LN (2017) Distribution of furanyl fentanyl and 4-ANPP in an accidental acute death: a case report. Forensic Sci Int 283. https://doi.org/10.1016/j.forsciint.2017.12.005
38. Poklis J, Poklis A, Wolf C, Hathaway C, Arbefeville E, Chrostowski L et al (2016) Two fatal intoxications involving butyryl fentanyl. J Anal Toxicol 40(8):703–708. https://doi.org/10.1093/jat/bkw048
39. McIntyre IM, Trochta A, Gary RD, Wright J, Mena O (2016) An acute Butyr-fentanyl fatality: a case report with postmortem concentrations. J Anal Toxicol 40(2):162–166. https://doi.org/10.1093/jat/bkv138
40. Bäckberg M, Beck O, Jönsson K-H, Helander A (2015) Opioid intoxications involving butyrfentanyl, 4-fluorobutyrfentanyl, and fentanyl from the Swedish STRIDA project. Clin Toxicol 53(7):609–617. https://doi.org/10.3109/15563650.2015.1054505
41. Baumann MH, Tocco G, Papsun DM, Mohr AL, Fogarty MF, Krotulski AJ (2020) U-47700 and its analogs: non-fentanyl synthetic opioids impacting the recreational drug market. Brain Sci 10(11):895
42. Armenian P, Olson A, Anaya A, Kurtz A, Ruegner R, Gerona RR (2017) Fentanyl and a

novel synthetic opioid U-47700 masquerading as street "Norco" in Central California: a case report. Ann Emerg Med 69(1):87–90. https://doi.org/10.1016/j.annemergmed.2016.06.014
43. Fleming SW, Cooley JC, Johnson L, Frazee CC, Domanski K, Kleinschmidt K et al (2016) Analysis of U-47700, a novel synthetic opioid, in human urine by LC–MS–MS and LC–QToF. J Anal Toxicol 41(3):173–180. https://doi.org/10.1093/jat/bkw131
44. Mohr ALA, Friscia M, Papsun D, Kacinko SL, Buzby D, Logan BK (2016) Analysis of novel synthetic opioids U-47700, U-50488 and furanyl fentanyl by LC–MS/MS in postmortem casework. J Anal Toxicol 40(9):709–717. https://doi.org/10.1093/jat/bkw086
45. Siczek M, Zawadzki M, Siczek M, Chłopaś-Konowałek A, Szpot P (2020) Etazene (N, N-diethyl-2-{[(4-ethoxyphenyl)methyl]-1H-benzimidazol-1-yl}-ethan-1-amine (dihydrochloride)): a novel benzimidazole opioid NPS identified in seized material: crystal structure and spectroscopic characterization. Forensic Toxicol 39(1):146–155. doi: https://doi.org/10.1007/s11419-020-00552-9
46. Hunger A, Kebrle J, Rossi A, Hoffmann K (1960) Benzimidazol-Derivate und verwandte Heterocyclen. II. Synthese von 1-Aminoalkyl-2-benzyl-benzimidazolen. Helv Chim Acta 43(3):800–809. https://doi.org/10.1002/hlca.19600430323
47. Vandeputte MM, Krotulski AJ, Papsun DM, Logan BK, Stove CP (2021) The rise and fall of isotonitazene and brorphine: two recent stars in the synthetic opioid firmament. J Anal Toxicol. https://doi.org/10.1093/jat/bkab082
48. NPS Discovery (2021) Flunitazene. Center for Forensic Science Research and Education (CFSRE)
49. Papsun DP, Isenschmid D, Logan BK (2017) Observed carfentanil concentrations in 355 blood specimens from forensic investigations. J Anal Toxicol
50. Solbeck P, Woodall KL, Martin TL (2021) Strategic decision-making by a forensic toxicology laboratory in response to an emerging NPS: detection, quantitation and interpretation of carfentanil in death investigations in Ontario, Canada, July 2017 to June 2018. J Anal Toxicol. https://doi.org/10.1093/jat/bkab079
51. Center for Forensic Science Research and Education (CFSRE) (2021) Recommended Scope for NPS Testing in the United States, Q2 2021
52. Salomone A, Palamar JJ (2021) Toxicosurveillance of novel opioids: just screening tests may not be enough. Am J Drug Alcohol Abuse 47(3):271–272. https://doi.org/10.1080/00952990.2021.1917588
53. Morrow JB, Ropero-Miller JD, Catlin ML, Winokur AD, Cadwallader AB, Staymates JL et al (2018) The opioid. J Anal Toxicol, Epidemic: moving toward an integrated, holistic analytical response. https://doi.org/10.1093/jat/bky049
54. Hamnett HJ, Jack RE (2019) The use of contextual information in forensic toxicology: an international survey of toxicologists' experiences. Sci Justice 59(4):380–389. https://doi.org/10.1016/j.scijus.2019.02.004
55. Chhabra N, Rizvanolli L, Rasin A, Marsden G, Hinami K, Aks SE (2021) A cross-sectional analysis of fentanyl analog exposures among living patients. Am J Drug Alcohol Abuse 47(3):344–349. https://doi.org/10.1080/00952990.2021.1891420
56. Gergov M, Nenonen T, Ojanpera I, Ketola RA (2015) Compensation of matrix effects in a standard addition method for metformin in postmortem blood using liquid chromatography-electrospray-tandem mass spectrometry. J Anal Toxicol 39(5):359–364. https://doi.org/10.1093/jat/bkv020
57. Ellison SLR, Thompson M (2008) Standard additions: myth and reality. Analyst 133(8): 992–997. https://doi.org/10.1039/B717660K
58. Mochizuki A, Nakazawa H, Adachi N, Takekawa K, Shojo H (2019) Postmortem distribution of mepirapim and acetyl fentanyl in biological fluid and solid tissue specimens measured by the standard addition method. Forensic Toxicol 37(1):27–33. https://doi.org/10.1007/s11419-018-0431-z
59. Bjorksten AR, Chan C, Crankshaw DP (2002) Determination of remifentanil in human blood by capillary gas chromatography with nitrogen-selective detection. J Chromatogr B 775(1): 97–101. https://doi.org/10.1016/S1570-0232(02)00178-2
60. Feldman PL, Brackeen MF (1990) A novel route to the 4-anilido-4-(methoxycarbonyl)-piperidine class of analgetics. J Org Chem 55(13):4207–4209. https://doi.org/10.1021/jo00300a047
61. Fort C, Curtis B, Nichols C, Niblo C (2016) Acetyl fentanyl toxicity: two case reports. J Anal Toxicol 40(9):754–757. https://doi.org/10.1093/jat/bkw068
62. Marchei E, Ferri MA, Torrens M, Farre M, Pacifici R, Pichini S et al (2021) Ultra-high

performance liquid chromatography-high resolution mass spectrometry and high-sensitivity gas chromatography-mass spectrometry screening of classic drugs and new psychoactive substances and metabolites in urine of consumers. Int J Mol Sci 22(8). https://doi.org/10.3390/ijms22084000

63. Stevens RA, Petty RH, Hill HF, Kao T-C, Schaffer R, Hahn MB et al (1993) Redistribution of sufentanil to cerebrospinal fluid and systemic circulation after epidural administration in dogs. Anesth Analg 76(2):323–327
64. Weldon Stephen T, Perry Dana F, Cork Randall C, Gandolfi AJ (1985) Detection of picogram levels of sufentanil by capillary gas chromatography. Anesthesiology 63(6): 684–687. https://doi.org/10.1097/00000542-198512000-00021
65. Strano-Rossi S, Bermejo AM, de la Torre X, Botrè F (2011) Fast GC-MS method for the simultaneous screening of THC-COOH, cocaine, opiates and analogues including buprenorphine and fentanyl, and their metabolites in urine. Anal Bioanal Chem 399(4): 1623–1630. https://doi.org/10.1007/s00216-010-4471-4
66. Palmquist KB, Swortwood MJ (2019) Data-independent screening method for 14 fentanyl analogs in whole blood and oral fluid using LC-QTOF-MS. Forensic Sci Int 297: 189–197. https://doi.org/10.1016/j.forsciint.2019.02.006
67. Noble C, Weihe Dalsgaard P, Stybe Johansen S, Linnet K (2018) Application of a screening method for fentanyl and its analogues using UHPLC-QTOF-MS with data-independent acquisition (DIA) in MSE mode and retrospective analysis of authentic forensic blood samples. Drug Test Anal 10(4):651–662. https://doi.org/10.1002/dta.2263
68. Butler DC, Shanks K, Behonick GS, Smith D, Presnell SE, Tormos LM (2017) Three cases of fatal acrylfentanyl toxicity in the United States and a review of literature. J Anal Toxicol 42(1): e6–e11. https://doi.org/10.1093/jat/bkx083
69. Boronat Ena MDM, Cowan DA, Abbate V (2021) Ambient ionization mass spectrometry applied to new psychoactive substance analysis. Mass Spectrom Rev. https://doi.org/10.1002/mas.21695
70. Kennedy JH, Palaty J, Gill CG, Wiseman JM (2018) Rapid analysis of fentanyls and other novel psychoactive substances in substance use disorder patient urine using paper spray mass spectrometry. Rapid Commun Mass Spectrom 32(15):1280–1286. https://doi.org/10.1002/rcm.8164
71. Wharton RE, Casbohm J, Hoffmaster R, Brewer BN, Finn MG, Johnson RC (2021) Detection of 30 fentanyl analogs by commercial immunoassay kits. J Anal Toxicol 45(2): 111–116. https://doi.org/10.1093/jat/bkaa181
72. Wang B-T, Colby JM, Wu AHB, Lynch KL (2014) Cross-reactivity of acetylfentanyl and risperidone with a fentanyl immunoassay. J Anal Toxicol 38(9):672–675. https://doi.org/10.1093/jat/bku103
73. Deo VS, Gilson TP, Kaspar C, Singer ME (2021) The fentanyl phase of the opioid epidemic in Cuyahoga County, Ohio, United States. J Forensic Sci 66(3):926–933. https://doi.org/10.1111/1556-4029.14665
74. Van Nimmen NFJ, Poels KLC, Veulemans HAF (2004) Highly sensitive gas chromatographic—mass spectrometric screening method for the determination of picogram levels of fentanyl, sufentanil and alfentanil and their major metabolites in urine of opioid exposed workers. J Chromatogr B 804(2): 375–387. https://doi.org/10.1016/j.jchromb.2004.01.044
75. Misailidi N, Athanaselis S, Nikolaou P, Katselou M, Dotsikas Y, Spiliopoulou C et al (2019) A GC-MS method for the determination of furanylfentanyl and ocfentanil in whole blood with full validation. Forensic Toxicol 37(1):238–244. https://doi.org/10.1007/s11419-018-0449-2
76. Valdez CA (2021) Gas chromatography-mass spectrometry analysis of synthetic opioids belonging to the fentanyl class: a review. Crit Rev Anal Chem:1–31. https://doi.org/10.1080/10408347.2021.1927668
77. Staeheli SN, Baumgartner MR, Gauthier S, Gascho D, Jarmer J, Kraemer T et al (2016) Time-dependent postmortem redistribution of butyrfentanyl and its metabolites in blood and alternative matrices in a case of butyrfentanyl intoxication. Forensic Sci Int 266:170–177
78. Garneau B, Desharnais B, Laquerre J, Côté C, Taillon M-P, Martin P-Y et al (2021) A comprehensive analytical process, from NPS threat identification to systematic screening: method validation and one-year prevalence study. Forensic Sci Int 318:110595. https://doi.org/10.1016/j.forsciint.2020.110595

Chapter 5

Target Analysis of Designer Benzodiazepines in Blood and Urine

Islam Amine Larabi and Jean-Claude Alvarez

Abstract

In recent years, the market for new psychoactive substances (NPS) has exploded, with a continuous stream of new compounds being introduced on a weekly basis. Designer benzodiazepines (DBZD) are sold as "legal" substitutes for prescribed benzodiazepines and are one of the most misused pharmaceuticals in the world. Up to 30 DBZD are now being monitored in Europe, 80% of which were detected for the first time between 2014 and 2020. Despite their chemical and structural similarity, these DBZD have distinct pharmacokinetic properties and metabolic pathways, making it difficult to analyze and interpret. In the literature, many analytical techniques for screening these substances in various biological specimens have been documented, some of which are similar to those used to detect traditional benzodiazepines. The aim of the chapter is to provide an update on the current analytical methods for target analysis of DBZD and their metabolites in human blood and urine.

Key words Designer benzodiazepines, New psychoactive substances, NPS, Mass spectrometry, Metabolism, Human liver microsomes

1 Introduction

Benzodiazepines (BZD) include more than 50 different types of pharmaceuticals that share a basic fundamental structure of diazepine ring bounded to a benzene ring and exhibit anxiolytic, anticonvulsant, muscle relaxant, sedative, and amnestic properties [1]. BZD interact with the gamma-aminobutyric acid A (GABAA) receptor and act as positive allosteric modulators, increasing the affinity of GABA to its receptor, GABA being the major inhibitor neurotransmitter in the central nervous system (CNS) [2]. In the mid-1950s, BZD were introduced as a substitute for barbiturates or meprobamate in the treatment of anxiety and sleep disorders [3]. Sixty years after their first clinical application, they have become a widely prescribed medication for various psychiatric disorders (e.g., anxiety, insomnia, and seizures), as well as in anesthesiology and critical care [3]. They are also used as

Marta Concheiro and Karl B. Scheidweiler (eds.), *Methods for Novel Psychoactive Substance Analysis*,
Methods in Pharmacology and Toxicology, https://doi.org/10.1007/978-1-0716-2605-4_5,

recreational drugs to increase opioid euphoric effects and reduce the "crash" that happens following stimulant intake or to commit drug-facilitated sexual assault (DFSA) by exploiting their sedative and amnestic properties [1, 4].

Despite their broad therapeutic index, BZD have a variety of adverse effects (e.g., respiratory depression, drowsiness, confusion, headache, syncope, nausea/vomiting, diarrhea, and tremor) and a serious risk of addiction and abuse [5]. Slow clearance of certain benzodiazepines causes accumulation in tissues, which have a high lipid content, and overdose signs may appear in case of chronic use. Additionally, benzodiazepines cause serious adverse effects when used with opioids. Both classes have a synergistic effect and cause respiratory depression [6]. As a result, BZD should be used by prescription, only for short-term use, and at a lowest possible dose, in agreement with the 1971 United Nations Convention on Psychotropic Substances, which currently includes 38 BZD under control [7].

2 Designer Benzodiazepines

Designer benzodiazepines (DBZD) are novel, but illegal, benzodiazepines that are increasingly becoming available as new psychoactive substances (NPS) outside of traditional prescription channels, such as counterfeit medication and the illegal market, which is on the rise [2, 7].

In Europe, phenazepam was the first novel benzodiazepine to hit the market in 2007. It was generally sold as a "legal high" or intentionally used to make counterfeit benzodiazepines. Phenazepam is still used as a medication in Russia and other countries of the former Soviet Union. In 2011, its identification was followed by that of etizolam, which was also authorized as a medicine in some countries. From mid-2012, pyrazolam was the first DBZD offered for sale online, followed by flubromazepam and diclazepam in 2013. As a result, an increasing number of DBZDs have reached the market, bringing the total number of monitored benzodiazepines in Europe to 30 as of February 28, 2021 [7] (Fig. 1).

3 Benzodiazepine Classification

Benzodiazepines can be classified in different manner, by the duration of action, the elimination half-life, or chemical structure [3, 7].

1. Duration of action
   - Short acting (1–12 h, e.g., midazolam)
   - Intermediate acting (12–40 h, e.g., lorazepam, oxazepam)
   - Long acting (>40 h, e.g., diazepam, flubromazepam)

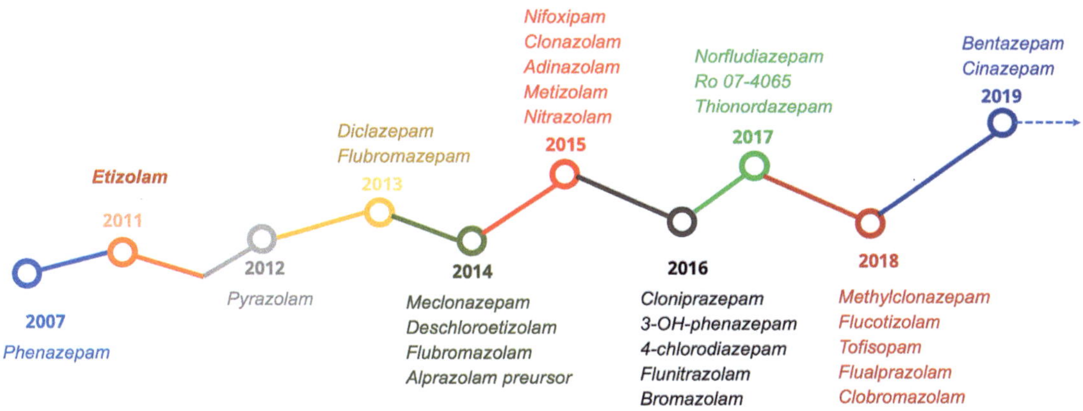

Fig. 1 Timeline of DBZD notified to the EU Early Warning System for the first time (2007–2019)

2. Elimination half-life ($t_{1/2}$)
   - Short (1–24 h, e.g., midazolam, pyrazolam)
   - Intermediate (24–48 h, e.g., oxazepam, diclazepam)
   - Long (>48 h, e.g., diazepam, flubromazepam)

In some circumstances, despite a relatively short elimination half-life, the duration of action might be rather extended. This is true for benzodiazepines that are metabolized to pharmacologically active metabolites with longer elimination half-lives than the parent drug, such as diclazepam metabolized to delorazepam [8]. Furthermore, potency varies widely, with the most potent (e.g., flunitrazolam, clonazolam, and flubromazolam) having strong effects at dosages far below 1 mg, whereas adinazolam, for example, generates equivalent effects at doses of around 30–50 mg. Another distinction is the onset of action, which varies from few minutes (e.g., deschloroetizolam ≈ 1–4 min) to hours (e.g., diclazepam ≈ 1.5 h) based on the route of administration and the absorption rate [3, 9].

3. Chemical structure

There are several BZD chemical structures, and those of DBZD can be classed in order of importance as follows (Fig. 2):

- **1,4-Benzodiazepines**: 3-hydroxyphenazepam, 4-chlorodiazepam, cinazepam, cloniprazepam, desalkylflurazepam (norflurazepam), fonazepam (norflunitrazepam), diclazepam, flubromazepam, meclonazepam, methyl-clonazepam, nifoxipam, nimetazepam, norfludiazepam, phenazepam, Ro07-4065 (difludiazepam), thionordazepam
- **Triazolobenzodiazepines**: adinazolam, alprazolam precursor, bromazolam, clobromazolam, clonazolam, flualprazolam, flubromazolam, flunitrazolam, nitrazolam, pyrazolam

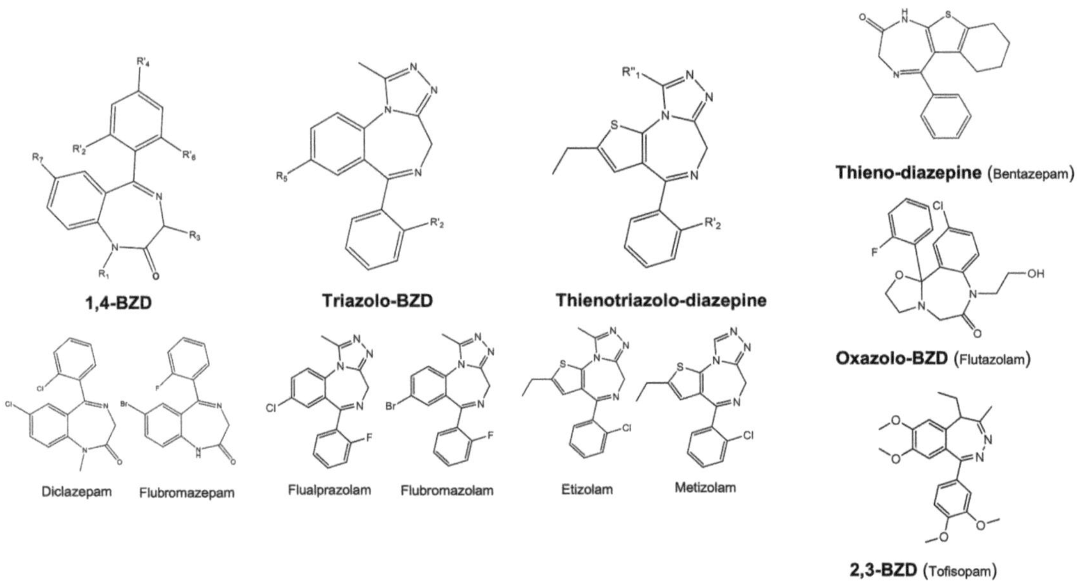

Fig. 2 DBZD chemical classes, along with some of the common compounds. (Adapted from: https:/www.emcdda.europa.eu/media-library/timeline-benzodiazepines-formally-notified-eu-early-warning-system-first-time-2007%E2%80%932020_en)

- **Thienotriazolodiazepines**: deschloroetizolam, etizolam, flucotizolam, metizolam
- **Thienodiazepines**: bentazepam
- **Oxazolobenzodiazepine**: flutazolam
- **2,3-Benzodiazepine**: tofisopam

Based on known DBZD structure-activity relationship, illegal producers appear to use a variety of strategies to keep flooding the market and avoid drug control legislation. Hundreds of 1,4-benzodiazepine derivatives were selected from scientific articles or patent filings, and various substituent combinations were created. Using active metabolites of traditional BZD or DBZD might also be considered as possibilities (e.g., flurazepam/norflurazepam, phenazepam/3-OH-phenazepam, flunitrazepam/fonazepam, or desmethylflunitrazepam). Lastly, synthesis of triazolo analogs of existing 1,4-benzodiazepines (e.g., clonazepam/clonazolam, flunitrazepam/flunitrazolam, nitrazepam/nitrazolam) is another approach to make novel molecules [3].

4 DBZD Toxicity

Although DBZD mono-intoxication is a rare cause of mortality, there have been several cases of DBZD-related deaths, particularly when alcohol and opioids/opiates are combined, resulting in profound unconsciousness and severe respiratory depression.

Flumazenil, a particular antidote that reversibly and competitively inhibits the binding site at GABA-receptors, is used in the therapeutic treatment of acute intoxications [7].

There is currently a wide range of data and studies on DBZD acute intoxications or deaths in the literature [10–14]. Between 2019 and April 2020, DBZD were detected in 48% of postmortem and 83% of driving under the influence of drugs (DUID) cases reported to the United Nation Office On Drugs and Crime (UNODC), with flualprazolam, flubromazolam, and etizolam being the most often encountered drugs [4]. In a review of 49 relevant reports in the worldwide early warning database involving 254 cases, Brunetti et al. reported that among 12 examined DBZD, etizolam, flualprazolam, flubromazolam, and phenazepam were the most implicated in adverse events and fatalities [4]. In Sweden and Finland, a study of 33 postmortem cases that tested positive for flualprazolam found that the cause of death was lethal poisoning by a drug in 23 (70%) of the cases, with flualprazolam being one of the substances implicated in 13 (40%) of the cases and the sole intoxicant in two poisoning cases [15]. Papsun et al. found flualprazolam in 197 blood samples from 171 postmortem and 22 DUID cases in the United States, most commonly (83%) in combination with opioids [16].

Etizolam has become the most often misused benzodiazepine in Scotland. In 2019, it may have played a role in 752 drug-related deaths (compared with 548 in 2018, 299 in 2017, and 223 in 2016) [17]. Finally, long-term benzodiazepines use has been associated to teratogenic effects, cognitive impairment, drug dependance, and neonatal withdrawal, but for DBZD, such data are sparse [7].

5 DBZD Metabolism

Once ingested, benzodiazepines are mainly metabolized by CYP3A4 enzymes and to a lesser extent by CYP3A5, CYP2B6, CYP2C18, and CYP2C19 [18–21]. The primary reactions involved in phase II metabolism are glucuronidation and acetylation by uridine 5′-diphosphoglucuronosyltransferase (UGT) and N-acetyltransferase 2 (NAT2), respectively. Benzodiazepines are mostly excreted in urine, with 1–20% of the parent drug excreted unchanged and glucuronide being the most frequent metabolite [22].

The pharmacokinetic characteristics and biotransformation patterns of the majority of DBZD remains scarce, and three main strategies are currently used to investigate their metabolism: (1) in vitro studies using human liver microsomes (HLM), S9 fraction, or human liver cells, followed by analysis of the incubates mainly by high-resolution mass spectrometry (HRMS), (2) analysis of urine samples from a large cohort of NPS users, and (3) analysis of biological samples after controlled self-administration studies [9].

Phase I metabolic reactions that have been reported in the literature for DBZD included hydroxylation, dealkylation, and reduction of the nitro moiety to the corresponding 7-amino derivative [7]. Debromination can also occur in compounds that include a bromine atom in their structure [9]. The phase II reactions involves O- or N-glucuronidation of the parent drug and/or its phase I metabolites, as well as the acetylation of the amine function of reduced nitro (triazolo-)benzodiazepines [7, 9]. Overall, the metabolism of some DBZD (e.g., flubromazolam, meclonazepam, clobromazolam, and bromazolam) appears to be similar to that of traditional benzodiazepines [23–26]. The principal synthetic benzodiazepines and their metabolites found in vitro or in human blood and urine samples are summarized in Table 1.

It is worth mentioning that the presence of some DBZD metabolites should be evaluated with caution since they may cause the analytical results to be misinterpreted. The following are some examples of DBZD metabolites that may be impacted by these interpretation issues:

- A prescribed benzodiazepine as metabolite of DBZD: e.g., estazolam is produced from adinazolam; clonazepam is produced from cloniprazepam and delorazepam; lormetazepam and lorazepam are all diclazepam metabolites; and nitrazepam is a metabolite of nimetazepam.
- A DBZD as metabolite of other benzodiazepines: e.g., 3-OH-phenazepam is derived from cinazepam; nifoxipam is derived from fonazepam (or norflunitrazepam); and norflurazepam is derived from flurazepam.
- A common metabolite of DBZD and other benzodiazepines: e.g., OH-alprazolam is a metabolite of both alprazolam and adinazolam; fonazepam (norflunitrazepam) is derived from flunitrazepam; 7-aminoclonazepam is a product of both clonazepam and cloniprazepam; while norflurazepam is a metabolite of both flurazepam and ethyl loflazepate (loflazepate prodrug).

6 Analytical Methods for DBZD Target Analysis in Blood and Urine

6.1 Methodology

The literature review of analytical methods dedicated to DBZD target analysis in blood and urine was based on a comprehensive search of PubMed (US National Library of Medicine) and Google Scholar using the keywords "designer benzodiazepines," "synthetic benzodiazepines," and each of the DBZD names mentioned in the 2021 EMCDDA report on DBZD associated with "mass spectrometry," "chromatography," "blood," "urine,"

Table 1
Target metabolites of major designer benzodiazepines identified in vitro and in human blood and urine

DBZD	Metabolites	In Vitro[a]	Blood	Urine	Authors (Year)	Ref.
3-OH-phenazepam	NA	ND	—	—	Moosmann et al. (2016)	[72]
4-Chlorodiazepam (Ro5-4864)	3-OH-4-chlorodiazepam	√	—	—	Moosmann et al. (2019)	[73]
	Nor-4-chlorodiazepam	√	—	—		
	3-OH-nor-4-chlorodiazepam	√	—	—		
Adinazolam	N-desmethyladinazolam	√	√	√	Lahti et al. (1983), Peng et al. (1984), Fraser et al. (1993), an Moosmann et al. (2016)	[72, 74–76]
	N, N-didesmethyladinazolam	√	—	—		
	α-OH-alprazolam			√		
	Estazolam			√		
Alprazolam triazolobenzophenone derivative	Alprazolam metabolite	√	—	—	Cho et al. (1986)	[77]
Bentazepam	NA	—	—	—	—	—
Bromazolam	α -OH-bromazolam	√	—	—	Wagmann et al. (2020)	[25]
	4-OH-bromazolam	√	√	√		
	Bromazolam glucuronide	√	—	√		
	Phenyl- OH-bromazolam	√	√	√		
	α-4-diOH-bromazolam	—	√	√		
	Phenyl-OH-bromazolam glucuronide	—	√	√		
	α -OH-bromazolam glucuronide	—	—	√		
	4-OH-bromazolam glucuronide	√	—	√		
Cinazepam	**3-OH-phenazepam**	—	—	√ (mice)	Schukin et al. (2011)	[78]

(continued)

Table 1 (continued)

DBZD	Metabolites	In Vitro[a]	Blood	Urine	Authors (Year)	Ref.
Clobromazolam	α -OH-clobromazolam	√	—	—	Wagmann et al. (2020)	[25]
	4-OH-clobromazolam	√	—	—		
	α-4-diOH-clobromazolam	√	—	—		
	Clobromazolam glucuronide	√	—	—		
	OH-clobromazolam glucuronide	√	—	—		
Clonazolam	7-Aminoclonazolam	√	—	√	Huppertz et al. (2015), Meyer et al. (2016), and El Belkhi et al. (2017)	[79–81]
	Desmethylclonazolam	√	—	—		
	OH-clonazolam	√	—	—		
	7-Acetaminoclonazolam	—	—	√		
	OH-clonazolam	—	—	√		
	7-Aminoclonazolam glucuronide	—	—	√		
	7-Acetaminoclonazolam glucuronide	—	—	√		
	OH-clonazolam glucuronide	—	—	√		
	8-Aminoclonazolam,	√	—	—		
	8-Acetamidoclonazolam	√	—	—		
Cloniprazepam	3-Ketocloniprazepam	√			Moosmann et al. (2016) and Mortelé et al. (2018)	[72, 82]
	3-OH-7-aminoclonazepam	√	—	—		
	7-Aminocloniprazepam	√	—	—		
	7-Aminoclonazepam	√	—	—		
	Clonazepam	√	—	—		
	DiOH-cloniprazepam	√	—	—		
	OH-cloniprazepam	√	—	—		
	OH-clonazepam	√	—	—		
	OH-cloniprazepam glucuronide	√	—	—		

Deschloroetizolam	OH-deschloroetizolam	√	—	—	Huppertz et al. (2015), El Belkhi et al. (2017), and Pettersson Bergstrand et al. (2019)	[26, 79, 81]
	DiOH-deschloroetizolam	√	—	—		
	Deschloroetizolam glucuronide	√	—	—		
Desmethylflurazepam (= norflurazepam or norfludiazepam)	OH-norflurazepam	√	—	—	Moosmann et al. (2018)	[73]
	DiOH-norflurazepam	√	—	—		
Diclazepam (2-chlorodiazepam)	**Delorazepam**	√	√	√	Moosmann et al. (2014), and El Belkhi et al. (2017)	[8, 81]
	Lormetazepam	√	—	√		
	Lorazepam	√	—	√		
Etizolam	α-OH-etizolam	√	√	—	Nakamae et al. (2008), El Belkhi et al. (2017), and Pettersson Bergstrand et al. (2019)	[26, 58, 81]
	8-OH-etizolam	√	√	√		
	keto-metabolite	√	—	√		
	Etizolam glucuronide	√	—	√		
Flualprazolam	α-OH-flualprazolam	√	√	√	Wagmann et al. (2020a)	[62]
	4-OH-flualprazolam	√	√	√		
	DiOH-flualprazolam	√	—	√		
	Flualprazolam glucuronide	√	—	√		
	α-OH-flualprazolam glucuronide	√	—	√		
	4-OH-flualprazolam glucuronide	√	—	√		
	diOH-flualprazolam glucuronide	—	—	√		
Flubromazepam	OH-flubromazepam	√	√	√	Moosmann et al. (2013)	[53]
	Debromoflubromazepam	√	√	—		
	Debromo-OH-flubromazepam	√	—	√		
Flubromazolam	4-OH-flubromazolam	√	—	√	Huppertz et al. (2015, 2018), El Balkhi et al. (2017), Noble et al. (2017), Wohlfarth et al. (2017), and Pettersson Bergstrand et al. (2018)	[23, 79, 81, 83, 84]
	α-OH-flubromazolam	√	—	√		
	DiOH-flubromazolam	√	—	√		
	Flubromazolam glucuronide	√				
	4-OH-flubromazolam glucuronide	—	—	√		
	α-OH-flubromazolam glucuronide	—	—	√		

(continued)

Table 1 (continued)

DBZD	Metabolites	In Vitro[a]	Blood	Urine	Authors (Year)	Ref.
Fluclotizolam	NA	—	—	—	—	—
Flunitrazolam	OH-flunitrazolam	√	—	√	Ameline et al. (2018), Moosmann et al. (2018), and Pettersson Bergstrand et al. (2019)	[26, 31, 73]
	(7)-Aminoflunitrazolam	√	—	√		
	DiOH-flunitrazolam	√	—			
	Flunitrazolam glucuronide	√	—	—		
	Desnitroflunitrazolam	—	—	√		
	7-Acetamidoflunitrazolam	—	—	√		
Fonazepam (norflunitrazepam)	3-OH-norflunitrazepam	√	—	—	Moosmann et al. (2016)	[72]
	7-Aminonorflunitrazepam	√	—	—		
	Nifoxipam	√	—	—		
Meclonazepam	7-Amino-meclonazepam	—	—	√	Huppertz et al. (2015), Meyer et al. (2016), El Balkhi et al. (2017), Richter et al. (2017), and Vikingsson et al. (2017)	[24, 79–81, 85]
	7-Acetaminomeclonazepam	—	—	√		
	OH-meclonazepam	√	—	—		
Methyl clonazepam	NA	—	—	—	—	—
Metizolam (desmethyl-etizolam)	2-OH-metizolam	√	—	√	Moosmann et al. (2016) and Kintz et al. (2017)	[50, 72]
	N-OH-metizolam	√	—	√		
	Metizolam glucuronide	√	—	—		
	2-OH-metizolam glucuronide	—	—	√		
Nifoxipam (3-OH-fonazepam)	7-Aminonifoxipam	√	—	√	Meyer et al. (2016), El Balkhi et al. (2017), and Pettersson Bergstrand et al. (2019)	[26, 80, 81]
	Denitro-nifoxipam	√	—	—		
	Nifoxipam glucuronide	√	—	√		
	7-Acetaminonifoxipam	—	—	√		
Nimetazepam	**Nitrazepam**	—	√	√	Lee et al. (2013)	[52]
	7-Aminonimetazepam	—	—	√		

Nitrazolam	8-Aminonitrazolam	√			Moosmann et al. (2016)	[72]
	4-OH-nitrazolam	√				
	α-OH-nitrazolam	√				
Phenazepam	3-OH-phenazepam	—	√	√	Zherdev et al. (1982), Maskell et al. (2012), WHO (2015), and Crichton et al. (2015)	[10, 11, 86, 87]
	5-Bromo-(2-chlorophenyl)-2-aminobenzophenone (ABPH)	—	√	√		
	6-Bromo-(2-chlorophenyl) quinazoline-2-one (QNZ)	—	—	√		
Pyrazolam	4-OH-pyrazolam	√	—	√	Moosmann et al. (2013) and Pettersson Bergstrand et al. (2018)	[88, 89]
	α-OH-pyrazolam		—			
	Pyrazolam glucuronide	√	—	√		
	4-OH-pyrazolam glucuronide	√	—	√		
	α-OH-pyrazolam glucuronide	√	—	√		

[a]Only human tissue-based experimental models were considered.

✓ present, — not currently reported, *ND* not detected

DBZD are ordered alphabetically, authors by publication date, and alphabetically for publications in the same year. Metabolites that might be classified as a conventional benzodiazepine, a designer benzodiazepine, or a common metabolite of several benzodiazepines are highlighted in bold.

"metabolites," and "metabolism" as search terms [7]. Only papers with full manuscripts were taken into consideration. Multidrug screening approaches were given priority, but when novel extraction or detection techniques are used, we also considered articles that presented an assay for one or a limited number of DBZD (which may include case reports and in vitro metabolism studies), as well as publications that revealed new data on DBZD concentrations. To complete this bibliographic search, three literature reviews [2, 4, 27], two book chapters [1, 3], and the most recent EMCDDA report on DBZD from 2021 were often reviewed [7]. Finally, analytical techniques that only allow for the screening of traditional benzodiazepines (e.g., norflurazepam, lorazepam, lormetazepam, and delorazepam) that may be considered metabolites of certain DBZD were excluded from this study.

6.2 Screening and Confirmation Methods

Immunoassays can be used successfully for BZD detection with sufficient sensitivity, even though these techniques are based on cross-reactivity with a BZD reference standard (e.g., oxazepam) and do not allow for specific characterization of the drug involved [28–30]. Furthermore, because the clinical cutoffs provided by the manufacturers do no not allow for the detection of a single intake, immunoassays should be used with caution when dealing with drug-facilitated crime (DFC) situations, particularly when extremely potent DBZD is involved (e.g., flunitrazolam) [31]. Finally, it is worth noting that an immunoassay-positive result should be confirmed with an updated method that includes DBZD screening to avoid being misinterpreted as a "false positive," as demonstrated by Pettersson Bergstrand et al., who found that 40% of BZD immunoassay-positive urine samples ($n = 156$) tested negative for traditional BZD by mass spectrometry but positive for DBZD [32].

Currently, the common DBZD targeted screening approaches are liquid [8, 11, 16, 23, 32–55] or gas [15, 48, 56–60] chromatography-tandem mass spectrometry-based techniques. High-resolution mass spectrometry often coupled with liquid chromatography (LC-HRMS) is also currently on the rise since it allows for the detection of a large variety of compounds without the need of reference standards, making it both practical and cost-effective [16, 24, 25, 48, 61–66]. The ability to reprocess raw data in a retrospective manner is the second feature of HRMS technology that is not shared with tandem mass spectrometry (so-called low-resolution mass spectrometry); this feature allows subsequent reprocessing of the data and thus the detection of compounds that were not present in the library at the time of the first analysis (this is the case of the newly appeared DBZD) [67]. Table 2 summarizes current analytical methods for detecting and quantifying DBZD in human blood and urine, as well as DBZD concentrations found in clinical and forensic settings.

7 DBZD Stability Studies

As a part of method validation process, most of the abovementioned analytical methods (Table 2) have undertaken DBZD stability studies in various stock solutions and/or biological matrices. For example, Pettersson Bergstrand et al. [68] investigated the stability of 28 DBZD in methanol stock solutions at −20 °C for 15 months, in urine at −80 °C for 4 months, at 4 °C for 1 week, at room temperature (20 °C) for 1 day, over three freeze-thaw cycles, and in an autosampler (12 °C) for 4 days. Supplementary Data 1 of this study provides an in-depth summary and can be found at (https://www.sciencedirect.com/science/article/pii/S2376999818300126#m0005). Except for clonazolam and nifoxipam, all analytes were stable under all conditions. Over the course of 4 months, the concentration of clonazolam at −80 °C fell by 21%, making it slightly beyond the acceptable limit (i.e., 20%). After 15 months at −20 °C in methanol, 1 week at 4 °C, and 4 months at −80 °C, the concentration of nifoxipam dropped marginally below the acceptability limit. As a result, storing patient urine samples at 4 °C for more than 1 week was not recommended. Nevertheless, previous studies from the same author showed that urine samples are stable for at least 2 months at −20 °C [32]. Over a 4-month period, urine standards and QC samples were stable at −80 °C and could be thawed at least three times without affecting the findings. In the autosampler, the methanol solutions were stable for 15 months at 20 °C, while prepared extracts were stable for 4 days at 12 °C in the autosampler [68].

Similarly, Degreef et al. [33] investigated the stability of 54 benzodiazepines and Z-drugs including 20 designer ones in spiked plasma at room temperature, at −4 °C, and at −20 °C. Also, benchtop and autosampler stabilities were tested after 3 h at room temperature and over 72 h, respectively. Overall, excellent stabilities were found (≥98% ± 6%) in different storage conditions for all but two compounds (cloxazolam and ethyl loflazepate) that showed concentrations decrease over the acceptability limit (±15%). After three freeze-thaw cycles, the acceptability criteria were met for all but ten compounds. Accurate concentrations of adinazolam, brotizolam, ethyl loflazepate, etizolam, loprazolam, medazepam, nifoxipam, and nitrazepam could be guaranteed for only two freeze-thaw cycles. Bentazepam showed reliable concentration up to one cycle only, but in a previous study, its in-serum benchtop stability for up to 10 h was excellent [69]. Surprisingly, cloxazolam showed inexplicable drop to 70% after one cycle at low concentration but an excellent freeze-thaw stability for the 3 cycles at high concentration. Finally, in-freezer stability at −20 °C showed the majority of compounds were stable for a minimum of 3 months. Adinazolam showed acceptable results only over 1 month, while long-term stability for nifoxipam seemed variable at −20 °C.

Table 2
Analytical methods for target analysis of designer benzodiazepines and their metabolites in human blood and urine

Method	Analyte	Matrix	Extraction	Chromatography conditions	LOD (ng/mL)	LLOQ (ng/mL)	ULOQ (ng/mL)	Concentrations found [min-max ng/mL] (n) or cases[a] (n)	Circumstances	Ref.
LC-MS/MS	3-OH-flubromazepam	Plasma	LLE (TBME, pH 9.5)	Zorbax Eclipse Plus C8 column (2.1 × 150 mm i.d., 3.5 μm) *Gradient* H_2O, 0.1% HCOOH (A)/ACN: H_2O (9:1), 0.1% HCOOH (B)	NP	2.5	1000	NA	NA	[33]
	Adinazolam				NP	2.5	1000			
	Bentazepam				NP	2.5	1000			
	Clonazolam				NP	2	800			
	Cloniprazepam				NP	0.5	200			
	Deschloro-etizolam				NP	1.25	500			
	Diclazepam				NP	1	400			
	Etizolam				NP	1.25	500			
	Flubromazepam				NP	2.5	1000			
	Flubromazolam				NP	2	800			
	Halazepam				NP	10	4000			
	Meclonazepam				NP	1	400			
	Metizolam				NP	1	400			
	Nifoxipam				NP	2.5	1000			
	Norflurazepam				NP	5	2000			
	Phenazepam				NP	2	800			
	Pivoxazepam				NP	2.5	1000			
	Pyrazolam				NP	2.5	1000			
LC-MS/MS	α-Hydroxyetizolam	Blood	LLE (MTBE: 1-chlorobutane, 60:40)	Acquity BEH C18 column *Gradient* NH_4HCOO (A)/MeOH(B)	0.5	1.98	200	2.1–510 (n = 159)	Retesting of samples with positive Immunoassay and negative traditional BZD results (n = 1262)	[34]
	Bromazepam				0.16	4.96	500	16–620 (n = 6)		
	Clonazolam				0.16	5.01	500	5–26 (n = 18)		
	Delorazepam				0.16	5.01	500	5.9–520 (n = 20)		
	Diclazepam				0.16	5.03	500	6.6–57 (n = 7)		
	Etizolam				0.06	1.98	200	2–900 (n = 607)		
	Flualprazolam				0.06	1.97	200	2–990 (n = 537)		
	Flubromazepam				0.63	20	2000	23–1100 (n = 2)		
	Flubromazolam				0.06	1.99	200	2.8–760 (n = 136)		
	Flunitrazolam				0.06	1.99	200	ND		
	Nitrazolam				0.16	4.93	500	ND		
	Phenazepam				0.63	1.88	NP	91–200 (n = 3)		

LC-MS/MS	Clonazolam Deschloroetizolam, Flubromazolam Meclonazepam Nifoxipam,	Plasma	Microextraction by packed sorbent (MEPS) (MeOH + SPE by polymeric sorbent) *Eluent*: CH_2Cl_2/2-propanol/NH_4OH (78:20:2, v/v/v)	ACQUITY BEH C18 column (× 100 × 2.1 mm i.d., 1.7 μm), 55 °C *Gradient* H_2O, 0.1% HCOOH (A)/ACN, 0.1% HCOOH (B)	5 0.5 5 5 1	10 1 10 10 2.5	500 500 500 500 500	ND ND ND ND ND	Overdose (n = 7)	[35]
LC-MS/MS	Etizolam Nimetazepam Phenazepam	Blood	LLE (butyl chloride isopropanol: 9:1)	Kinetex C18 column (50 × 4.6 mm, 2.6 μm), 60 °C *Gradient* 50 mM NH_4COOH pH 3.5 (A)/ACN, 0.1% HCOOH (B)	NP NP NP	10 10 10	1000 200 200	160 ND ND	Reanalysis of authentic casework (n = 90)	[36]
LC-MS/MS	Clonazolam Diclazepam Etizolam Flualprazolam Flubromazepam Flubromazolam Phenazepam	Blood	Protein precipitation (ACN)	Biphenyl column (50 × 2.1 mm i.d., 2.7 μm), 40 °C *Gradient* H_2O, 0.1% HCOOH (A)/CAN, 0.1% HCOOH (B)	5 1 5 1 1 5 5	5 5 25 5 NA 5 25	100 100 500 100 100 100 500	ND <25–1160 (n = 5) <2.5–45 (n = 2) ND ND ND	Reanalysis of DUID casework (n = 19)	[37]
LC-MS/MS (qualitative analysis)	Etizolam Flualprazolam Flubromazolam	Blood Urine	Protein precipitation (ACN-Acetone, 70: 30 v/v)	Zorbax Eclipse Plus C18 column (100 × 2.1 mm i.d., 3.5 μm) 50 °C *Gradient* 10 mM NH_4HCOO, MeOH (98:2) (A)/ACN(B)	0.05 0.01 0.05	NA NA NA	NA NA NA	43 cases 70 cases 79 cases	DUID, DFSA, death (n = 5079)	[38]
LC-MS/MS	Etizolam	Blood (femoral) Plasma (antemortem)	LLE (MTBE)	ZORBAX Eclipse Plus C18 (100 × 2.1 mm i.d., 1.8 μm) *Gradient* NH_4 CH_3COO, CH_3COOH (A)/MeOH (B)	1	1	NP	1–72 (n = 24) 4–44 (n = 4)	Postmortem cases (n = 24)	[39]
LC-TOF-MS (screening) + LC-MS/MS (quantification by standard addition protocol)	Flualprazolam	Blood	LLE (MTBE/1-chlorobutane) (60: 40, v/v) (pH 9.0)	Acquity UPLC BEH C18 (50 × 2.1 mm i.d., 1.7 μm), 55 °C *Gradient* 0.1% HCOOH, H_2O (A)/0.1% HCOOH, MeOH (B)	<1	2	200	<2–620 (death) 4.4–68 (DUID)	Death (n = 171) DUID (n = 22) Unknown (n = 4)	[16]

(continued)

Table 2
(continued)

Method	Analyte	Matrix	Extraction	Chromatography conditions	LOD (ng/mL)	LLOQ (ng/mL)	ULOQ (ng/mL)	Concentrations found [min-max ng/mL] (n) or cases[a] (n)	Circumstances	Ref.
LC-MS-MS	Flualprazolam	Blood (B) Urine	Protein precipitation (ACN)	Hypersil Gold column (100 × 2.1 mm i.d., 1.9 μm), 60 °C *Gradient* 0.1% HCOOH, 1 mM NH_4HCOO in H_2O (A)/.1% HCOOH, 1 mM NH_4HCOO in ACN (B)	NP	1	200	2.83–35.1 (blood, n = 9)	Death (mixed drug toxicity)	[40]
LC-QTOF-MS (qualitative analysis)	3-Hydroxyphenazepam	Blood	Protein precipitation (ACN)	Acquity HSS T3 column (100 mm 2.1 mm i.d., 1.8 μm), 50 °C *Gradient* 10 mM NH_4COOH, 0.05% HCOOH (A)/ACN, 0.05% HCOOH (B)	NP	NA	NA	ND	Retrospective data analysis (n = 1314)	[61]
	4-Chlorodiazepam				NP	NA	NA	ND		
	7-Aminoclonazolam				NP	NA	NA	ND		
	Adinazolam				NP	NA	NA	ND		
	Bentazepam				NP	NA	NA	ND		
	Bromazepam				NP	NA	NA	ND		
	Bromazolam				NP	NA	NA	ND		
	Clonazolam				NP	NA	NA	ND		
	Cloniprazepam				NP	NA	NA	ND		
	Deschloroetizolam				2	NA	NA	ND		
	Diclazepam				5	NA	NA	ND		
	Etizolam				2	NA	NA	ND		
	Flubromazepam				10	NA	NA	2 cases		
	Flubromazolam				5	NA	NA	ND		
	Meclonazepam				10	NA	NA	ND		
	Flubromazepam				10	NA	NA	ND		
	Flualprazolam				NP	NA	NA	ND		
	Flubromazolam				5	NA	NA	ND		
	Flunitrazolam				NP	NA	NA	ND		
	Meclonazepam				NP	NA	NA	ND		
	Metizolam				NP	NA	NA	ND		
	Nifoxipam				NP	NA	NA	ND		
	Nimetazepam				NP	NA	NA	ND		
	Norfludiazepam				NP	NA	NA	ND		
	Phenazepam				NP	NA	NA	ND		
	Pyrazolam				10	NA	NA	ND		

LC-MS/MS	Clonazolam	Blood	LLE (ethyl acetate:heptane 4:1) (pH 11) or TBME (pH 11)	*2016–2017* BEH Phenyl column *Gradient* 5 mM NH_4HCO_3 (pH 8) (A)/ MeOH (B) *Dec 2017–2019* BEH C18 column (100 mm × 2.1 i.d., 1.7 μm), 60 °C *Gradient* NH_4COOH (pH 5) (A)/ACN (B)	1.4	1.4	NP	1.7–53 (n = 22)	Apprehended drivers, drug offences, and autopsies (n = 575)	[41, 49]
	Diclazepam				1.6	1.6	NP	1.6–250 (n = 334)		
	Etizolam				14	14	NP	15–300 (n = 40)		
	Flualprazolam				1.6	1.6	NP	3.3–56 (n = 10)		
	Flubromazepam				3.3	3.3	NP	7–700 (n = 5)		
	Flubromazolam				0.37	0.37	NP	0.4–36 (n = 20)		
	Phenazepam				1.7	1.7	NP	1.8–850 (n = 138)		
GC-NCI-MS	Flualprazolam	Blood	LLE (butyl acetate)	ZB-DRUG-1 (15 m × 0.25 mm, 0.25 μm) column Carrier: helium5.6 160–320 °C (50 °C/min)	1 ng/g	2 ng/g	125 ng/g	3–68 ng/g	Postmortem cases (n = 33)	[15]
LC-MS/MS	Deschloroetizolam	Blood Urine	1) Protein precipitation (ice-cold 10:90 methanol:ACN (v/v) mixture) 2) SPE (eluent: hexane) (+ urine hydrolysis using β-glucuronidase)	Accucore Phenyl-Hexyl column (150 × 2.1 mm i.d., 2.6 μm), 40 °C *Gradient* 0.1% HCOOH in H_2O (A)/ 0.1% HCOOH in ACN (B)	1	2	80	NP	Forensic casework (n > 300)	[42]
	Nitrazolam				2.5	5	200	NP		
	Fonazepam				5	10	400	NP		
	Clonazolam				2.5	5	200	NP		
	2-OH-ethyl flurazepam				5	10	400	NP		
	Flualprazolam				2.5	5	200	NP		
	N-Desalkylflurazepam				5	10	400	NP		
	Flubromazolam				1	2	80	NP		
	Flubromazepam				10	20	800	NP		
	Bromazolam				5	10	400	NP		
	Delorazepam				2.5	5	200	NP		
	Meclonazepam				2.5	5	200	NP		
	Lormetazepam				5	10	400	NP		
	Etizolam				1	2	80	NP		
	Phenazepam				5	10	400	NP		
	3-OH-flunitrazepam				5	10	400	NP		
	Methylclonazepam				2.5	5	200	NP		
	Diclazepam				2.5	5	200	NP		
LC-HRMS/MS (quantification by standard addition method)	Flualprazolam	Plasma	Protein precipitation (ACN in 0.1% HCOOH)	Accucore Phenyl-Hexyl column (100 mm × 2.1 mm i.d., 2.6 μm) *Gradient* 0.1% HCOOH in 2 mM NH_4HCOO (A)/2 mM NH_4HCOO, 0.1% HCOOH in MeOH-ACN (50:50), 1% H_2O (B)	NP	NP	NP	8	Suspected flualprazolam ingestion (emergency department)	[62]

(continued)

Table 2
(continued)

Method	Analyte	Matrix	Extraction	Chromatography conditions	LOD (ng/mL)	LLOQ (ng/mL)	ULOQ (ng/mL)	Concentrations found [min-max ng/mL] (n) or cases[a] (n)	Circumstances	Ref.
LC-HRMS/MS	Bromazolam	Plasma (+ urine and HLM S9)	Plasma: protein precipitation (ACN in 0.1% HCOOH)	Accucore Phenyl-Hexyl column (100 mm × 2.1 mm i.d., 2.6 μm), 40 °C *Gradient* 0.1% HCOOH in 2 mM NH_4HCOO (A)/2 mM NH_4HCOO, 0.1% HCOOH in MeOH-ACN(50:50), 1% H_2O (B)	NP	NP	NP	6, 29	Drug abuse (n = 2)	[25]
	Clobromazolam				NP	NP	NP	—		
LC-MS/MS	Etizolam (and 7 traditional BZD)	Plasma (breast milk)	LLE (ethyl acetate)	C18 column (150 × 2.0 i.d., 3 μm), 40 °C *Gradient* 10 mM NH_4 CH_3COOH (A)/ CAN (B)	NP	0.5	100	—	—	[43]
LC-MS/MS (Standard addition calibration)	Diclazepam	Blood	ITSP-SPE (100 mM NH4HCOO, pH 4.2) QuEChERS (for delorazepam)	Biphenyl Core-Shell column (100 mm 2.1 mm, i.d., 2.6 μm) *Gradient* 2 mM NH_4HCOO, 0.1% HCOOH (A)/ACN, 0.1% HCOOH (B)	NP	NP	NP	1 (femoral, f) 1 (heart, h)	Postmortem (asphyxia with polydrug intoxication) (n = 1)	[44]
		Urine						1		
	Delorazepam	Blood						100 (f) 250 (h)		
		Urine						570		
	Lormetazepam	Blood						6 (f) 4 (h)		
		Urine						810		
	Lorazepam	Blood						22 (f) 22 (h)		
		Urine						820		
	Pyrazolam	Blood			NP	NP	NP	28 (f) 28 (h)		
		Urine						500		

LC-MS/MS	3-hydroxyphenazepam	Blood	SPE (ethyl acetate: ammonium hydroxide (28–30% NH_3 basis) (98:2)	Infinity Lab Poroshell 120 EC18 column (50 × 3.0 mm, i.d., 2.7 μm), 55 °C *Gradient* 0.1% HCOOH in H_2O (A)/ 0.1% HCOOH in ACN(B)	0.5	1	200	—	Postmortem reanalyzed samples (n = 33)	[45]
	Clonazolam				0.5	1	500	1.1 (n = 1)		
	Delorazepam				0.5	1	200	1.1–68 (n = 3)		
	Deschloroetizolam				0.5	1	200	—		
	Diclazepam				0.5	1	200	<1 (n = 1)		
	Flualprazolam				0.5	1	200	1.94 (n = 1)		
	Flubromazepam				0.5	1	500	—		
	Flubromazolam				0.5	1	200	40 (n = 1)		
	Flunitrazolam				0.5	1	200	—		
	Meclonazepam				0.5	1	200	—		
	Nifoxipam				0.5	1	200	—		
	Pyrazolam				0.5	1	500	—		
LC-MS/MS	Flubromazepam	Blood	LLE (1-chlorobutane)	Synergi Polar RP (150 × 2.0 mm i.d., 4.0 μm) *Gradient* 0.1% HCOOH, 1 mM NH_4HCOO in H_2O (A)/0.1 % HCOOH in MeOH (B)	NP	1	50	830 (42 min), 370 (9 h), 530 (24 h), 520 (33 h), 580 (41 h), 480 (57 h), 280 (81 h) (t1/2 ≈ 100 h)	Lethal intoxication with pharma cokinetic data (n = 1)	[46]
		Urine						13 (41 h)		
	OH-flubromazepam	Blood			NP	1	50	160 (42 min), 87 (9 h), 120 (24 h), 76 (33 h), 100 (41 h), 100 (57 h), 55 (81 h)		
		Urine						310		
	Desalkylflurazepam	Blood			NP	1	50	NP		
		Urine								
NACE-MS/MS (capillary electrophoresis-MS/MS)	Bentazepam	Serum	LLE (1-chlorobutne)	*Capillary electrophoresis* bare fused silica capillary (50 μm id, 365 μm od) Successive multiple ionic-polymer layer (SMIL) coating (60 cm × 50 μm i.d.), 25 °C (20 μL/min): hexadimethrine bromide (activation/rinse), dextran sulfate (reaction)	1.5	5	600	NA	Spiked samples	[90]
	Deschloroetizolam				1.5	5	600	NA		
	Diclazepam				1.5	5	600	NA		
	Etizolam				1.5	5	600	NA		
	Flubromazepam				3	10	600	NA		
	Flubromazolam				1.5	5	600	NA		
	Nimetazepam				1.5	5	600	NA		
	Phenazepam				3	10	600	NA		
	Pyrazolam				15	50	600	NA		

(continued)

Table 2
(continued)

Method	Analyte	Matrix	Extraction	Chromatography conditions	LOD (ng/mL)	LLOQ (ng/mL)	ULOQ (ng/mL)	Concentrations found [min-max ng/mL] (n) or cases[a] (n)	Circumstances	Ref.
LC-HRMS (retrospective data analysis)	Diclazepam	Blood Urine	LLE (butyl chloride) Urine hydrolysis using β-glucuronidase)	Acquity BEH C18 column (50 × 3.0 mm i.d., 1.7 μm), 30 °C *Gradient:* 0.1% HCOOH, H_2O (A)/ACN (B)	1	8	320	70	Death (n = 1)	[63]
LC-HRMS (screening and quantitative analysis)	3-OH-flubromazepam	Urine	Hydrolysis using β-glucuronidase + dilute and shoot	YMC UltraHT Hydrosphere C18 column (100 mm × 2.0 mm i.d., 2.0 μm), 60 °C *Gradient* 10 mM NH_4HCOO, 0.005% HCOOH (pH 4.8) (A)/ 10 mM NH_4HCOOin MeOH-H_2O (10:90), 0.005% HCOOH (B)	50	50	1000	90–11,580 (n = 12)	Routine drug testing and acute intoxication (STRIDA project) (Total n= 126)	[68]
	3-OH-phenazepam				50	50	1000	84–16,549 (n = 10)		
	Adinazolam				1	10	1000	ND		
	Bentazepam				1	5	1000	ND		
	Clonazolam				5	10	1000	10–132 (n = 8)		
	Cloniprazepam				5	10	1000	ND		
	Deschloroetizolam				5	10	1000	5 (n = 1)		
	Diclazepam				1	5	1000	8 (n = 1)		
	Etizolam				5	5	1000	40–1000 (n = 5)		
	Flubromazepam				5	5	1000	5–89 (n = 12)		
	Flubromazolam				5	5	1000	ND		
	Flunitrazolam				1	5	1000	5–121 (n = 24)		
	Ketazolam				700	NP	NP	NQ (n = 1)		
	Meclonazepam				10	10	1000	5–126 (n = 10)		
	Metizolam				5	5	1000	5–14 (n = 4)		
	N-desmethyl-flunitrazepam				5	10	1000	ND		
	Nifoxipam				50	50	1000	149–3635 (n = 3)		
	Nimetazepam				5	10	1000	ND		
	Nitrazolam				5	10	1000	ND		
	Phenazepam				10	10	1000	ND		
	Pivoxazepam				50	50	1000	ND		
	Pyrazolam				5	5	1000	6–1995 (n = 11)		

LC-HRMS (qualitative analysis)	3-OH-phenazepam	Serum (S) Urine (U)	Serum: not described Urine: dilute and shoot	Kinetex C18 column (50 × 3.00 mm i.d., 2.6 μm) *Gradient* 0.05% HCOOH in 5 mM NH_4HCOO (A)/0.05% HCOOH in MeOH-ACN (50:50) (B) gradient	50 (S) 50 (U)	NA	100 (S) 500 (u)	NA	Multidrug recreational ingestion (n = 1) (quantitative method not described)	[65]
	Adinazolam				25 (S) 25 (U	NA		NA		
	Bromazepam				25 (S) 25 (U	NA		NA		
	Clonazolam				2.5 (S) 250 (U)	NA		10.2 (S)		
	Delorazepam				5 (S) 25 (U)	NA		NA		
	Deschloroetizolam				2.5 (S) 5 (U)	NA		NA		
	Diclazepam				5 (S) 5 (U)	NA		NA		
	Etizolam				1 (S) 5 (U)	NA		281 (S)		
	Flubromazepam				25 (S) 50 (U)	NA		NA		
	Flubromazolam				5 (S) 25 (U)	NA		NA		
	Ketazolam				5 (S) 5 (U)	NA		NA		
	N-desalkyl-flurazepam				25 (S) 25 (U)	NA		NA		
	Nimetazepam				50 (S) 10 (U)	NA		NA		
	Phenazepam				1 (S) 25 (U)	NA		NA		
	Pyrazolam				25 (S) 50 (U)	NA		NA		
LC-MS/MS	Etizolam (and other traditional BZD)	Blood	Ionic liquid-dispersive liquid-liquid microextractions 5IL-DLLME- (1-butyl-3-methylimidazolium hexafluorophosphate BMIM PF6, pH = 8)	Kinetex® Biphenyl Column (100 mm × 2.1 mm i.d., 2.6 μm), 45 °C *Gradient* Aqueous buffer pH 8.0 (A)/ MeOH (B)	0.26	2	250	NA	NA	[47]
Review of Several methods	Fonazepam (desmethylflunitrazepam)	Blood Urine	LLE SPE Dilute and shoot	Different conditions according to the method	0.04–12	0.125–100	12.5–1000	Blood 0.125–1000 Urine 1–1000	Clinical and forensic tests, DUID, PK study, metabolism	[48]

(continued)

Table 2
(continued)

Method	Analyte	Matrix	Extraction	Chromatography conditions	LOD (ng/mL)	LLOQ (ng/mL)	ULOQ (ng/mL)	Concentrations found [min-max ng/mL] (n) or cases[a] (n)	Circumstances	Ref.
GC-MS/MS	Phenazepam Diclazepam Flubromazepam Etizolam	Urine	UA-LDS-DLLME (ultrasound-assisted low-density solvent dispersive liquid-liquid microextraction) (ethyl acetate)	DB-5MS capillary column (30 m × 0.32 mm i.d., 0.25 μm) 100–300 °C (20–10 °C/min) Carrier: helium	3 1 3 3	10 3.5 10 10	10000 10000 10000 10000	158 NA NA NA	Drug-trafficker suspect (n = 1)	[56]
LC-MS/MS	Flubromazolam	Blood Urine (+HLM and tissues)	Protein precipitation (ACN)	ACQUITY BEH C18 (100 × 2.1 mm i.d., 1.7 μm) *Gradient* HCOOH (A)/ACN (B)	NP	0.0005 mg/Kg	0.5 mg/Kg	0.0044–0.008 mg/kg blood	Death (n = 2)	[23]
LC-QTOF-MS (ddmS2) (qualitative analysis)	Meclonazepam	Urine (+ HLM)	Urine hydrolysis using β-glucuronidase arylsulfatase	Acquity HSS T3 column (150 mm × 2.1 mm i.d., 1.8 μm) *Gradient* 10 mM NH_4COO (A)/0.05% HCOOH, MeOH (B)	NP	NP	NP	NP	DUID, violent crimes (n = 4)	[24]
LC-MS/MS	Clonazolam, Diclazepam Etizolam Flubormazepam Flubromazolam Pyrazolam	Blood	LLE (ethyl acetate:heptane 4: 1)	BEH Phenyl column *Gradient* 5 mM NH_4HCO_3 (pH 8) (A)/ MeOH (B)	1.4 1.6 14 3.3 0.37 3.5	1.4 1.6 14 3.3 0.37 3.5	NP NP NP NP NP NP	1.9–11 (n = 7) 2.1–57 (n = 15) 19–170 (n = 14) 4.7–1200 (n = 24) 0.48–100 (n = 25) 74 (n = 1)	Forensic casework (DUID, traffic accident, criminal acts) (n = 22022)	[49]
LC-HRMS (qualitative analysis)	Phenazepam (among 19 traditional BZD and metabolites)	Blood Urine	SLE NH_4CH_3COO 0.5 M pH 5, H_2O, NaOH 4.5 M) *Eluent* CH_2Cl_2/ isopropanol (95:5) (+urine hydrolysis using β-glucuronidase)	C18 column (2.1 mm × 50 mm, 1.6 μm), 30 °C *Gradient* ACN, 0.1% HCOOH (A)/5 mM NH_4HCOO, 0.1% HCOOH (B)	3	NA	NA	NA	NA	[66]

LC-MS/MS	Metizolam	Urine	Hydrolysis using β-glucuronidase LLE (dichloromethane/ n-heptane/ isopropanol, 25/65/ 10, v/v)	Acquity BEH C18 column (100 × 2.1 mm i.d., 1.7 μm) *Gradient* H2O, HCOOH 0.1% (A)/ACN (B)	0.025	0.05	50	11 (12 h) 0.24 (35 h)	Self-administration of one 2 mg tablet, PK study	[50]
LC-MS/MS	Phenazepam (among 17 traditional BZD)	Blood Urine	Precipitation (ACN) Urine hydrolysis using β-glucuronidase)	BEH C18 column (100 × 2.1 × mm, i.d., 1.7 μm), 40 °C *Gradient* H_2O, 0.1% HCOOH (A)/ACN, 0.1% HCOOH (B)	1	1	NP	NP	Routine analysis since 2013	[51]
LC-HRMS	Clonazolam Meclonazepam Nifoxipam	Urine	Dilute and shoot	PepMap RSLC Acclaim C18 (150 mm × 50 μm, 2 μm), 40 °C *Gradient* H_2O, 0.1% HCOOH (A)/ACN, 0.1% HCOOH (B)	NP	NP	NP	NA	PK study	[80]
LC-MS/MS	Clonazolam	Urine	Hydrolysis using β-glucuronidase + dilute and shoot	BEH Phenyl column (50 × 1 mm i.d., 1.7 μm), 60 °C *Gradient* H_2O, 0.1% HCOOH (A)/CAN (B)	5	5	500	7.3–23 (n = 8)	DUID (n = 390)	[32]
	Deschloroetizolam				2	5	500	130 (n = 1)		
	Dicalzepam				2	2	200	ND		
	Etizolam				2	5	500	5.8–270 (n = 11)		
	Flubormazepam				2.5	2.5	250	2.7–30 (n = 14)		
	Flubromazolam				2	5	500	5.4–1500 (n = 96)		
	Flutazolam				5	5	500	ND		
	Meclonazepam				1	1	100	1.6–190 (n = 45)		
	Nifoxipam				10	10	1000	10–2800 (n = 4)		
	Phenazepam				5	5	500	26–70 (n = 3)		
	Pyrazolam				4	10	1000	32–920 (n = 9)		
LC-MS/MS (Q-trap)	Phenazepam	Blood (femoral f, cardiac h..) Urine	LLE (hexane:ethyl acetate, 70:30)	Gemini column (150 × 2.0 mm i.d., 5 μm), 40 °C *Gradient* H_2O, 0.1% HCOOH (A)/CAN, 0.1% HCOOH (B)	0.3	0.7	200	f: 90–370 (n = 22) h: 14–310 (n = 18) 7–49	Postmortem tissue distribution study (n = 27)	[11]
	3-OH-phenazepam	Blood (femoral f, cardiac h..) Urine			7	16	1000	f: 19–159 (n = 9) h: 19–161 (n = 9) 17–274 (n = 12)		

(continued)

Table 2 (continued)

Method	Analyte	Matrix	Extraction	Chromatography conditions	LOD (ng/mL)	LLOQ (ng/mL)	ULOQ (ng/mL)	Concentrations found [min-max ng/mL] (n) or cases[a] (n)	Circumstances	Ref.
GC-MS	Phenazepam	Blood	SPE (eluting solvent: ethyl acetate)	HP-5MS capillary column (30 m × 0.25 mm × 0.25 μm) 150–290 °C (5–30 °C/min) Carrier: helium (20 kPa)	NP	1	100	76	DUID (Vehicle crash) (n = 1)	[57]
LC-MS/MS	Nimetazepam 7-Amino-nimetazepam Nitrazepam	Urine	LLE (not described)	ACE5 C18 column (250 mm × 4.6 mm i.d., 5 μm) 5% CAN, 0.1% HCOOH (A)/ 95%ACN, 0.1% HCOOH (B) (isocratic with 2 mobile phases)	0.25 1 0.125	5 5 1	200 200 200	5.5–368 (n = 6) 9.1–9650 (n = 12) 9.4–433 (n = 7)	Confirmation of positive urine specimens by enzyme immunoassay (n = 87)	[52]
LC-MS/MS	Diclazepam Delorazepam Lorazepam Lormetazepam	Serum (S) Urine (U)	LLE (1-chlorobutane, borate buffer pH 9)	Synergi 4u Polar RP column (150 × 2 mm i.d., 4 μm), 40 °C *Gradient* 0.1% HCOOH, 1 mM NH_4HCOO (A)/0.1% HCOOH, MeOH	0.25	NA	NA	S: 3.4 (3 h) U: 0.1–0.5 ng/mg S: 2 (36 h) S: 0.4 (99 h) S: 0.34 (6 h)	Self-administration of 1 mg tablet, PK study	[8, 53]
LC-MS/MS	Flubromazepam				NP	1	100	S: 78 (6 h) U: 0.55–4 ng/mg	Self-administration of 4 mg tablet, PK study	
LC-MS/MS	Nimetazepam 7-Amino-nimetazepam Nitrazepam	Urine	SPE after β-glucuronidase hydrolysis Eluent: ethyl acetate/ NaOH (98: 2)	Synergy Polar-RP column (150 mm × 2.1 mm i.d.,), 40 °C *Gradient* 20 mM NH_4CH_3COO (A)/I(B)	NP	0.1 0.1 0.1	100 100 100	0.5–5.2 25.3–152 ND–3.9	DFSA (n = 7)	[55]
LC-MS/MS	Phenazepam	Plasma	SPE Eluting solvent: MeOH, 2% CH3COOH + Urine hydrolysis using β-glucuronidase/ arylsulfatase)	Luna phenyl-hexyl column (50 × 3 mm i.d., 3 μm), 40 °C *Gradient* I(A)/10 mM NH_4 CH_3COO, 0.1% HCOOH (B)	1.4	3.1	100	DUID: 4–3600 (n = 144) Autopsy: 7–1600 (n = 17)	DUID (n = 4007) Medicolegal autopsy (n = 17)	[54]
		Urine			5	14	2000	12–60 (n = 7)		
	OH-phenazepam	Urine			5	16	2000	37–2580 (n = 7)		

GC-MS/MS	Etizolam	Blood	SPE (eluent: CH3Cl2) Derivatization 70 °C (BSTFA 1% TMCS)	Fused silica capillary column (30 m × 0.25 mm i.d., 0.25 μm) Carrier: helium 150–270 °C (50 °C/min)–285 °C (1 °C/min)–300 (10 °C/min)	NP	5	50	25.8–263.7	Death (n = 2)	[58]
	α-OH-etizolam							7.2–9.4		
	8-OH-etizolam							9.3–11.5		
GC-MS	Nimetazepam	Blood	HLB SPE (eluent: CH2Cl2)	BPX5 capillary column (15 m30.32 mm I.D, film thickness 0.25 μm, SGE) Oven: 60–300 °C (20 °C/min) Carrier: helium (20 kPa)	2	5	500	NA	NA	[59]
	Etizolam				5	5	500	NA	NA	
GC-EC	3-methylclonazepam (Meclonazepam)	Plasma	LLE (butyl acetate)	CP-Sil 5 glass capillary column (25 m × 0.5 mm) 280–300 °C (2 °C/min) Carrier: helium N55 (0.75 bar) Auxiliary: argon-methane (95:5)	0.1	0.62	18.8	Intravenous (1 mg): approx. 1–10 Oral (1 mg): approx. 1–10 Oral (4 mg) approx. 1–20	Intravenous (IV) administration of 1 mg and oral (O) administrations of 1 and 4 mg	[60]
HPLC-UV	Adinazolam	Plasma	LLE (ethyl acetate, toluene)	RP column ACN-MeoH-phosphate buffer (30:20:85, v/v/v) UV absorbance 254 nm	NP	NP	NP	9–300	Self-administration of 20 mg tablet, PK study	[75]
	N-desmethyl-adinazolam							15–350		

The methods are sorted in descending order by publication date, and publications in the same year are listed alphabetically by author's name

NP not provided, *ND* not detected, *NA* not applied

[a]Number of cases are reported only for qualitative screening method

In a recent study conducted by Mastrovito et al., for the detection and quantification of DBZD panel in blood, etizolam, flubromazolam, flualprazolam, diclazepam, delorazepam, flubromazepam, and phenazepam were stable in the sample matrix refrigerated or frozen for 30 days. α-OH-etizolam was stable at room temperature for 14 days and when refrigerated or frozen for 30 days. Nitrazolam and flunitrazolam were stable at room temperature for 7 days and when refrigerated or frozen for 30 days. Clonazolam and bromazolam were found to be stable at room temperature for 48 h. They were stable when refrigerated for 14 days and frozen for 30 days [34].

In contrary, Mei et al. investigated the stability of 13 designer drugs in spiked blood samples in the autosampler at room temperature over 24 and 48 h. Concentrations of nifoxipam, flubromazepam, clonazolam, deschloroetizolam, diclazepam, flualprazolam, and flunitrazolam had dropped at least 25% after 24 h, and the authors concluded that no reinjections are allowed for the quantification of these DBZD. All other analytes had results within the acceptable criteria [45]. Overall, outcomes varied depending on the compound and storage conditions in different experiments [37, 38, 40, 42, 51, 55, 60, 70, 71].

8 Conclusion

The ever-changing DBZD situation continues to be a challenge and a cat-and-mouse game for public health authorities, clinicians, and forensic laboratories in terms of getting non-prohibited medicines on the market and scheduled as quickly as possible. Keeping targeted screening methods up to date in clinical and forensic laboratories is a near-impossible effort. To keep up with the rapidly arriving and fading DBZD, the option of using untargeted high-resolution screening is becoming more popular since it does not require reference standards and supports retrospective data reprocessing.

Another difficulty is the lack of understanding about DBZD metabolites, as well as the numerous interpretive errors associated with similar metabolic pathways between prescription and designer benzodiazepines. The information in this chapter is intended to provide an overview of current analytical techniques for target analysis of DBZD and their metabolites in blood and urine, as well as to point out the most prevalent interpretation errors to avoid.

References

1. Drummer OH, Di Rago M, Gerostamoulos D (2019) Analysis of benzodiazepines for drug-facilitated assaults and abuse settings (Urine). In: Langman LJ, Snozek CLH (eds) LC-MS in drug analysis. Springer, New York, pp 23–39
2. Manchester KR, Lomas EC, Waters L et al (2018) The emergence of new psychoactive substance (NPS) benzodiazepines: a review. Drug Test Anal 10:37–53
3. Moosmann B, Auwärter V (2018) Designer benzodiazepines: another class of new psychoactive substances. Handb Exp Pharmacol 252: 383–410
4. Brunetti P, Giorgetti R, Tagliabracci A et al (2021) Designer benzodiazepines: a review of toxicology and public health risks. Pharmaceuticals 14:560
5. Bounds CG, Nelson VL (2021) Benzodiazepines. In: StatPearls. StatPearls Publishing, Treasure Island
6. Griffin CE, Kaye AM, Bueno FR, Kaye AD (2013) Benzodiazepine pharmacology and central nervous system-mediated effects. Ochsner J 13:214–223
7. European Monitoring Centre for Drugs and Drug Addiction (2021) New benzodiazepines in Europe: a review. https://data.europa.eu/doi/10.2810/725973. Accessed 10 July 2021.
8. Moosmann B, Bisel P, Auwärter V (2014) Characterization of the designer benzodiazepine diclazepam and preliminary data on its metabolism and pharmacokinetics. Drug Test Anal 6:757–763
9. Zawilska JB, Wojcieszak J (2019) An expanding world of new psychoactive substances-designer benzodiazepines. Neurotoxicology 73:8–16
10. Maskell PD, De Paoli G, Nitin Seetohul L, Pounder DJ (2012) Phenazepam: the drug that came in from the cold. J Forensic Leg Med 19:122–125
11. Crichton ML, Shenton CF, Drummond G et al (2015) Analysis of phenazepam and 3-hydroxyphenazepam in post-mortem fluids and tissues. Drug Test Anal 7:926–936
12. Shearer K, Bryce C, Parsons M, Torrance H (2015) Phenazepam: A review of medico-legal deaths in South Scotland between 2010 and 2014. Forensic Sci Int 254:197–204
13. Bäckberg M, Pettersson Bergstrand M, Beck O, Helander A (2019) Occurrence and time course of NPS benzodiazepines in Sweden – results from intoxication cases in the STRIDA project. Clin Toxicol (Phila) 57: 203–212. https://doi.org/10.1080/15563650.2018.1506130
14. Carpenter JE, Murray BP, Dunkley C et al (2019) Designer benzodiazepines: a report of exposures recorded in the National Poison Data System, 2014–2017. Clin Toxicol (Phila) 57:282–286. https://doi.org/10.1080/15563650.2018.1510502
15. Kriikku P, Rasanen I, Ojanperä I et al (2020) Femoral blood concentrations of flualprazolam in 33 postmortem cases. Forensic Sci Int 307: 110101
16. Papsun DM, Krotulski AJ, Homan J et al (2021) Flualprazolam blood concentrations in 197 forensic investigation cases. J Anal Toxicol 45:226–232
17. National Records of Scotland. Drug-related Deaths in Scotland in 2019. https://www.nrscotland.gov.uk/files/statistics/drug-related-deaths/2019/drug-related-deaths-19-pub.pdf. Accessed 9 July 2021
18. Fukasawa T, Yasui–Furukori N, Suzuki A et al (2005) Pharmacokinetics and pharmacodynamics of etizolam are influenced by polymorphic CYP2C19 activity. Eur J Clin Pharmacol 61:791–795
19. Kilicarslan T, Haining RL, Rettie AE et al (2001) Flunitrazepam metabolism by cytochrome P450S 2C19 and 3A4. Drug Metab Dispos 29:460–465
20. Mizuno K, Katoh M, Okumura H et al (2009) Metabolic activation of benzodiazepines by CYP3A4. Drug Metab Dispos 37:345–351
21. Masica AL, Mayo G, Wilkinson GR (2004) In vivo comparisons of constitutive cytochrome P450 3A activity assessed by alprazolam, triazolam, and midazolam. Clin Pharmacol Ther 76:341–349
22. Drummer OH, Odell M (2001) The forensic pharmacology of drugs of abuse. Arnold, London
23. Noble C, Mardal M, Bjerre Holm N et al (2017) In vitro studies on flubromazolam metabolism and detection of its metabolites in authentic forensic samples. Drug Test Anal 9: 1182–1191
24. Vikingsson S, Wohlfarth A, Andersson M et al (2017) Identifying metabolites of meclonazepam by high-resolution mass spectrometry using human liver microsomes, hepatocytes, a mouse model, and authentic urine samples. AAPS J 19:736–742

25. Wagmann L, Manier SK, Felske C et al (2020) Flubromazolam-derived designer benzodiazepines: toxicokinetics and analytical toxicology of clobromazolam and bromazolam. J Anal Toxicol 45:bkaa161
26. Pettersson Bergstrand M, Richter LHJ, Maurer HH et al (2019) In vitro glucuronidation of designer benzodiazepines by human UDP-glucuronyltransferases. Drug Test Anal 11:45–50
27. Chetraru E, Ameline A, Gheddar L et al (2018) Les « designer benzodiazepines » : qu'en sait-on aujourd'hui ? Toxicol Anal Clin 30:5–18. https://doi.org/10.1016/j.toxac.2017.12.001
28. O'Connor LC, Torrance HJ, McKeown DA (2016) ELISA detection of phenazepam, etizolam, pyrazolam, flubromazepam, diclazepam and delorazepam in blood using immunalysis® benzodiazepine Kit. J Anal Toxicol 40: 159–161
29. Pettersson Bergstrand M, Helander A, Hansson T, Beck O (2017) Detectability of designer benzodiazepines in CEDIA, EMIT II Plus, HEIA, and KIMS II immunochemical screening assays. Drug Test Anal 9:640–645
30. Larabi I-A, Mayer C, Anderson V et al (2019) Rapid screening of NPS and conventional drugs of abuse (DoA) using a Biochip Array Technology (BAT) in whole blood and urine samples from forensic and clinical cases: a comparative study. Toxicol Anal Clin 31:S63–S64
31. Ameline A, Richeval C, Gaulier J-M et al (2019) Detection of the designer benzodiazepine flunitrazolam in urine and preliminary data on its metabolism. Drug Test Anal 11: 223–229
32. Pettersson Bergstrand M, Helander A, Beck O (2016) Development and application of a multi-component LC-MS/MS method for determination of designer benzodiazepines in urine. J Chromatogr B Analyt Technol Biomed Life Sci 1035:104–110
33. Degreef M, Vits L, Berry EM et al (2021) Quantification of 54 benzodiazepines and Z-drugs, including 20 designer ones, in plasma. J Anal Toxicol 45:141–153
34. Mastrovito RA, Papsun DM, Logan BK (2021) The development and validation of a novel designer benzodiazepines panel by LC–MS-MS. J Anal Toxicol 45:423–428. https://doi.org/10.1093/jat/bkab013
35. Ares-Fuentes AM, Lorenzo RA, Fernández P, Carro AM (2021) An analytical strategy for designer benzodiazepines and Z-hypnotics determination in plasma samples using ultra-high performance liquid chromatography/tandem mass spectrometry after microextraction by packed sorbent. J Pharm Biomed Anal 194:113779
36. Di Rago M, Pantatan S, Hargreaves M et al (2021) High throughput detection of 327 drugs in blood by LC-MS-MS with automated data processing. J Anal Toxicol 45: 154–183
37. Garcia L, Tiscione NB, Yeatman DT, Richards-Waugh L (2021) Novel and nonroutine benzodiazepines and suvorexant by LC–MS-MS. J Anal Toxicol 45:462–474
38. Garneau B, Desharnais B, Laquerre J et al (2021) A comprehensive analytical process, from NPS threat identification to systematic screening: method validation and one-year prevalence study. Forensic Sci Int 318:110595
39. Hikin LJ, Smith PR, Maskell PD et al (2021) Femoral blood concentrations of the designer benzodiazepine etizolam in post-mortem cases. Med Sci Law 61:122–129
40. Rice K, Hikin L, Lawson A et al (2021) Quantification of Flualprazolam in blood by LC–MS-MS: a case series of nine deaths. J Anal Toxicol 45:410–416
41. Heide G, Høiseth G, Middelkoop G, Øiestad ÅML (2020) Blood concentrations of designer benzodiazepines: relation to impairment and findings in forensic cases. J Anal Toxicol 44: 905–914
42. Sofalvi S, Lavins ES, Kaspar CK et al (2020) Development and validation of an LC-MS-MS method for the detection of 40 Benzodiazepines and three Z-drugs in blood and urine by solid-phase extraction. J Anal Toxicol 44: 708–717
43. Furugen A, Nishimura A, Kobayashi M et al (2019) Quantification of eight benzodiazepines in human breastmilk and plasma by liquid-liquid extraction and liquid-chromatography tandem mass spectrometry: application to evaluation of alprazolam transfer into breastmilk. J Pharm Biomed Anal 168: 83–93
44. Lehmann S, Sczyslo A, Froch-Cortis J et al (2019) Organ distribution of diclazepam, pyrazolam and 3-fluorophenmetrazine. Forensic Sci Int 303:109959
45. Mei V, Concheiro M, Pardi J, Cooper G (2019) Validation of an LC-MS/MS method for the quantification of 13 designer benzodiazepines in blood. J Anal Toxicol 43:688–695
46. Koch K, Auwärter V, Hermanns-Clausen M et al (2018) Mixed intoxication by the synthetic opioid U-47700 and the benzodiazepine flubromazepam with lethal outcome: pharmacokinetic data. Drug Test Anal 10:1336–1341

47. De Boeck M, Missotten S, Dehaen W et al (2017) Development and validation of a fast ionic liquid-based dispersive liquid-liquid microextraction procedure combined with LC-MS/MS analysis for the quantification of benzodiazepines and benzodiazepine-like hypnotics in whole blood. Forensic Sci Int 274: 44–54
48. Katselou M, Papoutsis I, Nikolaou P et al (2017) Metabolites replace the parent drug in the drug arena. The cases of fonazepam and nifoxipam. Forensic Toxicol 35:1–10
49. Høiseth G, Tuv SS, Karinen R (2016) Blood concentrations of new designer benzodiazepines in forensic cases. Forensic Sci Int 268: 35–38
50. Kintz P, Richeval C, Jamey C et al (2017) Detection of the designer benzodiazepine metizolam in urine and preliminary data on its metabolism. Drug Test Anal 9:1026–1033
51. Mata DC, Davis JF, Figueroa AK, Stanford MJ (2016) Ultra performance liquid chromatography with tandem mass spectrometry for the quantitation of seventeen sedative hypnotics in six common toxicological matrices. J Anal Toxicol 40:58–63
52. Lee HH, Lee JF, Lin SY et al (2013) Simultaneous quantification of urine flunitrazepam, nimetazepam and nitrazepam by using liquid chromatography tandem mass spectrometry. Clin Chim Acta 420:134–139
53. Moosmann B, Huppertz LM, Hutter M et al (2013) Detection and identification of the designer benzodiazepine flubromazepam and preliminary data on its metabolism and pharmacokinetics. J Mass Spectrom 48:1150–1159
54. Kriikku P, Wilhelm L, Rintatalo J et al (2012) Phenazepam abuse in Finland: findings from apprehended drivers, post-mortem cases and police confiscations. Forensic Sci Int 220: 111–117
55. Wang K-C, Cheng M-C, Hsieh C-L et al (2013) Determination of nimetazepam and 7-aminonimetazepam in human urine by using liquid chromatography-tandem mass spectrometry. Forensic Sci Int 224:84–89
56. Meng L, Zhu B, Zheng K, Fu S (2017) Ultrasound-assisted low-density solvent dispersive liquid-liquid microextraction for the determination of 4 designer benzodiazepines in urine samples by gas chromatography-triple quadrupole mass spectrometry. J Chromatogr B Analyt Technol Biomed Life Sci 1053:9–15
57. Kerrigan S, Mellon MB, Hinners P (2013) Detection of phenazepam in impaired driving. J Anal Toxicol 37:605–610
58. Nakamae T, Shinozuka T, Sasaki C et al (2008) Case report: Etizolam and its major metabolites in two unnatural death cases. Forensic Sci Int 182:e1–e6
59. Inoue H, Maeno Y, Iwasa M et al (2000) Screening and determination of benzodiazepines in whole blood using solid-phase extraction and gas chromatography/mass spectrometry. Forensic Sci Int 113:367–373
60. Coassolo P, Aubert C, Cano JP (1985) Plasma determination of 3-methylclonazepam by capillary gas chromatography. J Chromatogr 338: 347–355
61. Gundersen POM, Broecker S, Slørdal L et al (2020) Retrospective screening of synthetic cannabinoids, synthetic opioids and designer benzodiazepines in data files from forensic post mortem samples analysed by UHPLC-QTOF-MS from 2014 to 2018. Forensic Sci Int 311:110274
62. Wagmann L, Manier SK, Bambauer TP et al (2020) Toxicokinetics and analytical toxicology of Flualprazolam: metabolic fate, isozyme mapping, human plasma concentration and main urinary excretion products. J Anal Toxicol 44:549–558
63. Partridge E, Trobbiani S, Stockham P et al (2018) A case study involving U-47700, Diclazepam and Flubromazepam-application of retrospective analysis of HRMS data. J Anal Toxicol 42:655–660
64. Pettersson Bergstrand M, Meyer MR, Beck O, Helander A (2018) Human urinary metabolic patterns of the designer benzodiazepines flubromazolam and pyrazolam studied by liquid chromatography-high resolution mass spectrometry. Drug Test Anal 10:496–506
65. van Wijk XMR, Yun C, Hooshfar S et al (2019) A liquid-chromatography high-resolution mass spectrometry method for non-FDA approved benzodiazepines. J Anal Toxicol 43:316–320
66. Jagerdeo E, Schaff JE (2016) Rapid screening for drugs of abuse in biological fluids by ultra high performance liquid chromatography/Orbitrap mass spectrometry. J Chromatogr B Analyt Technol Biomed Life Sci 1027:11–18
67. Fabresse N, Larabi IA, Stratton T et al (2019) Development of a sensitive untargeted liquid chromatography-high resolution mass spectrometry screening devoted to hair analysis through a shared MS2 spectra database: a step

toward early detection of new psychoactive substances. Drug Test Anal 11:697–708
68. Pettersson Bergstrand M, Beck O, Helander A (2018) Urine analysis of 28 designer benzodiazepines by liquid chromatography–high-resolution mass spectrometry. Clin Mass Spectrom 10:25–32
69. Tomková J, Švidrnoch M, Maier V, Ondra P (2017) Analysis of selected designer benzodiazepines by ultra high performance liquid chromatography with high-resolution time-of-flight mass spectrometry and the estimation of their partition coefficients by micellar electrokinetic chromatography. J Sep Sci 40: 2037–2044
70. Papoutsis II, Athanaselis SA, Nikolaou PD et al (2010) Development and validation of an EI-GC-MS method for the determination of benzodiazepine drugs and their metabolites in blood: applications in clinical and forensic toxicology. J Pharm Biomed Anal 52:609–614
71. Mata DC (2016) Stability of 26 sedative hypnotics in six toxicological matrices at different storage conditions. J Anal Toxicol 40:663–668
72. Moosmann B, Bisel P, Franz F et al (2016) Characterization and in vitro phase I microsomal metabolism of designer benzodiazepines – an update comprising adinazolam, cloniprazepam, fonazepam, 3-hydroxyphenazepam, metizolam and nitrazolam. J Mass Spectrom 51:1080–1089
73. Moosmann B, Bisel P, Westphal F et al (2019) Characterization and in vitro phase I microsomal metabolism of designer benzodiazepines: an update comprising flunitrazolam, norflurazepam, and 4′-chlorodiazepam (Ro5-4864). Drug Test Anal 11:541–549
74. Lahti RA, Sethy VH, Barsuhn C, Hester JB (1983) Pharmacological profile of the antidepressant adinazolam, a triazolobenzodiazepine. Neuropharmacology 22:1277–1282
75. Peng GW (1984) Assay of adinazolam in plasma by liquid chromatography. J Pharm Sci 73:1173–1175
76. Fraser AD, Isner AF, Bryan W (1993) Urinary screening for adinazolam and its major metabolites by the Emit d.a.u. and FPIA benzodiazepine assays with confirmation by HPLC. J Anal Toxicol 17:427–431
77. Cho MJ, Sethy VH, Haynes LC (1986) Sequentially labile water-soluble prodrugs of alprazolam. J Med Chem 29:1346–1350
78. Schukin SI, Zinkovsky VG, Zhuk OV (2011) Elimination kinetics of the novel prodrug cinazepam possessing psychotropic activity in mice. Pharmacol Rep 63:1093–1100
79. Huppertz LM, Bisel P, Westphal F et al (2015) Characterization of the four designer benzodiazepines clonazolam, deschloroetizolam, flubromazolam, and meclonazepam, and identification of their in vitro metabolites. Forensic Toxicol 33:388–395
80. Meyer MR, Bergstrand MP, Helander A, Beck O (2016) Identification of main human urinary metabolites of the designer nitrobenzodiazepines clonazolam, meclonazepam, and nifoxipam by nano-liquid chromatography-high-resolution mass spectrometry for drug testing purposes. Anal Bioanal Chem 408:3571–3591
81. El Balkhi S, Chaslot M, Picard N et al (2017) Characterization and identification of eight designer benzodiazepine metabolites by incubation with human liver microsomes and analysis by a triple quadrupole mass spectrometer. Int J Legal Med 131:979–988
82. Mortelé O, Vervliet P, Gys C et al (2018) In vitro Phase I and Phase II metabolism of the new designer benzodiazepine cloniprazepam using liquid chromatography coupled to quadrupole time-of-flight mass spectrometry. J Pharm Biomed Anal 153:158–167
83. Wohlfarth A, Vikingsson S, Roman M et al (2017) Looking at flubromazolam metabolism from four different angles: metabolite profiling in human liver microsomes, human hepatocytes, mice and authentic human urine samples with liquid chromatography high-resolution mass spectrometry. Forensic Sci Int 274:55–63
84. Huppertz LM, Moosmann B, Auwärter V (2018) Flubromazolam – basic pharmacokinetic evaluation of a highly potent designer benzodiazepine. Drug Test Anal 10:206–211
85. Richter LHJ, Maurer HH, Meyer MR (2017) New psychoactive substances: studies on the metabolism of XLR-11, AB-PINACA, FUB-PB-22, 4-methoxy-α-PVP, 25-I-NBOMe, and meclonazepam using human liver preparations in comparison to primary human hepatocytes, and human urine. Toxicol Lett 280:142–150
86. Zherdev VP, Caccia S, Garattini S, Ekonomov AL (1982) Species differences in phenazepam kinetics and metabolism. Eur J Drug Metab Pharmacokinet 7:191–196
87. WHO (2015) Phenazepam. Pre-review Report. https://www.who.int/medicines/access/controlled-substances/5.8_Phenazepam_PreRev.pdf. Accessed 7 July 2021
88. Moosmann B, Hutter M, Huppertz LM et al (2013) Characterization of the designer benzodiazepine pyrazolam and its detectability in

human serum and urine. Forensic Toxicol 31: 263–271

89. Pettersson Bergstrand M, Meyer MR, Beck O, Helander A (2017) Human urinary metabolic patterns of the designer benzodiazepines flubromazolam and pyrazolam studied by liquid chromatography-high resolution mass spectrometry. Drug Test Anal 10:496–506

90. Švidrnoch M, Boráňová B, Tomková J et al (2018) Simultaneous determination of designer benzodiazepines in human serum using non-aqueous capillary electrophoresis – Tandem mass spectrometry with successive multiple ionic – polymer layer coated capillary. Talanta 176:69–76

Chapter 6

Target Analysis of NPS in Alternative Matrix Hair

Karen S. Scott and Stephanie R. VanLysebettens

Abstract

An overview of the detection of novel psychoactive substances (NPS) and their detection in hair is provided. The overview focuses on the major NPS groups, namely opioids, benzodiazepines, stimulants and hallucinogens, and synthetic cannabinoids with a focus on preparation techniques for each group followed by extraction and analysis information. Pre-analytical techniques for the preparation and extraction of NPS drugs from the hair matrix have remained somewhat constant over the last decade. Sample cleanup techniques remain focused on liquid–liquid and solid phase extraction, although with the increased availability of liquid chromatography mass spectrometric confirmation techniques many methods are now direct inject with no additional cleanup. Discussion of some of the challenges faced with this particular group of drugs is provided along with the specific challenges associated with the analysis of the hair matrix.

Key words Hair, NPS, NPO, Synthetic cannabinoids, Benzodiazepines, LC-MS/MS, Alternative matrix, Stimulants, Hallucinogens

1 Introduction

Hair is an important toxicological specimen as it provides mid- to long-term information on an individual's drug use history which cannot be provided from other biological samples without excessive sampling and analysis. It is therefore an important tool in drug-facilitated crimes, postmortem toxicology, child custody cases, and other toxicological cases where drug use history is important to the investigation. With the ever-expanding NPS drug market the detection of these drugs in hair can give us a detailed overview of an individual's historical NPS use and in some instances may provide the user with information on drugs they ingested but were not aware of.

This is a fast-changing market and what the user believes they are purchasing may not actually be what they are ingesting. Websites such as "NPS Discovery" likely provide the most up to date information on what drugs are being seen in casework in the United States [1]. Other countries worldwide may have similar

Marta Concheiro and Karl B. Scheidweiler (eds.), *Methods for Novel Psychoactive Substance Analysis*,
Methods in Pharmacology and Toxicology, https://doi.org/10.1007/978-1-0716-2605-4_6,

monitoring systems and laboratories need to make use of all available early warning statistics to keep on top of what to look for in casework.

The first Subheadings (1.1, 1.2, 1.3, and 1.4) of this chapter summarize current trends in NPS detection as it is important to know what drugs are currently being observed in casework, albeit briefly. It is interesting to note that while certain NPS drugs are short lived, many, e.g. mephedrone, recur, so continued monitoring in hair and other matrices is not futile. The information provided in this chapter is an overview on techniques for analysis of the major drug groups within the NPS categories and while some of the drugs listed will remain in use, some are already out of favor due to the nature of the NPS market. However, the broad chemistries of these drugs mean many of these methods are still useful starting points for future method development. In addition, there is no way to predict which of the drugs will come back into favor within the NPS using population in the future, and hair, unlike blood and urine, can be used to monitor long-term drug use, monitoring trends from previous months is useful.

1.1 Opioids

Opioids remain the largest subclass of new NPS to appear on the market since 2018 [2]. While fentanyl analogs (fentalogs), such as fluorofentanyl, are typically detected more frequently in casework, the most prevalent opioid analog group in 2021 was the nitazenes with nine analogs detected, compared to seven fentalogs. Table 1 includes the current list of novel psychoactive opioids (NPO) which have been detected in US casework since January 2021. The NPOs of most interest in Forensic Toxicology casework at the time of writing are fluorofentanyl, brophine, 2-methyl AP-237, metonitazene, N-pyrrolo-etonitazene, and isotonitazene. Even within this group, by late 2020 to early 2021, brophine has started to decline and fluorofentanyl cases showed a sharp increase.

1.2 Benzodiazepines

Not unexpectedly, the benzodiazepines account for the highest prevalence of detected NPS drugs in Forensic Toxicology casework [2]. They are frequently seen in polydrug cases alongside traditional and novel opioids. Many of the drugs are simple derivatives of traditional benzodiazepines with the addition or substitution of a halogen. For example, bromazolam is the brominated version of alprazolam. In 2021, the main novel psychoactive benzodiazepines detected were etizolam, flualprazolam, clonazolam, flubromazolam, bromazolam, and flubromazepam. While little is known of the pharmacology of these drugs, the pharmacology of benzodiazepines is well established. With the exception of bromazolam, all of these drugs have 2′ substitution with an electron withdrawing group indicating increased activity as well as a potential for amnesic effects associated with some of the benzodiazepines. Table 1 includes the current list of novel psychoactive benzodiazepines which were detected in US casework in 2021.

Table 1
Selection of NPS detected in US casework, 2021

Stimulants & Hallucinogens	Opioids	Benzodiazepines	Synthetic Cannabinoids
2-Fluorodeschloroketamine	**Nitazenes**	Adinazolam	ADB-HEXINACA
4-AcO-EPT	Etodesnitazene	Bromazepam	ADB-4en-PINACA
4-HO-DiPT	Flunitazene	Bromazolam	4F-MDMB-BINACA
α-D2PV	Isotonitazene	Clonazolam	4F-ABINACA
α-PBP	Metodesnitazene	Descholoroetizolam	4F-MDMB-BICA
α-PHiP/ α-PHP	Metonitazene	Desalkylflurazepam	5F-MDMB-PICA
α-PPP	N-Piperidinyl etonitazene	Diclazepam	ADB-BINACA
α-PVP	N-Pyrroloetonitazene	Estazolam	MDMB-4en-PINACA
Benzylone	Protonitazene	Etizolam	ACHIMINACA
Butylone		Flualprazolam	5F-EDMB-PICA
Cl-α-PVP	**Fentalogs**	Flubromazepam	5F-ADB (5F-MDMB-PINACA)
Cl-PCP	Bromofentanyl	Flubromazolam	
Deschloroketamine	Carfentanil	Fluclotizolam	
Dibutylone	Chlorofentanyl	Meclonazepam	
Dimethylone	Dimethoxyfentanyl	Metizolam	
Dimethylpentylone	Fluorofentanyl	Phenazepam	
Ethylphenidate	Methyltetrahydrofuranylfentanyl	Pyrazolam	
Eutylone	Valerylfentanyl		
Fluoromethamphetamine			
HO-PCP	**Other**		
HO-PCE	2-Methyl-AP-237		
Hydroxetamine	AP-238		
Methcathinone	Brophine		
MeO-PCP	Naphthyl-U-47700		
Methylbenzylpiperazine	Tianeptine		
Methylenedioxy-α-PHP	Trifluoromethyl-U-47700		

(continued)

Table 1 (continued)

Stimulants & Hallucinogens	Opioids	Benzodiazepines	Synthetic Cannabinoids
Methylenedioxy-α-PVP			
Methylenedioxy-PV8			
Methylone			
Methylmethcathinone			
N-Benzyl-dimethoxyamphetamine			
N-Butylpentylone			
N-Ethyldeschloroketamine			
N-Ethylhexedrone			
N-Ethylpentedrone			
N-Ethylpentylone			
Pentylone			

1.3 Stimulants and Hallucinogens

In terms of the number of individual drugs detected, stimulants and hallucinogens are the most diverse of the NPS categories, despite accounting for the smallest percentage of drugs detected in casework each year [2]. Thirty-seven individual drugs were detected 2021 (compared to 22 opioids, 17 benzodiazepines, and 15 synthetic cannabinoids). The vast majority of compounds are detected in only one or two cases with only N-ethyl pentylone, 2-fluorodeschloroketamine, and eutylone being prevalent in casework. Table 1 includes the current list of novel psychoactive stimulants and hallucinogens which were detected in US casework since January 2021. Some of the drugs, e.g. methylone, have been scheduled drugs for decades while others such as the PCP analog 3-chloro-PCP were only identified in Europe in 2020 [3]. Some drugs such as α-PVP and mephedrone are reappearing in casework making these useful drugs to include when developing methods for the analysis of drugs in hair.

1.4 Cannabinoids

Synthetic cannabinoids began to be manufactured and sold in the early 2000s. They are the largest group of novel substances monitored with 209 drugs reported between January 2001 and December 2020, and they are a structurally diverse class [4]. Fifteen synthetic cannabinoids were detected in the US in 2021 as shown in Table 1. The most prevalent drugs in forensic toxicology casework are currently all indazole-based synthetic cannabinoids; MDMB-4en-PINACA, ADB-BINACA, and 5F-MDMB-PINACA.

2 Pre-analytical Considerations

Regardless of the NPS being analyzed, all hair samples must undergo relatively complex pre-analytical considerations. A simplified schematic is shown in Fig. 1. The sample should be collected by trained personnel cutting as close to the skin as possible to allow establishment of an accurate time frame. Ideally hair from the posterior vertex of the scalp should be collected [5]. The collected sample should be carefully aligned in an approved hair collection kit to facilitate segmenting. By cutting the hair into segments of between 10 and 30 mm, it may be possible to obtain a historical representation of an individual's drug exposure. In cases of alleged drug-facilitated sexual assault, where a single drug exposure is expected, it may be necessary to cut into smaller segments. This also helps to narrow down the time window.

Following segmentation, each segment of hair should be washed, to remove hair care products, the body's natural fluids, and skin cells as these may interfere with the analysis. In addition, the wash procedure should be designed to remove any potential external contamination of drugs from the environment. As with liquid samples, it is important hair samples are homogenized prior to analysis. This can be achieved by either snipping with scissors or powdering in a ball mill or similar homogenizer unit. At this point, the hair is ready for analysis.

3 Analysis of NPS in Hair

Analysis of cut or powdered hair samples involves three stages [5]. The first stage involves an incubation or digestion to move the drugs from the hair matrix into solution. Once the drugs are in

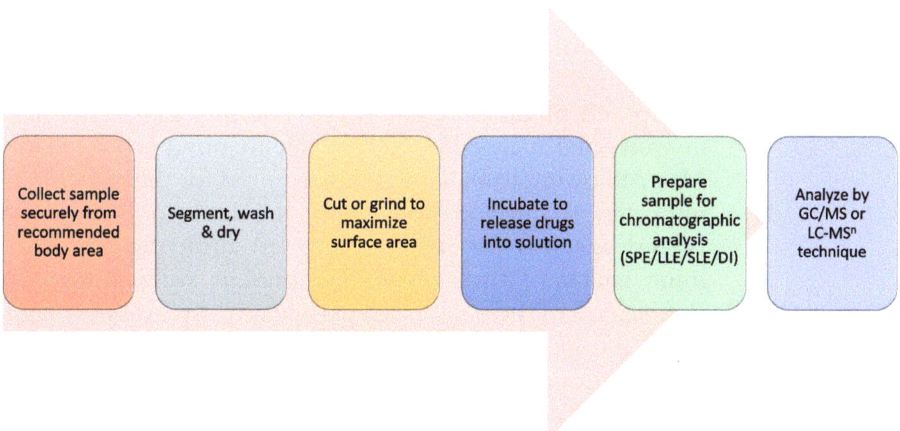

Fig. 1 Typical hair analysis workflow

solution, analysis proceeds similarly to other liquid matrix analysis. Traditionally liquid–liquid or solid phase extraction cleanup techniques are used prior to analysis by gas chromatography mass spectrometry (GC/MS). However, many of the newer NPS methods use liquid chromatography tandem mass spectrometry (LC-MS/MS), which does not require as rigorous a cleanup. While some labs still include SPE or LLE, others simply use a direct injection of the incubation medium and some others have combined the homogenization and incubation steps followed by a direct injection (see Table 2).

3.1 Screening for NPS in Hair

While the focus of this chapter is targeted analysis, it is worth mentioning preparation for screening. Screening for NPS drugs is challenging. Some groups of drugs, e.g. the benzodiazepines and fentanyl analogs, are relatively straight forward as all drugs have the same core structure and can be expected to have some cross reactivity with a standard immunoassay kit. Other groups however may have to rely on chromatographic screening methods such as quadrupole time of flight liquid chromatography mass spectrometry (QTOF-LC/MS). This technique is especially useful as data can be retrospectively probed once standards become available for new NPS drugs [2]. Chromatographic screening techniques such as GC/MS and LC/MS are more commonly used as these are still more readily available in most laboratories. For high-throughput laboratories using ELISA screening, it is essential to determine negative samples quickly and efficiently, and in these cases a portion of the sample may be removed from the bulk hair and not fully cut or pulverized. In general, post wash and cutting/powdering, and screening can be performed following a 2 h methanolic incubation, dry down and reconstitution in solvent suitable for the screening technique, e.g. ELISA buffer or LC mobile phase [6]. Following this, more specific, targeted conformational analysis can be completed.

3.2 Confirmation Analysis for NPS in Hair

As with other matrices, within the literature, there are papers describing both multianalyte, cross drug group analyses and single group and single drug analytical procedures [7]. The majority of instrumental analyses for NPS in hair are now performed by UPLC-MS/MS owing to the sensitivity required for this matrix. While it is important to develop methods for the detection of a wide range of NPS drugs, it is also important to test authentic samples to determine which of these drugs are present within the population. Publication of these studies is particularly useful as they help laboratories set upper and lower calibration limits when developing new methods, as well as set appropriate controls. Table 2 describes a selection of papers indicating which NPS have been detected in the hair of users. Details of the methods as well as the drugs detected in authentic hair samples along with the ranges of concentrations detected are also included.

Table 2
Analytical conditions of drugs detected in casework and brief analytical conditions

Drugs detected in casework	Hair concentrations (pg/mg)	Wash procedure	Sample preparation/incubation	Extraction	Chromatography/instrumentation	Reference
APP-FUBINACA Butyrylfentanyl Butyryl norfentanyl Mephedrone UR144 5-methoxy-methyltryptamine 6-(2-aminopropyl) benzofuran	50 380 160 240 100 70 70	Dichloromethane 2x	25 mg incubated with 500 μL M3® reagent at controlled temperature of 100 °C for 60 min	After cooling to room temperature supernatant was injected directly into system	UPLC Acquity I Class, Waters XEVO TQ-S Micro tandem quadrupole mass spectrometer, Oasis HLB (4.6 mm × 20 mm, 5 μm) column	[9]
AB-FUBINACA 5F-AB-PINACA 5F-APINACA AB-PINACA 5F-PB-22 PB-22 JWH-200 STS-135 MAM-2201 Pentedrone p-Methoxymethamphetamine Dimethoxyamphetamine Dimethyltryptamine Ethylone Butylone p-methoxy-methamphetamine α-pyrrolidinovalerophenone R/S Cathinone 2-methiopropanamine 3,4-methylenedioxy-α-pyrrolidinopropiophenone α-pyrrolidinobutiophenone Methylenedioxypyrovalerone Pentylone Ethylphenidate Pyrovalerone Dimethyl-tryptamine Methylone 25C-NBOMe		2 mL methanol 3x, 2 mL dichloromethane 3x, 2 mL methanol 3x, dried between washes	Homogenized using BeadRuptor™ in extraction medium (1 mL 0.1 M HCl in MeOH) before and after incubation, incubation for 3 h at 40 °C	Supernatant collected and injected directly into system	LC-MS/MS system consisting of a Shimadzu Nexera X2 30 AD, Sciex 4500 Q-Trap triple quadrupole mass spectrometer, Kinetex C18 (4.6 mm × 50 mm, 2.6 μm) column	[8]

(continued)

Table 2
(continued)

Drugs detected in casework	Hair concentrations (pg/mg)	Wash procedure	Sample preparation/incubation	Extraction	Chromatography/instrumentation	Reference
Methoxyphencyclidine						
U-47700						
Acetylfentanyl						
Isotonitazene	32–182	Shake for 5 min in Milli-Q Water, 5 min in acetone, 5 min in hexane, dry under nitrogen	1.2 mL methanol added to 50 mg hair, mixed with Retsch MM 400 for 90 min at 30 times per second, centrifuged for 10 min at 8000 rpm	Evolute CX Express extraction	UHPLC-MS/MS system consisting of an Acquity separations module, Acquity TQD mass detector, Accucore RP-MS C18 (50 mm × 2.1 mm, 2.6 μm) column	[16]
Diphenidine	4400	2 mL dichloromethane 2x, 2 mL methanol 2x	1.5 mL methanol added to 25 mg of hair followed by incubation at 55 °C for 15 h without stirring	Supernatant collected and injected directly into system	UHPLC-MS/MS system consisting of an Agilent 1290 Infinity LC system, QTRAP® 4500 mass spectrometer with a Zorbax Eclipse Plus C18 RRHD (100 mm × 2.1 mm, 1.8 μm) column	[10, 21]
Methylone	<LOQ-28					
Methoxetamine	7.7–27					
α-pyrrolidinovalerophenone	1040					
4-fluoroamphetamine	55					
Mephedrone	50–59					
4-methylethcathinone	330					
3,4-methylenedioxypyrovalerone	120					
Methoxetamine	3–19					
Methylone	<6–98					
α-pyrrolidinovalerophenone	6					
4-fluoroamphetamine	29					
5/6-(2-aminopropyl) benzofuran	82					
Furanyl-fentanyl	44	2 mL dichloromethane 2x, 2 mL methanol 2x	1.5 mL methanol added to 25 mg of hair followed by incubation at 55 °C for 15 h without stirring	Supernatant collected and injected directly into system	UHPLC-MS/MS system consisting of a Shimadzu LC-30A Series, API 5500 triple quadrupole mass spectrometer, CORTECS UPLC C18 plus (100 mm × 2.1 mm, 1.6 μm) column	[14]

m-chlorophenylpiperazine	6578	Not provided	30 mg incubated with 2 mL methanol with 0.1% HCl for 1 h at 60 °C	Solid phase extraction with Strata-XC cartridges	LC-MS/MS system with an Atlantis T_3 (50 mm × 2.1 mm, 3 μm) column	[26]
JWH-018 JWH-018 N-5-OH JWH-018 N-COOH JWH-073 JWH-073 N-COOH AM-2201 AM-2201 N-4-OH AM-2201 N-6-OHindole JWH-122 JWH-122 N-5-OH MAM-2201	0.4–38.9 0.3–37.2 0.2–1.1 0.1–0.8 0.3 1.7–739.0 0.4 0.2–3.1 0.1–402.0 0.1–3.5 0.2–276.0	Washed in fritted reservoirs, 2 mL of methanol, distilled water, and methanol	2 mL methanol added to samples followed by mild stirring at 38 °C for 16 h	Supernatant collected and injected directly into system	Agilent 1290 Infinity UHPLC, AB Sciex 5500Qtrap tandem mass spectrometer, Zorbax Eclipse plus C_{18} (RRHD 2.1 mm × 100 mm, 1.8 μm)	[28]
Mephedrone	21.11	2 mL dichloromethane for 2 min 2x	Enzymatic digestion in glass vial containing Cleland's reagent (100 mg) and enzyme proteinase K (15 mg) Incubation with Tris buffer (1 mL) for 2 h at 37.5 °C with constant stirring	Hexane (3 mL) vortexed, centrifuged at 5 °C for 5 min at 1750 × *g*, dry down using nitrogen gas, reconstituted in 200 μL acetonitrile	Accela UPLC system, Triple Quadrupole TSQTM mass spectrometer, Agilent SB-C_{18} (2.1 μm × 150 mm × 1.8 μm) column	[25]
4-bromo-2,5 dimethoxyphenethylamine N-ethylamphetamine N-ethylbuphedrone N-ethyldeschloroketamine 3-trifluoromethylphenylpiperazine 3,4-dimethylmetcathinone 4-fluoroamphetamine 4-fluoromethcathinone 4-methoxyphencyclidine 4-methoxy-α-pyrrolidinopropiophenone 5-methoxy-N,n-dimethyltryptamine Benzedrone Benzylpiperazine Brephedrone	14.3–121 10.1–633 32–158 40 10.3–99 10–11 11–113,600 42 647 19 96–945 11.2–152 78.4–196 2730	Water and acetone	Cut into small snippets. 50 mg added to vial with extraction buffer, (methanol/acetonitrile/H_2O with ammonium formate; 25/25/50), samples shaken twice at 40 °C for 18 h at 900 rpm. Supernatant removed and second extraction performed with 500 μL of extraction buffer,	Supernatants combined and filtered through a regenerated cellulose membrane filter	LC Infinity 1290, 6460 triple quadrupole mass spectrometer, Kinetex C_{18} (2.1 mm × 150 mm, 1.7 μm) column	[11]

(continued)

Table 2
(continued)

Drugs detected in casework	Hair concentrations (pg/mg)	Wash procedure	Sample preparation/incubation	Extraction	Chromatography/instrumentation	Reference
Bupropion	11.9–3470					
Butylone	312					
Cathinone	19.8–1,270					
m-chlorophenylpiperazine	12–580					
Deschloroketamine	13–44					
Dimethyltryptamine	13.3–609					
Ethylcathinone	12–52					
Etizolam	53					
Eutylone	65					
Heliomethylamine	22–28					
Methylenedioxypyrovalerone	15.8–8790					
Mephedrone	10.8–3500					
Metamfepramone	10.3–72.1					
Methcathinone	10–13					
Pentedrone	15–61					
Pentylone	24					
α-PPP	11–12					
Pyrazolam	56					
Pyrovalerone	20					
U-47700	10–333					
Etizolam	24–107 (n = 1)	Water, detergent, organic solvents		Ultrasonic extraction using n-chlorobutane	LC-MS	[19]
Valeryl fentanyl Carfentanil *Para/ortho*-Fluorofentanyl	None detected	Water 2x, Acetone 3x, airdry at room temperature	Cryogenic Grinding	MeOH, ACN, 2 mmol/L ammonium acetate (pH 5.3)	Waters Acquity UPLC with Acquity HSS T3 (100 mm × 2.1 mm, 1.8 μm) column: (A) 20 mmol NH_4COO w 0.1% FA, (B) ACN gradient, Sciex 6500+ QTRAP	[12]
Bromazepam N-Desalkylflurazepam Phenazepam	None detected	Water, acetone, and hexane, 2 min each	Cut into small snippets then pulverized in a mixer mill	Methanol followed by 5 mM ammonium formate buffer adjusted to pH 3.5 with formic acid) and methanol (1:1, v/v) in a mixer mill	LC-MS system with AB Sciex 3200 QTRAP and AB Sciex 5500 QTRAP, Phenomenex Kinetex, (50 mm × 2.1 mm, 2.6 μm) column	[20]

AB-001	0.8–40	Deionized water, acetone, petroleum ether, 4 mL each shaken for 4 min, air-dry at room temperature	50 mg of hair weighed with 1.5 mL MeOH added, cut into 1–2 mm pieces	3 h ultrasonication with 1 mL of methanolic extract transferred and evaporated to dryness under gentle nitrogen stream at 40 °C, reconstitution in 100 µL mobile phase A/B (50/50, *v/v*)	LC-MS system with Sciex QTRAP 4000 triple-quadrupole mass spectrometer, Shimadzu Prominence HPLC system, Kinetex C18 (100 mm × 2.1 mm, 2.6 µm) column	[29]
AB-CHMINACA	0.8–1000					
AB-FUBINACA	1.0–1700					
ABICA-5F	1.8–31					
AB-PINACA	1.5–220					
AB-PINACA-5F	0.8–46					
ADB-5F	1.3–2400					
ADB-CHMINACA	3.5–210					
ADB-PINACA	25–31					
AKB48	0.9–54					
AKB48-5F	1.1–1500					
AMB-CHMICA	140					
AMB-PICA-5F	2.1–7.5					
APICA	1.2-16					
MDMB-CHMICA	1.5-1,300					
MDMB-FUBICA	1.4-120					
MDMB-PICA-5F	2.2-29					
para-Methoxyamphetamine	870	2x with acetonitrile then Milli-Q water, both 5 mL for 2 min, dry at room temperature with gentle nitrogen stream	Cut into ~2 mm pieces with scissors and incubated with 1% HCl in methanol at 45 °C for 24 h with continuous stirring	Reconstitution in 100 µL IS solution and 5 µL directly injected	LC-MS/MS system with Shimadzu LC-$20AD_{XR}$, Applied Biosystem MDS Sciex API 2000 triple quadrupole, Kinetex PFP (75 mm × 2.1 mm, 2.6 µm) column	[22]
para-Methoxy-N-methylamphetamine	<LOQ					
Trimethoxyamphetamine	720					
Dimethoxyamphetamine	1300					
2,5-Dimethoxy-4-methylamphetamine	700					
2,5-Dimethoxy-4-ethylamphetamine	900					
2,5-Dimethoxy-4-bromoamphetamine	850					
2,5-Dimethoxy-4-bromophenethylamine	370					
2,5-Dimethoxy-4-iodophenethylamine	450					
2,5-Dimethoxy-4-ethylthiophenethylamine	250					
2,5-Dimethoxy-4-*n*-propylthiophenethylamine	300					

(continued)

Table 2 (continued)

Drugs detected in casework	Hair concentrations (pg/mg)	Wash procedure	Sample preparation/incubation	Extraction	Chromatography/instrumentation	Reference
Etizolam	1.7–5.2 0.44–0.57[a]	Gently wiped with tissue paper moistened with distilled water	Pulverized (3 min, 1500 strokes/min), simultaneously extracted with 50 mM borate buffer (pH 8.4, 0.2 mL), ultrasonication for 30 min	Liquid–liquid extraction with 600 μL of chloroform-isopropanol (3:1, v/v), hydrochloric acid ethanol solution (1:20, v/v; 20 μL) added, dry down under nitrogen, reconstitution with 100 μL of 10% methanol, centrifuge at 7000 g for 10 min, 20 μL aliquot of supernatant injected onto system	Shimadzu Prominence Series LC System, AB Sciex QTRAP 5500, InertSustain C18 semimicro (50 mm × 1.0 mm, 3 μm) column	[18]

[a] After 1 month and 3 months of ingestion (n = 3)

3.3 Polydrug Methods

In 2017, Boumba et al. successfully validated a method for 132 NPS including stimulants, hallucinogens, and synthetic cannabinoids in human hair by LC-MS/MS [8]. The method was successfully applied to 23 medicolegal cases where NPS had been detected in blood or cases where use had been suspected. The majority of the cases (20/23) had at least one NPS present in the hair samples. The methodology used is summarized in Table 2. Drugs detected in casework included a variety of synthetic cannabinoids including AB-FUBINACA, 5F-APINACA, and 5F-PB-22; stimulant/hallucinogenic drugs including pentedrone, para-methoxymethamphetamine, dimethoxyamphetamine, and methoxyphencyclidine as well as opioids including U-47700 and acetylfentanyl.

Mannocchi et al. developed a versatile method for both hair and nails for a range of 87 NPS drugs from the cannabinoid, opioid, benzodiazepine, and stimulant/hallucinogenic drug categories by LC-MS/MS [9]. The method was applied to ten hair samples from polyconsumers of classic and NPS drugs. Six of the targeted NPS were detected in authentic hair samples, namely 5-methoxy-α-methyltryptamine, 6-(2-aminopropyl)benzofuran, UR-144, APP-FUBINACA, mephedrone, and butyrylfentanyl. In addition, the nor-metabolite of butyrylfentanyl was detected. An additional four drugs were detected in nail, three of which were JWH synthetic cannabinoids and one cathinone. The drugs detected in hair are shown in Table 2 along with a summary of the methods used.

In a 2015 study of attendees of night-club and dance music venues, 80 hair samples were collected, 26 samples tested positive for at least one NPS [10]. Not surprisingly, all NPS detected were stimulant, dissociative, empathogenic, or hallucinogenic although the analytical method also evaluated synthetic cannabinoids. The most frequently detected drug was butylone ($n = 25$) followed by methylone ($n = 5$), methoxetamine ($n = 4$) and 5/6-(2-aminopropyl)benzofuran ($n = 1$), α-pyrrolidinovalerophenone ($n = 1$), and 4-fluoroamphetamine ($n = 1$). The concentrations of the drugs along with the method used are shown in Table 2.

More recently in 2022, a range of novel psychoactive stimulants, hallucinogens, opioids, and benzodiazepines were evaluated in 1203 cases from 2008 to 2020 in Germany [11]. NPS were detected in 381 cases, with polydrug cases being common. The most frequently detected drug was N-ethylamphetamine ($n = 81$) followed by α-PVP ($n = 38$) and mephedrone ($n = 38$). Other commonly detected drugs were benzedrone ($n = 30$), metamfepramone ($n = 28$), and 4-fluoroamphetamine ($n = 26$). Over 80% of the drugs detected were in the stimulant/hallucinogenic class. In addition, six novel benzodiazepines and two opioids were detected. It should however be noted that the only opioids the method was validated for were U-47700 and bupropion. No fentalogs were included.

3.4 Single Drug Groups Methods

Most of the scientific literature on NPS in hair focuses on the analysis of a single group of drugs. The following section provides examples from the literature describing cases when NPS in each drug category were detected in human hair. Most of the authors develop multianalyte as opposed to single target methods, and not all of the papers reviewed for the purpose of writing this chapter resulted in the detection of NPS in hair; however, several papers have been included in Table 2 as useful methodologies as they include target drugs being observed in blood (fluorofentanyl, carfentanil, etizolam, etc.).

3.4.1 Opioids

The opioid pandemic has resulted in many methods for fentalogs appearing in the scientific literature mainly by liquid chromatography – tandem mass spectrometry techniques. The majority of papers evaluating NPOs detect fentalogs with a few also looking at other compounds such as U-47700, AH-7921, W-18, U-49900, and MT-45. Polyfentalog papers are common [12–15]. The newest NPO trends reflect the 2-benzylbenzimidazole analogs (nitazene) drugs being popular; however, to date only one paper has been reported on the detection of these drugs in human hair [16]. Three deaths were reported in Europe with hair collected in each case. Hair was collected from the head in two cases and the thorax in one case. Isotonitazene in hair was a targeted analysis as it had been determined in blood. The hair analysis was complementary as it indicated past use of isotonitazene in the prior months leading up to death in addition to the time of death. The authors used the same extraction and instrumentation for the analysis of isotonitazene in hair as for blood and urine. Methodology is starting to appear for other nitazenes in matrices such as blood, urine, and tissue with liquid–liquid extraction followed by LC-MS/MS, and these methods provide useful starting points for development for hair methods [17].

As noted above, and not surprisingly, as they still represent the most prevalent drugs within this category of casework, the majority of published NPO papers deal with the analysis of fentalogs [12–15]. While most of these publications include case studies, not all cases test positive for NPOs. That being said, fentanyl and other fentanyl analogs, e.g. sufentanil, have the ability to be diverted to the black market so these methods are shown to be fit for purpose through the detection of these substances [12]. *p/o*-Fluoro-fentanyl, carfentanil and valeryl fentanyl while not detected in authentic hair samples ($n = 2$) remain prevalent in casework and are worth continued monitoring. For this reason, this method is included in Table 2.

By targeting known users, it is possible to obtain better data on fentalog concentrations in hair including information on polydrug use [13]. Hair from 17 users was reported in a 2020 publication with acetyl fentanyl (n = 7) being the most prevalent fentalog after fentanyl itself (n = 15). Other fentalogs detected were furanyl fentanyl (n = 5), tetrahydrofuranylfentanyl (n = 1) and valerylfentanyl (n = 1). Poly fentalog use was prevalent (n = 9) with one case showing fentanyl, norfentanyl, acetylfentanyl, tetrahydrofentanyl, and valerylfentanyl in multiple hair segments. In only one case was a NPO (furanylfentanyl) detected in the absence of fentanyl.

Highlighting the value of multianalyte assays, in a 2020 case report, a 43-year-old male was admitted to the emergency room in respiratory depression after snorting a white powder labeled butyrylfentanyl [15]. Blood and urine analysis were not conducted. Hair samples were collected 1 month after the incident and showed repeated exposure to meta-fluorofentanyl, furanylfentanyl and fentanyl. Additional cases discussed by the same authors showed use of methoxyfentanyl, carfentanil, and acetylfentayl. All of the above described methods were conducted by LC-MS/MS techniques and are described in Table 2.

3.4.2 Benzodiazepines

The benzodiazepines are probably the least reported NPS group in terms of positive casework in single drug group studies. In order to simulate a DFSA case, Sasaki et al., dosed three subjects with sedative hypnotic drugs including the novel psychoactive benzodiazepine etizolam (1 mg) [18]. Samples were collected after 1 month and 3 months. Etizolam was detected in all tested segments of hair following LLE and LC-MS/MS. Etizolam was also detected in the hair of a 49-year-old male found dead at home following a suspected alprazolam overdose [19]. Unfortunately, while the concentrations of etizolam in multiple sections of hair were reported, the exact details of the method were not reported for this case. A method for 20 traditional benzodiazepines and 3 Z-drugs plus the NPS phenazepam was developed in 2012 when this drug was prevalent in Europe. While phenazepam was not detected in any of the authentic hair samples, the method was found to be fit for purpose and as such has been included in Table 2 [20].

3.4.3 Stimulants and Hallucinogens

Currently the most divergent class of NPS, it is not surprising that there are many papers focusing on the analysis of stimulant and hallucinogenic type NPS. Due to the fact that parent drugs preferentially incorporate into hair over metabolites and even when the pharmacokinetics of NPS are known, metabolites may not be available, Salomone et al. proposed hair as a useful alternative sample for driver license recovery where NPS use was suspected [21]. Thirty-two samples were collected from ecstasy and ketamine abusers along with 54 samples that had previously tested negative during driver license recovery in Italy. Six samples tested positive for at least one of 26 targeted analytes (methoxetamine (n = 3), α-PVP (n = 2), 4-MEC, MDPV, and diphenidine).

In vivo studies in the rat model can be useful in determining both the incorporation of different NPS into hair as well as the detection of metabolites. In a 2015 study, Nieddu et al. evaluated 11 phenethylamine derivatives in dark gray and white rat hair [22]. None of the drugs studied were detected in white hair which is not unexpected for weakly basic drugs. All 11 drugs were detected in dark gray hair with 2,5-dimethoxyamphetamine (DMA) being incorporated to the highest degree; however, the authors did not note if this was significant or not. *p*-methoxy-methamphetamine (PMMA) was the least incorporated drug, being detected below the LOQ of the method; however, its metabolite *p*-methoxyampetamine (PMA) was also detected. The concentrations of all 10 drugs detected at measurable levels in rat hair are shown in Table 2.

The dance scene is an obvious place to obtain biological samples containing stimulant drugs including NPS stimulants [23]. Hair samples were collected following a survey of NYC nightclub attendees. Of the 48 acceptable samples, 47.9% were found to be positive for butylone, and 10.4% for methylone. In the survey, participants were asked about lifetime use of "bath salts," stimulant NPS, and unknown pills or powders. Butylone, methylone, α-PVP, 5/6-APB, and 4-FA were found in 41.2% of the hair of those who reported no known use of these products. Unfortunately, this chapter did not disclose the concentrations of drugs in hair and so has not been included in Table 2. However, the authors used a UHPLC-MS/MS method to screen 11 common drugs of abuse, 24 NPD stimulant/hallucinogenic drugs, 3 NPS dissociative drugs and 2 dissociative non-NPS drugs.

Mephedrone (4-methylmethcathinone) has been evaluated on its own by both GC and LC methods yielding a total of three quantitative results [24, 25]. In a letter to the editor in 2010, Torrance and Cooper reported mephedrone in fatalities in Scotland for the first time. In one case mephedrone was detected in two separate segments of hair at 4.2 and 4.7 ng/mg. In contrast, Shah et al. evaluated mephedrone and two of its metabolites, 4-methylephedrine and 4-methylnorephedrine in 154 samples from healthy volunteers age 18–56 years old. Four samples were positive for mephedrone but no metabolites were detected in any of the samples. Only one of the samples was able to be quantitated. The concentration detected was much lower than that reported by Torrance and Cooper (see Table 2). Mephedrone has been reported by several authors in hair [9, 11, 20, 24, 25], and it has been observed in casework from 2010 until present day highlighting the importance of continued monitoring of some NPS drugs in hair. The concentrations have varied widely from around 20 to

>4000 pg/mg. NPS drugs are typically analyzed alongside traditional drugs of abuse with the traditional drugs being detected more frequently. Meta-chlorophenylpiperazine (mCPP) was the only NPS of 8 cathinones and piperazines which tested positive ($n = 1$) in a 2014 mixed traditional/NPS method when applied to 7 authentic hair samples [26].

Finally, in a fatal case study, a 39-year-old woman, known as a multidrug addict, was murdered by her partner by ligature strangulation [27]. A comprehensive toxicological screening by gas chromatography and ultra-high performance liquid chromatography with tandem mass spectrometry showed polydrug intoxication at the time of death, including the presence of the dissociative hallucinogen anesthetic drug (3-methoxy-(1-(1-phenylcyclohexyl) piperidine) or 3-MeO-PCP. Concentrations of 3-MeO-PCP in 3 × 2-cm segments at 731, 893, and 846 pg/mg indicated long-term abuse of the drug. This chapter is the first ever to report hair concentrations of 3-MeO-PCP in the scientific literature.

3.4.4 Synthetic Cannabinoids

Many of the papers including case examples for synthetic cannabinoids include naphthoylindole based compounds which are less frequently seen in casework in the 2020s; however, as these drugs are older, more is known regarding their pharmacology. AM-2201, JWH-122, MAM-2201, and JWH-018 have been evaluated in authentic human hair samples along with their monohydroxylated metabolites [28]. Synthetic cannabinoids tend to be smoked, and by using parent drug to metabolite ratios, it is possible to establish criteria on external contamination. A rat model was also used to establish that hair pigmentation did not significantly affect the ratio of metabolite to parent drug for AM-2201. Wide variation in the concentrations of synthetic cannabinoids in the hair of 9 human subjects by the same authors. All subjects were in their 20's, and a mix of hair colors and varying lengths were tested. Both parent drugs and metabolites were able to be detected. A comprehensive method validation for 72 synthetic cannabinoids in hair was completed by Franz et al. in 2018 [29]. A large number of the drugs ($n = 52$) were detected in the hair of authentic users ($n = 306$) and a selection of these drugs along with a summary of the method is provided in Table 2. Poly synthetic cannabinoid use was prevalent within the study population with up to 17 individual drugs being detected in some cases.

4 Interpretation Challenges

Analytically, best practices for method validation should be adhered to according to current regulations when developing instrumental methods. In addition, careful consideration of calibration models, and the application of appropriate statistics are critical for reliable and reproducible results [30, 31]. The major challenges faced by all laboratories who conduct hair testing are compounded when considering NPS testing. These are outlined below.

- Lack of information about metabolites: In order to prove ingestion of drug, it is recommended that metabolites are evaluated in hair samples. Some metabolites, e.g. α-hydroxyetizolam, are now available and it is recommended that these are incorporated into testing methods to prove ingestion as opposed to only testing for parent compounds. This is a requirement by the Society of Hair Testing.
- As NPS are not as well studied as traditional drugs, their metabolic pathways are not as well established, if at all. In addition, as previously noted, metabolites may not be commercially available. By using either animal models or human liver microsomal studies, headway is being made into the metabolic pathways for some of these drugs. Reference standard manufactures are quick to respond to the ever changing drug market, including metabolites where demand is made apparent, thus this research is a crucial part of the NPS scene. [32, 33]
- Drugs that can be smoked are especially problematic in hair testing. Within the drugs discussed above, this is of particular importance to the synthetic cannabinoids. Parent drugs and pyrolysis products can deposit onto the surface of the hair resulting in external contamination. Efficient wash procedures are therefore critically important especially where the use of metabolites to indicate ingestion may not be possible. Analysis of wash solutions can give guidance as to whether external contamination is likely to be of concern.

Hair analysis is a powerful tool in determining NPS prevalence especially as intake information is often unclear. Unlike with prescription drugs, the user cannot be sure of what they are ingesting. On a monthly and even weekly basis, the drugs that are being sold are changing. What is contained in the products they are taking may not be what they think they are, and many users may not even be aware they are consuming NPS drugs [23]. Continued development of techniques to monitor these drugs in hair provides a way to monitor NPS consumption in the community both through targeted analysis such as LC-MS/MS or through use of techniques such as QTOF-LC/MS, which can retrospectively probe data and obtain accurate mass determinations on new drugs. NPS are appearing across the board in DUID, postmortem, and drug-facilitated crime casework, so monitoring for these drugs routinely is now crucial for labs. Hair testing, provides complementary information to conventional matrices in all areas of forensic toxicology where longer-term drug use history is required to aid interpretation of results. This chapter shows that NPS drugs in all categories can be detected in human hair and that LC-MS/MS is the preferred instrumental technique.

References

1. https://www.npsdiscovery.org/reports/dashboard/
2. https://www.npsdiscovery.org/reports/trend-reports/
3. "Analytical Report. 3Cl-PCP (C17H24ClN). 1-[1-(3-chlorophenyl)cyclohexyl]piperidine" (PDF). National Forensic Laboratory. Slovenia.
4. Synthetic cannabinoids in Europe– a review. EMCDDA technical report. European Monitoring Centre for Drugs and Drug Addiction, 2021. https://www.emcdda.europa.eu/publications/rapid-communications/synthetic-cannabinoids-europe-review_en
5. Cooper GAA, Kronstrand R, Kintz P (2012) Society of hair testing guidelines for drug testing in hair. Forensic Sci Int 218(1–3):20–24
6. NEOGEN®'s Hair Extraction Procedure for ELISA. https://www.youtube.com/watch?v=JPQjY0ki3bI
7. Farley M, Tran H, Towler S, Gevorkyan J, Pearring S, Rodda LN (2021) A single method for 127 recommended and additional DUID drugs in blood and urine by LC-MS/MS. J Anal Toxicol 00:1–12
8. Bouma VA, Di Rago M, Peka M, Drummer OH (2017) The analysis of 132 novel psychoactive substances in human hair using a single step extraction by tandem LC/MS. Forensic Sci Int 279:192–202
9. Mannocchi G, Di Trana A, Tini A, Zaami S, Gottardi S, Pichini S, Busardò SP (2020) Development and validation of fast UHPLC-MS/MS screening method for 87 NPS and 32 other drugs of abuse in hair and nails: application to real cases. Anal Bioanal Chem 412: 5125–5145
10. Salomone A, Palamar JJ, Gerace E, Di Corcia D, Vincenti M (2017) Hair testing for drugs of abuse and new psychoactive substances in a high-risk population. J Anal Toxicol 41:376–381
11. Niebel A, Westendorf L, Krumbiegel F, Hartwig S, Parr MK, Tsokos M (2022) Prevalence and concentrations of new designer stimulants, synthetic opioids, benzodiazepines, and hallucinogens in postmortem hair samples: a 13-year retrospective hair study. Drug Test Anal 14:110–121
12. Qin N, Shen M, Wen D, Shen B, Deng H, Qiang H, Song F, Shi Y (2020) Determination of 37 fentanyl analogues and novel synthetic opioids in hair by UHPLC-MS/MS and its application to authentic cases. Nature 10: 11569
13. del Mar Ramírez Fernández M, SMR W, Janowski D (2020) Development of an UPLC-MS/MS method for the analysis of 16 synthetic opioids in segmented hair, and evaluation of polydrug history in fentanyl analog users. Forensic Sci Int 307:110137
14. Salomone A, Palamar JJ, Bigiarini R, Gerace E, Di Corcia D, Vincenti M (2019) Detection of fentanyl analogs and synthetic opioids in real hair samples. J Anal Toxicol 43:259–265
15. Labari IA, Martin M, Fabresse N, Etting I, Edel Y, Pfau G, Alvarez JC (2020) Hair testing for 3-fluorofentayl, furanofentanyl, methoxyacetylfentanyl, carfentanil, acetylfentanyl and fentanyl by LC-MS/MS after unintenional overdose. Forensic Toxicol 38:277–286
16. Mueller F, Bogdal C, Pfeiffer B, Andrello L, Ceschi A, Thomas A, Grata E (2021) Isotonitazene: fatal intoxication in three cases involving this unreported novel psychoactive substance in Switzerland. Forensic Sci Int 320(110636):1–13
17. Walton SE, Krotulski AJ, Logan BK (2021) A forward-thinking approach to addressing the new synthetic opioid 2-benzylbenzimidazole nitazene analogs by liquid chromatography--tandem quadrupole mass spectrometry (LC–QQQ-MS). J Anal Toxicol 00:1–11
18. Sasaki K, Shima N, Kamata T, Ishikawa A, Nitta A, Wada M, Nakano-Fujii S, Sato T, Katagi M (2021) Incorporation of five common hypnotics into hair after a single dose and application to a forensic case of drug facilitated crimes. Forensic Sci Int 325:110881
19. Kolbe V, Rentsch D, Boy D, Schmidt B, Kegler R, Büttner A (2020) The adulterated XANAX pill: a fatal intoxication with etizolam and caffeine. Int J Legal Med 134:1727–1731
20. Rust KY, Baumgartner MR, Meggiolaro N, Kraemer T (2012) Detection and validated quantification of 21 benzodiazepines and 3 "z-drugs" in human hair by LC–MS/MS. Forensic Sci Int 215:64–72
21. Salomone A, Gazzilli G, Di Corcia D, Gerace E, Vincenti M (2016) Determination of cathinones and other stimulant, psychedelic, and dissociative designer drugs in real hair samples. Anal Bioanal Chem 408:2035–2042
22. Nieddu M, Burrai L, Demontis MP, Varoni MV, Baralla E, Trignano C, Boatto G (2015) Simultaneous determination of 11 illicit

phenethylamines in hair by LC–MS-MS: *In Vivo* application. J Anal Toxicol 39:532–537

23. Palamara JJ, Salomone A, Vincenti M, Cleland CM (2016) Detection of "bath salts" and other novel psychoactive substances in hair samples of ecstasy/MDMA/"Molly" users. Drug Alcohol Depend 161:200–205
24. Torrance H, Cooper G (2010) The detection of mephedrone (4-methylmethcathinone) in 4 fatalities in Scotland. Forensic Sci Int 202: e62–e63
25. Shah SAB, Deshmukh NIK, Barker J, Petróczi A, Cross P, Archer R, Naughton DP (2012) Quantitative analysis of mephedrone using liquid chromatography tandem mass spectrometry: application to human hair. J Pharm Biomed Anal 61:64–69
26. Lendoiro E, Jiménez-Morigosa C, Cruz A, de Castro A (2014) Hair analysis of amphetamine-type stimulant drugs (ATS), including synthetic cathinones and piperazines, by LC-MSMS. Toxicologie Analitique et Clinique 26(2):S13
27. Kintz P, Ameline A, Walch A, Farrugia RJ-S (2019) Murdered while under the influence of 3-MeO-PCP. Int J Legal Med 133:475–478
28. Kim J, Park Y, Park M, Kim E, Yang W (2015) Simultaneous determination of five naphthoylindole-based synthetic cannabinoids and matabolites and their deposition in human and rat hair. J Pharm Biomed Anal 102: 162–175
29. Franz F, Jechle H, Angerer V, Pegoro M, Auwarter V, Neukamm MA (2018) Synthetic cannabinoids in hair – pragmatic approach for method updates, compound prevalences and concentration ranges in authentic hair samples. Analytica Chimica Acta 1006:61–73
30. Standard Practices for Method Validation in Forensic Toxicology, ANSI/ASB Standard 036, 1st Ed. (2019). https://www.aafs.org/asb-standard/standard-practices-method-validation-forensic-toxicology
31. Desharnais B, Camirand-Lemyre F, Mireault F, Skinner CD (2017) Procedure for the selection and validation of a calibration model I—description and application. J Anal Toxicol 41(4):261–268
32. Krotulski AJ, Mohr ALA, Papsun DM, Logan BK (2018) Metabolism of novel opioid agonists U-47700 and U-49900 using human liver microsomes with confirmation in authentic urine specimens from drug users. Drug Test Anal 10(1):127–1363
33. Temporal KH, Scott KS, Mohr ALA, Logan BK (2017) Metabolic profile determination of NBOMe compounds using human liver microsomes and comparison with findings in authentic human blood and urine. J Anal Toxicol 3: 1–12

Chapter 7

Target Analysis of NPS in Oral Fluid

Maria Rosaria Varì, Simona Pichini, Sara Malaca, Annagiulia Di Trana, Nunzia La Maida, Silvia Graziano, Giulio Mannocchi, Pietro Brunetti, Alfredo Fabrizio Lo Faro, and Francesco Paolo Busardò

Abstract

The need to identify and quantify a growing number of NPS represents a crucial challenge for toxicological and forensic purposes. Among biological specimens investigated for pharmaco-toxicological analyses, oral fluid (OF) provides greater advantages compared to blood to prove drug current use. OF is regarded with particular interest for drug screening at workplace and roadside testing, among others. This chapter contains an overview of analytical methodologies applied to determine the most used NPS (synthetic cannabinoids, cathinones, phenethylamines, designer piperazines, tryptamines, synthetic opioids, and designer benzodiazepines), and metabolites in OF by hyphenated techniques, reporting details of specimen extraction, separation, and detection with sensitivity, specificity, and accuracy details. The majority of the published methods used the Intercept® or the Quantisal® device for OF collection, and analytes were mostly extracted by solid-phase or liquid–liquid extraction. Liquid chromatography coupled to (tandem) mass spectrometry was the main applied technique for target analysis of NPS in OF.

Key words Oral fluid, New psychoactive substances, Liquid chromatography tandem mass spectrometry, Gas chromatography tandem mass spectrometry, High resolution mass spectrometry

Abbreviations

OF	Oral fluid
NR	not reported
SPE	solid-phase extraction
LLE	liquid–liquid extraction
SPME	solid-phase microextraction;
SLE	solid liquid extraction
LFIA	lateral flow development & manufacturing
ELISA	enzyme-linked immunosorbent assay
MEPS	microextraction packed sorbent
GC-MS	gas chromatography-mass spectrometry
UHPLC-HRMS	ultra-high performance liquid chromatography high resolution mass spectrometry

Marta Concheiro and Karl B. Scheidweiler (eds.), *Methods for Novel Psychoactive Substance Analysis*, Methods in Pharmacology and Toxicology, https://doi.org/10.1007/978-1-0716-2605-4_7,

UHPLC-MS	ultra-high-pressure liquid chromatography mass spectrometry
LC-HRMS	liquid chromatography high resolution mass spectrometry
LC-MS/MS	liquid chromatography-tandem mass spectrometry
LC-QqQ-MS/MS	liquid chromatography triple quadrupole tandem mass spectrometry
UPLC-MS/MS	ultraperformance liquid chromatography-tandem mass spectrometry
US DLLME	Ultrasonic-assisted dispersive liquid–liquid microextraction
GC/EI-MS	gas chromatography/electron ionization-mass spectrometry
CE-FD	capillary electrophoresis-fluorescence detector
IMS	ion mobility spectrometry
MIP-SPE	Molecularly Imprinted Polymer-solid-phase extraction
FSF	fluorescence spectral fingerprinting
SERS	surface-enhanced Raman spectroscopy
MSTFA	N-methyl-N(trimethylsily) trifluoroacetamide
ISTD	internal standard
TPP	triphenyl phosphate
SWGTOX	Scientific Working Group for Forensic Toxicology
NH_4HCO_2	Ammonium formate
FA	Formic acid (CH_2O_2)
ACN	Acetonitrile
CH_2Cl_2	Dichloromethane
EtOAc	Ethyl acetate
MeOH	Methanol
NH_4Cl	Ammonium chloride
$(NH_4)_2CO_3$	Ammonium carbonate
H3CCOONH4	Ammonium acetate
H_2O	water
NaH2PO4•H2O	phosphate buffer
C_6H_{14}	hexane
C_7H_{16}	heptane
$Na_2B_4O_7$	sodium tetraborate
$CH_3CH(OH)CH_3$	2-propanol
CH_3COOH	acetic acid
H_3PO_4	phosphoric acid
BBS	borate buffer
NH_4OH	ammonium hydroxide
THF	tetrahydrofuran
$C_6H_{14}O$	1-hexanol
PTFE	poly(tetrafluoroethylene)
$(C_2H_5)_2O$	ether
$(CH_3)_2CHCH_2CH_2OH$	isoamyl alcohol
JWH-122	(4-methyl-1-naphthalenyl)(1-pentyl-1H-indol-3-yl)-methanone

JWH-210	(4-ethyl-1-naphthalenyl)(1-pentyl-1H-indol-3-yl)-methanone
UR-144	(1-pentyl-1H-indol-3-yl)(2,2,3,3-tetramethylcyclopropyl)-methanone
JWH-122 N-(4-OH)	(1-(4-hydroxypentyl)-1H-indol-3-yl)(4-methyl-naphthalen-1-yl)methanone
JWH-210 N-(4-OH)	(4-ethylnaphthalen-1-yl)(1-(4-hydroxypentyl)-1H-indol-3-yl)methanone
JWH-210 N-(5-OH)	(4-ethylnaphthalen-1-yl)(1-(5-hydroxypentyl)-1H-indol-3-yl)methanone
UR-144 N-(5-OH)	[1-(5-hydroxypentyl)-1H-indol-3-yl](2,2,3,3-tetramethylcyclopropyl)-methanone
(+/−) JWH 018 N-(2-hydroxypentyl)	[1-(2-hydroxypentyl)-1H-indol-3-yl]-1-naphthalenyl-methanone
(+/−) JWH 018 N-(3-hydroxypentyl)	(1-(3-hydroxypentyl)-1H-indol-3-yl)(naphthalen-1-yl)methanone
(+/−) JWH 018 N-(4-hydroxypentyl)	(1-(4-hydroxypentyl)-1H-indol-3-yl)(naphthalen-1-yl)-methanone
(+/−) WIN 55,212 (mesylate)	(R)-(5-Methyl-3-(morpholinomethyl)-2,3-dihydro-[1,4]oxazino[2,3,4-hi]indol-6-yl)(naphthalen-1-yl)methanone methanesulfonate
5-chloro AB-PINACA	N-[(2S)-1-amino-3-methyl-1-oxobutan-2-yl]-1-(5-chloropentyl)indazole-3-carboxamide
5-Fluoro ABICA	N-[(1S)-1-(aminocarbonyl)-2-methylpropyl]-1-(5-fluoropentyl)-1H-indole-3-carboxamide
5-Fluoro AB-PINACA	N-[(2S)-1-amino-3-methyl-1-oxobutan-2-yl]-1-(5-fluoropentyl)indazole-3-carboxamide
5-Fluoro ADBICA	N-[1-(aminocarbonyl)-2,2-dimethylpropyl]-1-(5-fluoropentyl)-1H-indole-3-carboxamide
5-Fluoro ADB-PINACA	N-[1-(aminocarbonyl)-2,2-dimethylpropyl]-1-(5-fluoropentyl)-1H-indazole-3-carboxamide
5-Fluoro AMB	N-[[1-(5-fluoropentyl)-1H-indazol-3-yl]carbonyl]-L-valine, methyl ester
5-Fluoro PB-22	1-(5-fluoropentyl)-8-quinolinyl ester-1H-indole-3-carboxylic acid
5-Fluoro PB-22 3-carboxyindole	1-(5-fluoropentyl)-1H-indole-3-carboxylic acid
AB-CHMINACA	N-[(2S)-1-amino-3-methyl-1-oxobutan-2-yl]-1-(cyclohexylmethyl)indazole-3-carboxamide
AB-FUBINACA	N-[(2S)-1-amino-3-methyl-1-oxobutan-2-yl]-1-[(4-fluorophenyl)methyl]indazole-3-carboxamide
AB-PINACA	N-[(2S)-1-amino-3-methyl-1-oxobutan-2-yl]-1-pentylindazole-3-carboxamide
AB-PINACA N-(4-hydroxypentyl)	N-(1-amino-3-methyl-1-oxobutan-2-yl)-1-(4-hydroxypentyl)-1H-indazole-3-carboxamide
AB-PINACA N-(5-hydroxypentyl)	N-(1-amino-3-methyl-1-oxobutan-2-yl)-1-(5-hydroxypentyl)-1H-indazole-3-carboxamide
AB-PINACA pentanoic acid metabolite	5-[3-[(1-amino-3-methyl-1-oxobutan-2-yl)carbamoyl]indazol-1-yl]pentanoic acid
ADB-FUBINACA	N-(1-amino-3,3-dimethyl-1-oxobutan-2-yl)-1-[(4-fluorophenyl)methyl]indazole-3-carboxamide

ADBICA	N-(1-amino-3,3-dimethyl-1-oxobutan-2-yl)-1-pentylindole-3-carboxamide
ADB-PINACA	N-(1-amino-3,3-dimethyl-1-oxobutan-2-yl)-1-pentylindazole-3-carboxamide
AM-1220	(1-((1-Methylpiperidin-2-yl)methyl)-1H-indol-3-yl)(naphthalen-1-yl)methanone
AM2201	1-[(5-Fluoropentyl)-1H-indol-3-yl]-(naphthalen-1-yl)methanone
AM2201 2-hydroxyindole	(1-(5-fluoropentyl)-2-hydroxy-1H-indol-3-yl)(naphthalen-1-yl)methanone
AM2201 5-hydroxyindole	(1-(5-fluoropentyl)-5-hydroxy-1H-indol-3-yl)(naphthalen-1-yl)methanone
AM2233	(2-Iodophenyl){1-[(1-methyl-2-piperidinyl)methyl]-1H-indol-3-yl}methanone
AM694	[1-(5-fluoropentyl)indol-3-yl]-(2-iodophenyl)methanone
HU-211	(6a,10a)-9-(hydroxymethyl)-6,6-dimethyl-3-(2-methyloctan-2-yl)-6a,7,10,10a-tetrahydro-6H-benzo[c]chromen-1-ol
JWH 007	(2-methyl-1-pentylindol-3-yl)-naphthalen-1-ylmethanone
JWH 015	(2-methyl-1-propylindol-3-yl)-naphthalen-1-ylmethanone
JWH 018	naphthalen-1-yl-(1-pentylindol-3-yl)methanone
JWH 018 2-hydroxyindole	(2-hydroxy-1-pentyl-1H-indol-3-yl)(naphthalen-1-yl)-methanone
JWH 018 4-hydroxyindole	(4-hydroxy-1-pentyl-1H-indol-3-yl)(naphthalen-1-yl)-methanone
JWH 018 5-hydroxyindole	(5-hydroxy-1-pentyl-1H-indol-3-yl)(naphthalen-1-yl)-methanone
JWH 018 6-hydroxyindole	(6-hydroxy-1-pentyl-1H-indol-3-yl)(naphthalen-1-yl)-methanone
JWH 018 7-hydroxyindole	(7-hydroxy-1-pentyl-1H-indol-3-yl)(naphthalen-1-yl)-methanone
JWH 018 adamantyl analog	(1s,3s)-adamantan-1-yl(1-pentyl-1H-indol-3-yl)methanone
JWH 018 N-(4-oxo-pentyl)	5-(3-(1-naphthoyl)-1H-indol-1-yl)pentan-2-one
JWH 018 N-(5-hydroxypentyl)	(1-(5-hydroxypentyl)-1H-indol-3-yl)(naphthalen-1-yl)methanone
JWH 018 N-propanoic acid	3-(1-naphthalenylcarbonyl)-1H-indole-1-propanoic acid
JWH 019	1-Hexyl-3-(naphthalen-1-oyl)indole
JWH 020	(1-heptyl-1H-indol-3-yl)-1-naphthalenyl-methanone
JWH 073	(1-butylindol-3-yl)-naphthalen-1-ylmethanone
JWH 081	(4-methoxynaphthalen-1-yl)-(1-pentylindol-3-yl)methanone
JWH 098	(4-Methoxy-1-naphthyl)(2-methyl-1-pentyl-1H-indol-3-yl)methanone

JWH 182	(1-pentyl-1H-indol-3-yl)(4-propylnaphthalen-1-yl)methanone
JWH 200	[1-(2-morpholin-4-ylethyl)indol-3-yl]-naphthalen-1-ylmethanone
JWH 203	2-(2-chlorophenyl)-1-(1-pentylindol-3-yl) ethanone
JWH 210	(4-ethylnaphthalen-1-yl)-(1-pentylindol-3-yl) methanone
JWH 249	2-(2-Bromophenyl)-1-(1-pentylindol-3-yl) ethanone
JWH 251	2-(2-methylphenyl)-1-(1-pentylindol-3-yl) ethanone
JWH 307	(5-(2-Fluorophenyl)-1-pentylpyrrol-3-yl)-naphthalen-1-ylmethanone
JWH 307 5′-isomer	(2-(2-Fluorophenyl)-1-pentyl-1H-pyrrol-3-yl) (naphthalen-2-yl)methanone
JWH 387	(4-bromonaphthalen-1-yl)(1-pentyl-1H-indol-3-yl)methanone
JWH 398	1-Pentyl-3-(4-chloro-1-naphthoyl)indole
JWH 412	(4-fluoronaphthalen-1-yl)(1-pentyl-1H-indol-3-yl)methanone
MAM2201 N-(3-fluoropentyl) isomer	(1-(3-fluoropentyl)-1H-indol-3-yl)(4-methyl-naphthalen-1-yl)methanone
MAM2201 N-(5-chloropentyl) analog	(1-(5-chloropentyl)-1H-indol-3-yl)(4-methyl-naphthalen-1-yl)methanone
PB-22	1-pentyl-8-quinolinyl ester-1H-indole-3-carboxylic acid
PB-22 3-carboxyindole	1-pentyl-1H-indole-3-carboxylic acid
PB-22 N-(4-hydroxypentyl)	quinolin-8-yl 1-(4-hydroxypentyl)-1H-indole-3-carboxylate
PB-22 N-(5-hydroxypentyl)	quinolin-8-yl 1-(5-hydroxypentyl)-1H-indole-3-carboxylate
PB-22 N-pentanoic acid	5-(3-((quinolin-8-yloxy)carbonyl)-1H-indol-1-yl) pentanoic acid
RCS-4	2-(4-methoxyphenyl)-1-(1-pentyl-indol-3-yl) methanone
RCS-4-C4 homolog	(1-butylindol-3-yl)-(4-methoxyphenyl) methanone
RCS-8	1-(2-Cyclohexylethyl)-3-(2-methoxyphenylacetyl) indole
THJ 018	naphthalen-1-yl-(1-pentylindazol-3-yl)methanone
THJ 2201	[1-(5-fluoropentyl)indazol-3-yl]-naphthalen-1-ylmethanone
XLR11	[1-(5-fluoropentyl)indol-3-yl]-(2,2,3,3-tetramethylcyclopropyl)methanone
XLR11 6-hydroxyindole	1-(5-fluoropentyl)-6-hydroxy-1H-indol-3-yl] (2,2,3,3-tetramethylcyclopropyl)-methanone

XLR11 N-(4-hydroxypentyl)	[1-(5-fluoro-4-hydroxypentyl)indol-3-yl]-(2,2,3,3-tetramethylcyclopropyl)methanone
5F-ADB	(methyl (s)-2-[1-(5-fluoropentyl)-1H-indazole-3-carboxamido]-3,3-dimethylbutanoate)
MMB-CHMICA	(methyl(n-{[1-(cyclohexylmethyl)-1H-indol-3-yl] carbonyl}-l-valinate))
THJ-2201	([1-(5-fluoropentyl)-1H-indazol-3-yl] (1-naphthyl)methanone)
CUMYL-4CN-BINACA	(1-(4-cyanobutyl)-n-(2-phenylpropan-2-yl)-1H-indazole-3-carboxamide)
MDMB-CHMCZCA	(methyl (s)-2-(9-(cyclohexylmethyl)-9H-carbazole-3-carboxamido)-3,3-dimethylbutanoate)
APINACA (AKB-48)	N-(adamantan-1-yl)-1-pentyl-1H-indazole-3-carboxamide
MMB-FUBINACA	Methyl (2S)-2-{[1-[(4-fluorophenyl)methyl]indazole-3-carbonyl]amino}-3-methylbutanoate
JWH-176	1-([(1E)-3-Pentylinden-1-ylidine]methyl) naphthalene
JWH-250	2-(2-Methoxyphenyl)-1-(1-pentylindol-3-yl) ethanone
AM-694	1-[(5-Fluoropentyl)-1H-indol-3-yl]-(2-iodophenyl)methanone
5F-PB-22	1-Pentyfluoro-1H-indole-3-carboxylic acid 8-quinolinyl ester
CP47,497	2-[(1R,3S)-3-hydroxycyclohexyl]-5-(2-methyloctan-2-yl)phenol
CP47,497(C8)	2-[(1R,3S)-3-hydroxycyclohexyl]-5-(2-methylnonan-2-yl)phenol
5F-AKB48	N-(adamantan-1-yl)-1-(5-fluoropentyl)-1H-indazole-3-carboxamide
AB-005	[1-[(1-Methylpiperidin-2-yl)methyl]-1H-indol-3-yl](2,2,3,3-tetramethylcyclopropyl)-methanone
AB-CHMINACA	N-[(2S)-1-Amino-3-methyl-1-oxobutan-2-yl]-1-(cyclohexylmethyl)indazole-3-carboxamide
STS-135	N-(Adamantan-1-yl)-1-(5-fluoropentyl)-1H-indole-3-carboxamide
A-796,260	[1-(2-morpholin-4-ylethyl)-1H-indol-3-yl]-(2,2,3,3-tetramethylcyclopropyl)methanone
MAM2201	(1-(5-Fluoropentyl)-1H-indol-3-yl)(4-methyl-1-naphthalenyl)methanone
5-Fluoro PB-22	1-Pentyfluoro-1H-indole-3-carboxylic acid 8-quinolinyl ester
AB-PINACA 5-pentanoic acid metabolite	5-[3-[(1-Carbamoyl-2-methyl-propyl)carbamoyl] indazol-1-yl]pentanoic acid
APINACA 5-Hydroxypentyl	1-(5-hydroxypentyl)-N-tricyclo[3.3.1.13,7]dec-1-yl-1H-indazole-3-carboxamide
AM2201 4-Hydroxypentyl	(1-(5-fluoro-4-hydroxypentyl)-1H-indol-3-yl) (naphthalen-1-yl)methanone
JWH-018 5-Pentanoic acid	5-(3-(1-naphthoyl)-1H-indol-1-yl)-pentanoic acid

JWH-019 5- Hydroxyhexyl	[1-(5-hydroxyhexyl)-1H-indol-3-yl]-1-naphthalenyl-methanone
JWH-073 4-Butanoic acid	4-(3-(1-naphthoyl)-1H-indol-1-yl)-butanoic acid
JWH-122 4-Hydroxypentyl	(1-(4-hydroxypentyl)-1H-indol-3-yl)(4-methylnaphthalen-1-yl)methanone
JWH-210 4-Hydroxypentyl	(4-ethylnaphthalen-1-yl)(1-(4-hydroxypentyl)-1H-indol-3-yl)methanone
JWH-250 4-Hydroxypentyl	1-(1-(4-hydroxypentyl)-1H-indol-3-yl)-2-(2-methoxyphenyl)ethanone
MAM2201 4-Hydroxypentyl	(1-(5-fluoro-4-hydroxypentyl)-1H-indol-3-yl)(4-methylnaphthalen-1-yl)methanone
UR-144 4-Hydroxypentyl	(1-(4-hydroxypentyl)-1H-indol-3-yl)(2,2,3,3-tetramethylcyclopropyl)methanone
UR-144 5-Pentanoic acid	5-(3-(2,2,3,3-tetramethylcyclopropanecarbonyl)-1H-indol-1-yl)pentanoic acid
XLR-11 4-Hydroxypentyl	[1-(5-fluoro-4-hydroxypentyl)-1H-indol-3-yl](2,2,3,3-tetramethylcyclopropyl)-methanone
MAM-2201 N-pentanoic acid	5-(3-(4-methyl-1-naphthoyl)-1H-indol-1-yl)pentanoic acid
UR-144 N-5-hydroxypentyl	[1-(5-hydroxypentyl)-1H-indol-3-yl](2,2,3,3-tetramethylcyclopropyl)-methanone
WIN-55,212-2	(R)-(+)-[2,3-Diidro-5-metil-3-(4-morfolinilmetil)pirrolo[1,2,3-de]-1,4-benzossazin-6-il]-1-naftalenilmetanone
JWH-018 N-5-hydroxypentyl	(1-(5-hydroxypentyl)-1H-indol-3-yl)(naphthalen-1-yl)methanone;
UR-144-PYR	3,3,4-trimethyl-1-(1-pentyl-1H-indol-3-yl)pent-4-en-1-one
XLR11-PYR	1-(1-(5-fluoropentyl)-1H-indol-3-yl)-3,3,4-trimethylpent-4-en-1-one
4-hydroxy-UR-144	[1-(5-hydroxypentyl)-1H-indol-3-yl](2,2,3,3-tetramethylcyclopropyl)methanone
4-hydroxy-XLR11	(1-(5-fluoro-4-hydroxypentyl)-1H-indol-3-yl)(2,2,3,3-tetramethylcyclopropyl)-methanone
AM-694	1-[(5-Fluoropentyl)-1H-indol-3-yl]-(2-iodophenyl)methanone
HU-210	(6aR,10aR)-9-(Hydroxymethyl)-6,6-dimethyl-3-(2-methyloctan-2-yl)-6a,7,10,10a-tetrahydrobenzo[c]chromen-1-ol
AM-2232	5-[3-(naphthalene-1-carbonyl)indol-1-yl]pentanenitrile
Methanandamide	(5Z,8Z,11Z,14Z)-N-[(2R)-1-Hydroxypropan-2-yl]icosa-5,8,11,14-tetraenamide
RCS ortho isomers	(2-methoxyphenyl)(1-pentyl-1H-indol-3-yl)methanone
WIN 48,098	(4-Methoxyphenyl)-[2-methyl-1-(2-morpholin-4-ylethyl)indol-3-yl]methanone
CP 47497	2-[(1RS,3SR)-3-Hydroxycyclohexyl]-5-(1,1-dimethylheptyl)phenol

CP 47497 C8	rel-5-(1,1-dimethyloctyl)-2-[(1R,3S)-3-hydroxy-cyclohexyl]-phenol
4-MMC	mephedrone
4-MEC	4-methylethcathinone
ETH-CAT	ethcatinone
MDPV	3,4-methylenedioxypyrovalerone
4-FMC	flephedrone
PMMC	methedrone
bk-MBDB	butylone
bk-MDEA	ethylone
bk-MBDP	pentylone
N,N-DMC	dimethylcathinone
α-PVP	α-pyrrolidinopentiophenone
MPBP	4′-Methyl-α-pyrrolidinobutiophenone
MABP	buphedrone
MIP	molecularly imprinted polymer
3-MMC	3-methylmethcathinone
3-FMC	3-fluoromethcathinone
4-MeMABP	4-methylbuphedrone
4-CMC	4-chloromethcathinone
4-MMC-MeO	mexedrone
α-PHP	α-pyrrolidinohexiophenone
MDPPP	3′,4′-methylenedioxy-α-pyrrolidinopropiophenone
4-CEC	4-chloroethcathinone
NEP	N-ethylpentylone
MPHP	4′-methyl-α-pyrrolidinohexiophenone
4Cl-PVP	4-chloro-α-pyrrolidinopentiophenone
TH-PVP	3′,4′-tetramethylene-α-pyrrolidinopentiophenone
PV9	1-phenyl-2-(pyrrolidin-1-yl)octan-1-one
3,4-MDPHP	3′,4′-Methylenedioxy-α-pyrrolidinohexiophenone
DMBDP	dipentylone
bk-MDDMA	dimethylone
3,4-DMMC	3,4-Dimethylmethcathinone
25B-NBOH	2-({[2-(4-bromo-2,5-dimethoxyphenyl)ethyl] amino}methyl)phenol
25B-NBOMe	2-(4-bromo-2,5-dimethoxyphenyl)-N-[(2-methoxyphenyl)methyl]ethanamine
25C-NBOH	2-({[2-(4-chloro-2,5-dimethoxyphenyl)ethyl] amino}methyl)phenol
25C-NBOMe	2-(4-Chloro-2,5-dimethoxyphenyl)-N-[(2-methoxyphenyl)methyl]ethan-1-amine
25D-NBOMe	2-(2,5-dimethoxy-4-methylphenyl)-N-(2-methoxybenzyl)ethanamine
25E-NBOH	2-({[2-(4-ethyl-2,5-dimethoxyphenyl)ethyl] amino}methyl)phenol
25E-NBOMe	2-(4-ethyl-2,5-dimethoxyphenyl)-N-(2-methoxy-benzyl)ethanamine
25G-NBOMe	2-(2,5-dimethoxy-3,4-dimethylphenyl)-N-[(2-methoxyphenyl)methyl]ethanamine

25H-NBOMe	2-(2,5-Dimethoxyphenyl)-N-[(2-methoxyphenyl) methyl]ethanamine
25I-NBOH	2-((2-(4-Iodo-2,5-dimethoxyphenyl)ethylamino) methyl)phenol
25I-NBOMe	2-(4-iodo-2,5-dimethoxyphenyl)-N-[(2-methoxyphenyl)methyl]ethanamine
25N-NBOMe	2-(2,5-Dimethoxy-4-nitrophenyl)-N-[(2-methoxyphenyl)methyl]ethan-1-amine
25T2-NBOMe	2-(4-trifluoromethyl-2,5-dimethoxyphenyl)-N-[(2-methoxyphenyl)methyl]ethanamine
2C-B	2-(4-bromo-2,5-dimethoxyphenyl)ethanamine
2C-C	2-(4-Chloro-2,5-dimethoxyphenyl)ethan-1-amine
2C-D	2-(2,5-Dimethoxy-4-methylphenyl)ethan-1-amine
2C-E	2-(4-Ethyl-2,5-dimethoxyphenyl)ethanamine
2C-G	2-(2,5-Dimethoxy-3,4-dimethylphenyl)ethan-1-amine
2C-I	2-(4-Iodo-2,5-dimethoxyphenyl)ethan-1-amine
2C-T	2-[2,5-Dimethoxy-4-(methylsulfanyl)phenyl] ethan-1-amine
2C-T-2	2-[4-(Ethylsulfanyl)-2,5-dimethoxyphenyl]ethan-1-amine
2C-T-4	2-{2,5-Dimethoxy-4-[(propan-2-yl)sulfanyl]phenyl}ethan-1-amine
PMA	1-(4-methoxyphenyl)propan-2-amine
PMMA	1-(4-Methoxyphenyl)-N-methylpropan-2-amine
TMA	1-(3,4,5-Trimethoxyphenyl)propan-2-amine,3,4,5-trimethoxyamphetamine
2C-T-7	2-[2,5-Dimethoxy-4-(propylsulfanyl)phenyl] ethan-1-amine
DOET	1-(4-Ethyl-2,5-dimethoxyphenyl)propan-2-amine
DOM	1-(2,5-Dimethoxy-4-methylphenyl)propan-2-amine
DOB	1-(4-Bromo-2,5-dimethoxyphenyl)propan-2-amine
DMA	N,N-Dimethylacetamide
4-FA	4-Fluoroamphetamine
MDEA	3,4-Methylenedioxy-N-ethylamphetamine
MDA	1-(2H-1,3-Benzodioxol-5-yl)propan-2-amine
MBDB	1-(1,3-Benzodioxol-5-yl)-N-methylbutan-2-amine
MTA	alpha-Methyl-p-(methylthio)phenethylamine
MDPV	1-(1,3-Benzodioxol-5-yl)-2-(pyrrolidin-1-yl)pentan-1-one
25T4-NBOMe	2-(2,5-Dimethoxy-4-propan-2-ylsulfanylphenyl)-N-[(2-methoxyphenyl)methyl]ethanamine
Ethylphenidate	(RS)-Ethyl 2-phenyl-2-piperidin-2-ylacetate
BZP	1-Benzylpiperazine
TFMPP	1-[3-(trifluoromethyl)phenyl]piperazine
mCPP	1-(3-chlorophenyl)piperazine
Piperonyl piperazine	1-(Benzo[d][1,3]dioxol-5-ylmethyl)piperazine
4-MeOPP	1-(4-methoxyphenyl)-piperazine

DMT	2-(1H-Indol-3-yl)-N,N-dimethylethanamine
5-MeO-DMT	2-(5-Methoxy-1H-indol-3-yl)-N,N-dimethylethanamine
AMT	1-(1H-Indol-3-yl)propan-2-amine
AcO-DMT	[3-[2-(dimethylamino)ethyl]-1H-indol-4-yl] acetate
5-MeO-AMT	1-(5-methoxy-1H-indol-3-yl)propan-2-amine
4-OH-DET	3-(2-Diethylaminoethyl)-1H-indol-4-ol
5-MeO-MIPT	N-[2-(5-methoxy-1H-indol-3-yl)ethyl]-N-methylpropan-2-amine
4-AcO-DIPT	3-{2-[Di(propan-2-yl)amino]ethyl}-1H-indol-4-yl acetate
5-MeO-DPT	N-[2-(5-methoxy-1H-indol-3-yl)ethyl]-N-propylpropan-1-amine
5-MeO-DALT	N-[2-(5-methoxy-1H-indol-3-yl)ethyl]-N-(prop-2-en-1-yl)prop-2-en-1-amine
U-47700	3,4-Dichloro-N-[(1R,2R)-2-(dimethylamino)-cyclohexyl]-N-methylbenzamide
AH-7921	3,4-dichloro-N-{[1-(dimethylamino)cyclohexyl] methyl}benzamide
U-49900	trans-3,4-dichloro-N-[2-(diethylamino)-cyclohexyl]-N-methyl-benzamide
U-50488	2-(3,4-dichlorophenyl)-N-methyl-N-[(1R,2R)-2-pyrrolidin-1-ylcyclohexyl]acetamide
MT-45	1-cyclohexyl-4-(1,2-diphenylethyl)piperazine
W-18	4-chloro-N-[(2Z)-1-[2-(4-nitrophenyl)ethyl] piperidin-2-ylidene]benzene-1-sulfonamide
W-15	(NE)-4-chloro-N-[1-(2-phenylethyl)piperidin-2-ylidene]benzenesulfonamide
Fentanyl	N-phenyl-N-[1-(2-phenylethyl)piperidin-4-yl] propanamide
Carfentanil	methyl 1-(2-phenylethyl)-4-[phenyl(propanoyl)-amino]piperidine-4-carboxylate
Norfentanyl	N-phenyl-N-(piperidin-4-yl)propionamide
Norcarfentanyl	4-[(1-oxopropyl)phenylamino]-4-piperidinecarboxylic acid, methyl ester
Remifentanil	methyl 1-(3-methoxy-3-oxopropyl)-4-(N-phenyl-propanamido)piperidine-4-carboxylate
Acetyl Fentanyl	N-(1-Phenethylpiperidin-4-yl)-N-phenylacetamide
Alfentanyl	N-{1-[2-(4-ethyl-5-oxo-4,5-dihydro-1H-1,2,3,4-tetrazol-1-yl)ethyl]-4-(methoxymethyl)piperidin-4-yl}-N-phenylpropanamide
4-ANPP	N-Phenyl-1-(2-phenylethyl)piperidin-4-amine
Sufentanil	N-[4-(Methoxymethyl)-1-(2-thiofuran-2-ylethyl)-4-piperidyl]-N-phenylpropanamide
Valeryl fentanyl	N-(1-(2-Phenylethyl)-4-piperidinyl)-N-phenylpentylamide
Furanyl fentanyl	N-Phenyl-N-[1-(2-phenylethyl)piperidin-4-yl] furan-2-carboxamide
Remifentanil	methyl 1-(3-methoxy-3-oxopropyl)-4-(N-phenyl-propanamido)piperidine-4-carboxylate

Methoxyacetyl norfentanyl	2-methoxy-N-phenyl-N-piperidin-4-ylacetamide
Acetyl norfentanyl	N-phenyl-N-(piperidin-4-yl)acetamide
Carboxy butyryl fentanyl	4-oxo-4-(N-[1-(2-phenylethyl)piperidin-4-yl]anilino)butanoic acid
Carboxy valeryl fentanyl	5-oxo-5-[phenyl[1-(2-phenylethyl)-4-piperidinyl] amino]-pentanoic acid
Methoxyacetyl fentanyl	2-Methoxy-N-(1-phenethylpiperidin-4-yl)-N-phenylacetamide
Furanyl norfentanyl	N-phenyl-N-4-piperidinyl-2-furancarboxamide
Cis-3-methyl norfentanyl	cis-N-(3-methyl-4-piperidinyl)-N-phenylpropanamide
Trans-3-methyl norfentanyl	trans-N-(3-methyl-4-piperidinyl)-N-phenylpropanamide
Butyryl norfentanyl	N-Phenyl-N-(4-piperidinyl)butanamide
Cycloproyl norfentanyl	N-phenyl-N-4-piperidinyl-cyclopropanecarboxamide
β-hydroxyfentanyl	N-[1-(2-Hydroxy-2-phenylethyl)piperidin-4-yl]-N-phenylpropanamide
Despropionil-p-F-fentanyl	N-(4-fluorophenyl)-1-(2-phenylethyl)-4-piperidinamine
Phenyl acetyl fentanyl	N-(1-phenethylpiperidin-4-yl)-N,2-diphenylacetamide
Bentazepam	5-phenyl-3,4,6,7,8,9-hexahydro-[1]benzothiolo [2,3-e][1,4]diazepin-2-one
Diclazepam	7-Chloro-5-(2-chlorophenyl)-1-methyl-1,3-dihydro-2H-1,4-benzodiazepin-2-one
Etizolam	4-(2-Chlorophenyl)-2-ethyl-9-methyl-6H-thieno [3,2-f][1,2,4]triazolo[4,3-a][1,4]diazepine
Flubromazepam	7-Bromo-5-(2-fluorophenyl)-1,3-dihydro-1,4-benzodiazepin-2-one
Phenazepam	7-Bromo-5-(2-chlorophenyl)-1,3-dihydro-2H-1,4-benzodiazepin-2-one
Pyrazolam	8-Bromo-1-methyl-6-(pyridin-2-yl)-4H-[1,2,4] triazolo[4,3-a][1,4]benzodiazepine
Metizolam	4-(2-Chlorophenyl)-2-ethyl-6H-thieno[3,2-f] [1,2,4]triazolo[4,3-a][1,4]diazepine
Flunitrazolam	1-methyl-8-nitro-6-(2-fluorophenyl)-4H-[1,2,4] triazolo[4,3-a][1,4]benzodiazepine
Deschloroetizolam	4-phenyl-2-ethyl-9-methyl-6H-thieno[3,2-f] [1,2,4]triazolo[4,3-a][1,4]diazepine

1 Introduction

Since the 1970s, oral fluid (OF) has been studied as a biological matrix alternative to blood to disclose current consumption of psychotropic drugs [1–6]. OF is the biological fluid that has successfully been used as an alternative to blood in clinical and forensic toxicology [7]. It is evident that when a given pharmacologically active drug is detected in OF specimens, there is a high probability that the tested individual is under the effects of the drug itself [8]. Published scientific studies have documented OF testing reliability for psychoactive drugs in drug treatment, workplace drug testing, pain management, and driving under the influence of drugs (DUID) programs [9].

OF collection is easy and noninvasive, with low biohazard risk, and is gender-neutral during direct observation in ways that other matrices are not, thereby reducing the opportunity for adulteration [7, 10]. As only unbound or free drugs are excreted into OF, there is evidence for some drugs that OF concentrations correlate with free drug plasma concentrations. This makes OF might better reflect recent drug use, providing a better correlation with pharmacodynamics effects, such as impaired performances. Disadvantages for OF tests include small sample volume, oral cavity contamination after oral or smoked drug consumption, dry mouth [6, 11], reduced salivation following intake of drugs with sympathomimetic properties (e.g., *Cannabis* or amphetamines), and lower concentrations compared to that in blood or than in urine [1, 7].

Although in terms of window of detection, OF shows in general a shorter window (hours) than that of blood or urine (days) [7], the overlap between detection times in OF and blood or urine can be ameliorated by using sensitive methods for OF analysis [7, 12]. Further, the detection window for drugs in OF may be expanded after repeated ingestion of high doses [7, 12, 13].

Drug transfer from blood into OF is affected by OF composition, flow rate, and pH, as well as the drug's pKa, protein binding, lipophilicity, spatial configuration, and molecular weight [9, 14, 15]. Free, unbound parent drugs can enter OF by passive diffusion from blood. Glucuronidated metabolites can be also detected, although in lower concentrations than free metabolites [16, 17]. Following smoking, insufflation, sublingual and/or oral drug administration, the oral mucosa is directly exposed to drug (s) resulting in relatively high OF drug concentrations [16, 18–20]. Conversely, capsule ingestion typically does not contaminate the oral mucosa unless the capsule is adulterated or chewed [21, Different OF collection methods have been described [14, 23–26]. A detailed review article is available explaining the different techniques in full details but, generally speaking, OF can be collected by passive drool and expectoration (with or without stimulation) and via a wide variety of devices. As expected, each collection method has advantages and disadvantages (e.g., passive drool is unpleasant for donors and collectors and stimulated OF samples are diluted due to increased salivary flow) [9]. Certainly, OF collection, sampling, and proper storage are crucial to correctly interpret the results and the outcomes of clinical and forensic cases. OF samples should be rapidly analysed upon collection and diluted with stabilizing/preservative buffers. It is proven that OF drug stability depends on different factors such as: individual OF composition, sample handling, stabilizing buffer composition, transportation, and storage duration and conditions [25, 27].

New psychoactive substance (NPS) is defined as "a new narcotic or psychotropic drug, in pure form or in preparation, that is not controlled by the United Nations drug conventions, but which may pose a public health threat comparable to that posed by substances listed in these conventions" [22].

Since the beginning of twenty-first century, more than 800 new psychoactive substances have entered the illicit street and web markets, posing continuous new serious health threats for drug consumers [22]. Hence, the objective assessment of their consumption through the analysis of biological matrices of presumptive consumers provide a tool in the above-reported clinical and forensic scenarios. In this chapter, we present the methodological and analytical details concerning the determination of principal classes of NPS in OF.

2 Synthetic Cannabinoids

Together with synthetic cathinones (SCath), synthetic cannabinoids (SC) are the class of NPS mostly abused to mimic and enhance the subjective effects of phytocannabinoids contained in the cannabis plant [22]. For this reason, several different analytical methods have been developed to investigate the presence of SC and their metabolites in OF (Table 1). Most of them are based on chromatography coupled to spectrometry assays, e.g., ultra-high-performance liquid chromatography-tandem mass spectrometry (UHPLC-MS/MS), liquid chromatography-tandem mass spectrometry (LC-MS/MS), and gas chromatography tandem mass spectrometry (GC-MS/MS).

Our group published the most recent methodology for SC in OF. Three different SC (JWH-122, JWH-210, UR-144) have been identified in the consumers' OF by last generation GC-MS screening method and then quantified with their respective metabolites JWH-122 N-(4-OH), JWH-210 N-(4-OH), JWH-210 N-(5-OH), and UR-144 N-(5-OH) using UHPLC-HRMS. The method showed matrix effects lower than 15%, extraction recovery higher than 70% and process efficiency better than 70%. After smoking 1 mg JWH-122 or UR-144 and 3 mg JWH-210, maximum SC concentrations were obtained 20 min after inhalation with disappearance hours post administration [28].

In the year 2020, two investigation groups validated comprehensive analytical methods to measure a great number of SC in OF. The first one by Mulet et al. sought to screen 72 SC (among

Table 1
Analytical methods for the screening and determination of synthetic cannabinoids in oral fluid

Synthetic Cannabinoids	Instrumental method	Collection Device	Sample preparation (including derivatization)	Column	Mobile phase	Run time (min)	LOD (ng/mL)	LOQ (ng/mL)	Authentic samples' concentrations (ng/mL)	References
JWH-122, JWH-210, UR-144, JWH-122 N-(4-OH), JWH-210 N-(4-OH), JWH-210 N-(5-OH), UR-144 N-(5-OH)	GC-MS (screening)	Salivette	a. 400 μL OF extracted + 100 μL 0.1 M NaH2PO4•H2O + 5 μL of IS pH 3.0 b. Extracted twice 3.0 mL C_6H_{14}:EtOAc (9:1, v/v) c. Centrifuged 4000 rpm/ 4 min d. Derivatized + 50 μL BSTFA +1% TMCS + ACN (1:1, v/v) 70 °C – 30 min	Ultra-inert (HP-5MS UI, 30 m × 250 μm i.d, film thickness 0.25 μm)	Helium	17	0.3–0.7	0.5–2.30	3.14–8.10	[28]
	UHPLC-HRMS	Salivette	a. 400 μL OF extracted + 100 μL 0.1 M NaH2PO4•H2O + 20 μL of IS pH 3.0 b. Extracted twice 3.0 mL C_6H_{14}:EtOAc (9:1, v/v) c. Centrifuged 4000 rpm/ 4 min	Kinetex biphenyl 100A (100 × 2.1 mm, 2.6 μm)	A: NH_4HCO_2 2 mM in H_2O + 0.1% FA B: NH_4HCO_2 2 mM + MeOH/ ACN 50:50 (v/v) + 0.1% FA	18	0.02–0.07	0.07–0.25	0.29–3.14 JWH-122 N-(5-OH) < LOQ	
33 from JWH-family, 17 from PINACA-family, AM-1220, AM-2201, AM2201 2-hydroxyindole metabolite, AM2201 5-hydroxyindole metabolite, AM2233, AM6, MAM2201 N-(3-fluoropentyl) isomer 94, HU-211, MAM2201 N-(5-chloropentyl), PB-22, PB-22 3-carboxyindole, PB-22 N-(4-hydroxypentyl), PB-22 N-(5-hydroxypentyl),	LC-QqQ-MS/MS	NR	Online SPE	C18 Rapid resolution HD (3.0 × 100 mm × 1.8 μm)	A: 5 mM NH_4HCO_2 + 0.1% FA for H_2O phase B: MeOH +0.1% FA for organic phase	12	0.4–3.8	1.1–11.6	9–40	[29]

PB-22 N-pentanoic acid, RCS-4, RCS-4-C4 homolog, RCS-8, THJ 018, THJ 2201, UR-144, XLR11, XLR11 6-hydroxyindole, XLR11 N-(4-hydroxypentyl)										
AB-FUBINACA, AB-PINACA, APINACA, AM2201, AM2233, HU-211, 12 from JWH-family, RCS-4, RCS-8, XLR11, PB22, MMB-FUBINACA	UHPLC-MS	Quantisal	a. 500-μL OF Quantisal elution buffer + 25 μL IS + 500 μL saturated $Na_2B_4O_7$ aqueous solution + 2 mL MTBE b. Vortexed 2 min c. Centrifuged 4500 rpm/ 5 min + 1.7-mL organic phase + nitrogen (40 °C) d. Reconstituted + 100-μL MeOH	Raptor™ biphenyl (100 mm × 2.1 mm, 2.7 μm)	A: H_2O + 0.1% FA (v/v) + 2 mmol/L NH_4HCO_2 B: MeOH +0.1% FA (v/v) + 2 mmol/L NH_4HCO_2	13.5	0.05–1	NR	NR	[30]
5F-ADB, MMB-CHMICA, THJ-2201, CUMYL-4CN-BINACA, MDMB-CHMCZCA	GC–MS	Eppendorf tube	a. OF centrifuged at 10000 rpm/4 min b. pH adjusted 480 μL sample + 20 μL buffer (pH 7, 2.5 M) c. Extracted by MEPS-C18 sorbent + 100 μL $CH_3CH(OH)CH_3$ + 100 μL deionized H_2O d. 5 μL TPP at 5 mg/ L + MEPS extract	HP-5 MS (30 m, 0.25 mm, 0.25 μm)	Helium	22	10–20	30–60	60–1000	[31]
JWH-018	LC-MS/MS	Intercept	1 mL OF + 2 mL of the stabilizing solution b. Centrifuged 3500 rpm/ 10 min c. 16.7 μL IS + 500 μL diluted OF + 2 mL C_6H_{14} d. Shook + centrifuged + evaporated + dryness under nitrogen - 40 °C e. Reconstituted in 1 mL of mobile phase A/B (90/10, v/v)	Acquity UPLC C18 1.7 μm (2.1 mm × 100 mm)	A: H_2O + 1 mM NH_4HCO_2 B: MeOH	5	NR	0.25	< LLOQ-100	[32]
5F-ADB, AM-694, 5F-PB-22, AB-FUBINACA, JWH-018, 5F-AKB48	FSF	NR	a. OF dissolved in either deionized H_2O b. Filtered 0.44 μm syringe-driven filter	NR	NR	NR	NR	NR	NR	[33]

(continued)

Table 1
(continued)

Synthetic Cannabinoids	Instrumental method	Collection Device	Sample preparation (including derivatization)	Column	Mobile phase	Run time (min)	LOD (ng/mL)	LOQ (ng/mL)	Authentic samples' concentrations (ng/mL)	References
UR 144, CP47,497, CP47497(C8), JWH-250, JWH-019, AM-2201, JWH-122, JWH-081, JWH-200	GC-MS	NR	a. OF thawed b. Centrifuged 4400 rpm/ 5 min c. 1 mL supernatant + spiked 10 μL of deuterated IS (final concentration: 100 ng/mL) d. PDMS 100 μm fiber dipped 15 min – room temperature e. Desorbed 270 °C – 15 min	Capillary column (30 m × 0.25 mm. i.d., 0.25 μm)	Helium	30	1–10	10	NR	[34]
5F-AKB48 (5F-APINACA)	UHPLC	No device	a. 0.5 mL OF + 10 μL ISs (5 μg/mL) + 1.0 mL H2O b. Vortexed 30 s c. Centrifuged 3500 rpm/3 min d. 2 μL supernatant injected into UHPLC–MS/MS	Acquity UPLC BEH C18 column (2.1 mm × 75 mm, 1.7 μm)	A: 0.1% FA in H_2O B 0.1% FA in ACN	6	0.6	2	NR	[35]
JWH-018	SERS	Polypropylene tube	a. OF + JWH-018 MeOH (37.5% v/v) b. 400 μL OF centrifuged 9.3 rcf /10 min c. 49.5 μL supernatant - μ-SPE (0.6 μL reversed phase C18) d. Wet, equilibrated, bound, washed, eluted	NR	NR	NR	0.8	2	2.0–7.8	[36]
AM2233, JWH-200, AB-005, AB-FUBINACA, AB-PINACA, AB-CHMINACA, AM2201, RCS-4, JWH-250, STS-135, JWH-73, XLR-11, JWH-251, JWH-18, JWH-122, JWH-19,	LC-MS/MS	Sterile tube	a. 100 μL spiked OF + 300 μL H_2O + 200 μL ACN + IS b. Centrifuged 5000 rpm/ 5 min c. 100 μL supernatant in autosampler vials d. 1 μL injected into LC–MS-MS	Kinetex biphenyl column (50 mm × 3 mm × 2.6 μm)	A: 0.1% FA in H_2O B: ACN	6	1	2.5	NR	[37]

UR-144, JWH-20, AKB-48										
JWH-200	LFIA ELISA UHPLC -MS/MS	NR	a. Freshly OF spiked with JWH-200 b. Two sets of prepared OF analyzed immunomethods and UHPLC–MS/MS	Zorbax eclipse plus C18 column (2.1 × 50 mm; 1.8 μm)	A: 10 mM FA B: MeOH	6	LOD_{LFIA} 0.08 LOD_{ELISA} 0.04	NR	0–1000	[38]
A-796,260, AB-FUBINACA, AB-PINACA, APINACA (AKB-48), JWH-018, JWH-073, MAM2201, PB-22, 5-Fluoro PB-22, UR-144, XLR-11, AB-PINACA 5-Pentanoic acid, APINACA 5- Hydroxypentyl, AM2201 4-Hydroxypentyl, JWH-018 5-Pentanoic acid,JWH-019 5-Hydroxyhexyl, JWH-073 4-Butanoic acid, JWH-122 4-Hydroxypentyl, JWH-210 4-Hydroxypentyl, JWH-250 4-Hydroxypentyl, MAM2201 4-Hydroxypentyl, UR-144 4-Hydroxypentyl, UR-144 5-Pentanoic acid, XLR-11 4-Hydroxypentyl	UHPLC -MS/MS	Quantisal	a. 0.2 mL OF + 0.6 mL Quantisal buffer + 50 μL IS +200 μL + 1 M CH_3COOH solution b. Vortexed 10 seconds c. Loaded onto Isolute 1 mL SLE + cartridges d. Equilibrated 15 minutes e. Eluted 2 × 3 mL of ethyl acetate f. Evaporated g. Reconstituted 150 μL mobile phase A:B, 85:15 (v/v) h. Transferred to LC-MS autosampler vials	C18 column (2.1 × 100 mm, 1.7 μm)	A: 0.1% FA in H_2O B: 0.1% FA in ACN	9	NR	0.5	0.8–38.9 (AB-FUBINACA)	[7]
AM-1220, JWH-200, AB-005, JWH-018 *N*-pentanoic acid, JWH-018 *N*-5-hydroxypentyl, WIN-55, XLR-11 *N*-4-hydroxypentyl, MAM-2201 N-pentanoic acid, JWH-073, UR-144 N-5-hydroxypentyl, JWH-250, MAM-2201, XLR-11, JWH-018, JWH-081, JWH-122, UR-144	UHPLC -MS/MS	Plastic disposable containers (Salimetrics, Carlsbad, CA, USA)	a. Ninety μL OF + 20 μL MeOH + IS (100 ng/mL) b. Diluted 30 μL MeOH + 60 μL ultrapure H_2O (total volume 200 μL) c. Vortexed + mixed 1 minute d. Sonicated 25 ° C 6 minutes e. Centrifuged 15000 rpm/ 5 min Room temperature f. Supernatant Clean-up - MEPS syringe (C18 sorbent inside BIN)	Reverse-phase C18 column (1.7 μm, 2.1 × 100 mm)	A: H_2O with 5 mM FA B: MeOH/ACN (80: 20 v/v) with 5 mM FA	8.3	0.005–0.850	0.015–0.820	LLOQ-200	[39]

(continued)

Table 1
(continued)

Synthetic Cannabinoids	Instrumental method	Collection Device	Sample preparation (including derivatization)	Column	Mobile phase	Run time (min)	LOD (ng/mL)	LOQ (ng/mL)	Authentic samples' concentrations (ng/mL)	References
XLR11, UR-144, XLR11-PYR, UR-144-PYR, 4-hydroxy-XLR11, 4-hydroxy-UR-144	UHPLC – MS-MS	Quantisal	a. 10 mL IS + 300 mL 4% H_3PO_4 in water + 400 mL human subject samples mixed on a rotary shaker b. Sample loaded into SPE c. Eluted with 5% NH_4 in H_2O:ACN (60:40, v/v)	BEH (C-18, 100 × 2.1 mm, 1.7 mm)	A: 0.1% FA in H_2O: ACN (95: 5, v/v) B: 0.1% FA in ACN	4.5	0.35–1.93	5.0	5–60	[40]
AM-694, AM-2201, HU-210, 11 from JWH-family, RCS-4, RCS-4-C4, RCS-8, WIN55,212–2	UPLC-MS/MS	Intercept	a. 0.5 ml collected samples at 4 °C extracted with LLE using EtOAc:C_7H_{16} 4:1	HSS T3 C18 column (100 mm × 2.1 mm, 1.8 μm	A: 10 mM NH_4HCO_2 buffer pH 3.1 B: MeOH	6	0.05–1.2	NR	0.2–115	[41]
JWH-200, JWH-018, JWH-073, JWH-022, AM-2201, AM-2232, AM-1220	ELISA & LC-MS/MS	Quantisal	a. OF (100 mL) diluted 1:1 IS (5 ng/mL) b. Centrifuged 15000 rpm/ 2 min c. 20 mL prepared sample injected into the LC system	DB biphenyl column (5 mm, 50 × 2.1 mm)	A: H_2O 0.1% FA + 0.2% NH_4HCO_2 B: ACN + 0.1% FA + 0.2% NH_4HCO_2	7.5	0.1	0.25	0.25–25	[42]
19 from JWH-family, 4 from AM-family, CRA-13, Methanandamide, RCS-4, RCS ortho isomers, RCS-8, WIN 48,098, WIN 55,212–2	LC-ESI-MS/MS	Tube	a. 200 μL nOF + 10 μL of ISTDs + 600 μL ice-cold ACN centrifuged at 3400 g/20 min 0 °C b. 600 μL supernatant evaporated to dryness at 60 °C under nitrogen c. Reconstituted A/B, 50: 50 V/V	Luna phenyl hexyl (50 × 2 mm, 5 μm)	A: H_2O + 0.2% FA + 2 mmol/L NH_4HCO_2 B: MeOH	8	0.015–0.6	0.15–3	<0.1–7.2	[43]
JWH-200, AM 694, JWH-250, JWH-073, JWH-018, JWH-019, JWH-122, HU 210, CP 47497	UHPLC-ESI-MS/MS	DCD5000	a. OF centrifuged 4000 rpm/ 5 min, and stored −20 ° C b. 250 μL OF sample + 25 μL ISs + 250 μL mobile phase A c. 10 μL solutions injected in UHPLC–MS/MS equipment	Kinetex (C18, 100 mm × 2.1 mm, 2.6 μm)	A: 5 mM NH_4HCO_2 + 0.05% FA B: MeOH/ACN 1:1 + 0.1% FA	9	1–20	NR	NR	[44]

JWH-018, JWH-250, JWH-073, CP 47497 C8, CP47497, HU-210	LC-MS-MS	Quantisal	a. OF vortex mixed 15 min prior to extraction b. N solid-phase extraction columns (SPE) + C_6H_{14}/ glacial CH_3COOH (98:2, v/v, 2 mL) c. Eluent evaporated to dryness under nitrogen at 40 °C d. Reconstituted MeOH (50 μL) e. Injected into LC– MS–MS system	Extend C18 (2.1 × 50 mm × 1.8 μm)	A: 0.2% CH_3COOH B: ACN	9.2	NR	0.5	0.11–16.9	[46]

which 33 from JWH-family and 17 from PINACA-family) and metabolites in OF by LC- MS/MS, using online solid-phase extraction (SPE) for sample pre-treatment. The method was regarded as a reliable screening tool in identifying SC exposure in forensic toxicological settings, even no real cases were reported [29]. The second one by da Cunha et al. screened 104 NPS, among which 23 SC in OF samples by UHPLC–MS-MS. The Quantisal™ OF device (Immunalysis Corporation; Pomona, CA, USA) was used to collect OF samples. Matrix effects were generally less than 20% and overall extraction recoveries greater than 60%. The highest matrix effect was observed within the SC group (PB-22, −55.5%) [30]. No real cases were presented also in the case of the GC-MS assay that Sorribes-Soriano et al. developed to detect of third-generation synthetic cannabinoids (5F-ADB, MMB-CHMICA, THJ-2201, CUMYL-4CN-BINACA, and MDMB-CHMCZCA) in OF. Extraction of oral fluid samples by microextraction by packed sorbent (MEPS) was performed [31].

Cannaert et al. [32] compared a semi-quantitative activity-based detection and quantitative LC-MS/MS method for JWH-018 in OF after vaping of 23 healthy *Cannabis*-experienced participants in this placebo-controlled, crossover study. Participants inhaled the smoke of 75 μg JWH-018/kg bodyweight (average 5.5 mg) or 50 mg Knaster Hemp (placebo) on two separate days. JWH-018 powder was heated via a vaporizer pen, which reaches temperatures of approximately 380 °C. Participants inhaled the vapor in five intakes, according to a strict inhalation regimen. OF samples were obtained using the Intercept i2 OF Collection Device at several time points after inhalation. A total of 175 OF samples were collected and analysed by LC-MS/MS method, 75 were below the limit of quantification (LOQ) of 0.25 ng/mL, whereas 18 were above the ULOQ (upper limit of quantification) of 100 ng/mL [32].

A further new method was developed and validated by May et al., using fluorescence spectral fingerprints. Six SC (5F-ADB, AM-694, 5F-PB-22, AB-FUBINACA, JWH-018, 5F-AKB48) were detected in real OF samples collected from volunteers who confirmed no drug use in the preceding month. The obtained results, suggest that the method proposed could be a useful tool in patient triage to manage overdoses and with application in harm reduction strategies for users of Spice/K2 [33]. In the same year, 2019, Anzillotti et al. described a pilot study by solid-phase microextraction (SPME) coupled to gas chromatography/mass spectrometry (GC-MS) to determine synthetic and natural cannabinoids in OF. The method was applied to 106 young student volunteers with an average age of 25.0 ± 9.8 years showing a positivity rate of 3.8%. The method was validated according to the FDA guidelines for linearity, LOD, LOQ, accuracy, and repeatability. This study showed to have a number of advantages such as simplicity, rapidity, low cost, and easy applicability to a wide variety of matrices [34].

In 2019, a fast UHPLC–MS/MS method was presented by Malaca et al. for the simultaneous determination of 13 different psychoactive drugs, belonging to different chemical classes, 6-monoacetylmorphine, morphine, codeine, cocaine, benzoylecgonine, amphetamine, methamphetamine, 3,4-methylenedioxymethamphetamine, delta-9-tetrahydrocannabinol, cannabidiol, mephedrone, ketamine, and 5F-AKB48 (5F-APINACA), in OF. Real specimens were diluted and inject directly onto LC system. Intra-assay and inter-assay precision and accuracy were better than 15% [35].

The first screening test for the detection of the synthetic cannabinoid JWH-018 in OF by micro solid-phase extraction surface-enhanced Raman spectroscopy (μ-SPE/SERS) was proposed by Deriu et al. in 2019. SERS was proposed as an alternative method to current immunoassays as a screening tool in OF samples in the emergency room settings [36].

In 2018, Williams et al. reported a validated method to detect the presence of 19 SC in OF (AM2233, JWH-200, AB-005, AB-FUBINACA, AB-PINACA, AB-CHMINACA, AM2201, RCS-4, JWH-250, STS-135, JWH-73, XLR-11, JWH-251, JWH-18, JWH-122, JWH-19, UR-144, JWH-20, and AKB-48). A rapid LC–MS-MS analysis with minimal sample preparation was proposed and validated in accordance with National Association of Testing Authorities (NATA) guidelines, where LOD, LOQ, selectivity, linearity of calibration, precision, repeatability, ion suppression, and measurement of uncertainty (MOU) were evaluated. This method was applied to 12 authentic samples previously submitted for routine testing; however, no cannabinoids were detected [37].

In 2017, Fojtíková et al. developed a lateral flow immunoassay (LFIA) and an enzyme linked immunosorbent assay (ELISA) for the detection of JWH-200 in OF. For this purpose, they reported a synthesis of a new hapten structurally derived from JWH-200 for the development of immuno-methods. The hapten was prepared using a ten-step synthetic sequence from commercially available naphthalen-1-ol. LFIA method was optimized using artificial saliva while the ELISA method was developed as an indirect competitive format using polyclonal antibody targeted at JWH-200 [38].

In the same year, Blandino et al. set up a LC-MS/MS method for the determination of 11 synthetic cannabinoids and correlated urine and OF as matrices to monitor SC recent exposure among HIV-infected homeless individuals. OF samples were collected by Quantisal devices and stored in the freezer at −20 ° C until analysis. For synthetic cannabinoids, parent compounds AB-FUBINACA, PB-22, and 5-Fluoro-PB-22 were detected in urine at low concentrations (1.4–3 ng/mL). In OF samples, the only synthetic cannabinoid detected was AB-FUBINACA (n = 5). Although PB-22,

5-Fluoro-PB-22, or UR-144 were detected in urine samples, these compounds were negative in paired OF samples [7]. Rocchi et al. determined 31 NPS standards, (synthetic cathinones, piperazines, phenethylamines, and synthetic cannabinoids) in OF without any real sample to prove methodology robustness by means of micro-extraction by packed sorbent followed by UHPLC-MS/MS. The SC included in this study were 17 (AM-1220, JWH-200, AB-005, JWH-018 N-pentanoic acid, JWH-018 N-5-hydroxypentyl, WIN-55, XLR-11 N-4-hydroxypentyl, MAM-2201 N-pentanoic acid, JWH-073, UR-144 N-5-hydroxypentyl, JWH-250, MAM-2201, XLR-11, JWH-018, JWH-081, JWH-122, UR-144). OF samples cleanup as performed by using a MEPS syringe with a C18 sorbent inside the barrel-in-needle (BIN) [39].

Before all the methods described so far, Amaratunga et al. reported in 2014 a LC-MS/MS method to quantify XLR11, UR-144, their metabolites (4-OH-UR-144 and 4-OH-XLR11), and their pyrolysis product (UR-144-PYR and XLR11-PYR) in OF. The samples were collected from drug treatment facilities, criminal justice institutions, and child and family welfare agencies with a Quantisal collection device and then were diluted in an OF-extraction buffer mixture. In the case of UR-144, the parent drug, the metabolite, and the pyrolysis product were present in OF; however, in the case of XLR11, the metabolite was absent [40].

The year before, Øiestad et al. described a screening method to detect SC in preserved OF by UPLC-MS/MS. The compounds under investigation were 18 SC (JWH-018, JWH-073, RCS-4, AM-2201, HU-210, JWH-019, JWH-081, JWH-122, JWH-200, JWH-210, JWH-250, AM-694, JWH-015, JWH-020, JWH-251, WIN55,212-2, RCS-4-C4, and RCS-8). AM-2201 and JWH-018 were found in authentic samples [41]. In the same year, Rodriguez et al. suggested an immunoassay (ELISA) as a powerful tool for the detection of seven SC (JWH-200, JWH-018, JWH-073, JWH-022, AM-2201, AM-2232, and AM-1220) in OF. The validated method was applied to 32 real samples previously confirmed by LC-MS-MS and shown to contain AM-2201 (0.25–80.5 ng/mL), along with either JWH-018 (0.36–12.5 ng/mL) or JWH-210 (0.36–27.7 ng/mL) [42].

A fully LC/ESI-MS/MS method for the analysis of 30 SC (16 naphthoylindoles, 4 phenylacetylindoles, 5 benzoylindoles, JWH-018 adamantyl derivative (AB-001), JWH-307, WIN 55,212–2, CRA-13, and methanandamide) in OF samples was presented by Kneisel et al. The method was successfully applied to 264 real samples collected from forensic psychiatry clinics (82%), withdrawal therapy (16%) as well as socio-pedagogical centers (2%), during routine analysis. A total of 31 samples (12%) was confirmed positive for at least one of the reported synthetic cannabinoids (AM-694 (3%), AM-2201 (20%), JWH-018 (23%), JWH-019 (7%), JWH-081 (26%), JWH-122 (55%), JWH-203 (7%),

JWH-210 (23%), JWH-250 (3%), JWH-307 (3%), MAM-2201 (3%), and RCS-4 (7%)) [43].

An UHPLC–ESI–MS/MS screening method for the direct analysis of 24 drugs in OF, including SC (JWH-200, AM 694, JWH-250, JWH-073, JWH-018, JWH-019, JWH-122, HU 210, and CP 47497) has been proposed by Strano-Rossi et al.in 2012 [44]. This screening method was validated in accordance with Peters et al. [45] for qualitative purpose. The method was applied to 400 authentic OF samples collected during random traffic controls. No synthetic cannabinoids were detected. Finally, Coulter et al. reported the first LC-MS/MS method for the simultaneous quantification of JWH-018, JWH-073, JWH-200, CP-47497, and cannabicyclohexanol (CP-47497 C8) in OF. The procedure was validated and applied to real samples taken from two naïve individuals who had purchased the compounds while still legally available in the U.S. [46].

3 Synthetic Cathinones

UHPLC-MS/MS has been the most used analytical technique for the determination of synthetic cathinones (SCath) in OF (Table 2). The first UHPLC-MS/MS method for OF detection of four SCath (Methylone, 4-MMC, 4-MEC, and MDPV) among 24 designer drugs appeared in the literature in 2012. The method was applied to 400 OF samples collected during Italian Police road checks and only two specimens resulted positive to MDPV a 4-MEC [44].

Another specific method for the quantification of 10 SCath (methylone, methcathinone, pyrovalerone, 4-MMC, MDPV, 4-FMC, PMMC, bk-MBDB, bk-MDEA, and α-PVP) in OF was developed, validated, and applied to authentic unknown OF samples by Amaratunga et al. in 2013. Five subjects resulted positive to the SCaths MDPV and α-PVP) [47].

Almost the same SCath as before, methylone, PMMC, 4-MMC, 4-FMC, and MDPV, were investigated in OF by HPLC-MS/MS in DUID cases by de Castro et al. in 2014. Ten OF samples from drivers tested positive to on-site methamphetamine screening tests were analyzed through the new method and one resulted positive for MDPV [48].

In 2016, Mercolini et al. used a volumetric absorptive microsampling device (VAMS) for fast sampling of small OF volume (10 μL) to quantify methylone, bk-MBDB, bk-MDEA, 4-MMC, and MDPV by a LC-MS/MS method. The dried saliva was extracted from the VAMS tips with methanol then the dried extracts were reconstituted with the LC mobile phase [49].

In the same year, the stability of 10 SCath in neat OF and OF preserved with two types of buffers, Quantisal and Oral-Eze™ (Thermo Fisher Scientific, Waltham, MA, USA), stored at room

Table 2
Analytical methods for the screening and determination of synthetic cathinones in oral fluid

Synthetic Cathinones	Instrumental method	Collection Device	Sample preparation (including derivatization)	Column	Mobile phase	Run time (min)	LOD (ng/mL)	LOQ (ng/mL)	Authentic samples' concentrations (ng/mL)	References
Methylone, 4-MMC, 4-MEC, MDPV	UHPLC-MS/MS	DCD5000	a. OF centrifuged 4000 rpm/5 min, and stored −20 ° C b. 250 μL OF sample + 25 μL ISs + 250 μL mobile phase A c. 10 μL solutions injected in UHPLC–MS/MS equipment	Kinetex (C18, 100 mm × 2.1 mm, 2.6 μm)	A: 5 mM NH_4HCO_2 + 0.05% FA B: MeOH:ACN (1:1, v/v) + 0.1% FA	9	1–5	NR	2–200 (MDPV, 4-MEC)	[44]
Methylone, Methcathinone, Pyrovalerone 4-MMC, MDPV, 4-FMC, PMMC, bk-MBDB, bk-MDEA, α-PVP	UHPLC-MS/MS	Quantisal	a. 10 mL IS + 300 mL 4% H_3PO_4 in H_2O + 400 mL human subject samples mixed on a rotary shaker b. Sample loaded into SPE c. Eluted with 5% NH_4 in H2O:ACN (60:40, v/v)	Acquity BEH (C-18, 100 × 2.1 mm, 1.7 mm)	A: 0.1% FA:ACN (95:5, v/v) B: 0.1% FA in ACN	3	NR	1	20.5–935 (MDPV, α-PVP)	[47]
Methylone, PMMC, 4-MMC, 4-FMC, MDPV	LC-MS/MS	Salivette Quantisal	a. 25 μL IStd + 2 mL BBS pH 9 + 0.5 mL OF b. Loaded + 2 mL MeOH + 2 mL H_2O c. Washed steps + 2 mL H_2O:MeOH (95:5, v/v) + 2 mL H_2O:MeOH: NH_4OH (70:29.5:0.5, v/v/v) d. Eluted with 3 mL CH_2Cl_2:$CH_3CH(OH)CH_3$ (75:25) e. Added 100 μL HCl 0.1% f. Evaporated nitrogen 35 °C g. Reconstituted 75 μL FA 0.1%:ACN (90:10, v/v) h. Injected into LC–MS/MS system	Atlantis® T3 (100mm × 2.1 mm, 3 μm) reversed-phase	A: 0.1% FA B: ACN	10	0.025–0.1	0.2–0.5	31.8 (MDPV)	[48]
Methylone, bk-MBDB, bk-MDEA, 4-MMC, 4-MEC, MDPV	LC-MS/MS	VAMS absorptive microsampling device	a. 10 μL OF extracted + CH_3OH b. Dried extracts reconstituted with LC mobile phase	Zorbax C18 (2.1 mm × 50 mm, 3.5 μm)	A: 0.1% FA in ACN B: 0.1% FA in H_2O	15	3	10	18–40 (4-MMC)	[49]
Methylone, Pyrovalerone, Pentedrone, ETH-CAT,	UHPLC-MS/MS	Salivette	a. 300 μL OF spiked + 200 μL MeOH shaken by hand 30 s b. Centrifuged 15596 g/5 min c. 300 μL supernatant	Acquity UPLC™ BEH shield RP 18 (100 mm × 2.1 mm, 1.7 μm)	A: 0.1% FA in water B: 0.1% FA in ACN	7.5	0.25	0.5	NR	[52]

4-MMC, 4-FMC, bk-MBDB, bk-MDEA, MDPV			+ 200 μL NaH2PO4•H2O (50 mM, pH 9) centrifuged 15596 g/5min d. MEPS procedure by automated analytical syringes (500 and 100 μL) e. Extracted evaporated under nitrogen 40 °C f. Reconstituted 50 μL mobile phase g. 5 μL injected into UPLC–MS/MS instrument							
Cathinone, Methcathinone, Methylone, Naphyrone, ETH-CAT, MABP, 4-MMC, 4-MEC, MDPV, α-PVP	UHPLC-MS/MS	Quantisal Oral-Eze	2 ml 1M CH_3COOH + 25 uL IS (100 ng/mL) + 0.4 mL OF-Quantisal (0.3 mL OF-Oral-Eze, or 0.1 mL neat OF) vortexed b. Loaded Strata Drug-X B cartridges + 2 mL MeOH + 2 mL deionized H_2O c. Eluted + 2 mL 2% NH_4OH + CH_2Cl_2: $CH_3CH(OH)CH_3$ (95:5, v/v) d. Dried + reconstituted 100 mL 0.1% FA + H_2O	Kinetex C18 (100 mm × 2.1 mm, 1.7 μm)	A: 0.1% FA in H_2O B: 0.1% FA in ACN	14	0.75–1	1	NR	[50]
440 NPS	LC-HRMS LC-MS/MS	Drugwipe-5S®	a. Centrifuged OF + 300 μL MeOH 1 hour - room temperature b. 100 μL supernatant + 50 μL IS dried c. Reconstituted 100 μL MeOH/FA 0.1% + NH_4HCO_2 5 mM (80:20, v:v) pH 3	Acquity™ UPLC HSS C18	NR	NR	NR	NR	NR	[51]
Cathinone, Methcathinone, Methylone, Pentedrone, Naphyrone, 4-FMC, PMMC, bk-MBDB, 4-MMC, 4-MEC, α-PVP, MPBP, MDPV	UHPLC-MS/MS	Sterile tube	a. 100 μL OF + 300 μL H_2O + 200 μL ACN + ISs centrifuged 5000 rpm/5 min b. 1 μL supernatant injected into LC–MS-MS system	Kinetex F5 (50 mm × 3 mm × 2.6 μm)	A: 0.1% FA B: ACN	8.5	1.00	2.5	4 (MPBP)	[54]
Methylone, ETH-CAT, N,N-DMC,	UHPLC-MS/MS	Plastic disposable containers	a. 90 μL OF + 20 μL MeOH + IS (100 ng/mL) b. Diluted 30 μL MeOH + 60 μL	Reverse-phase Kinetex C18	A: 5 mM FA + H_2O B: MeOH:ACN (80:20, v/v) + 5 mM FA	8.3	0.045–0.830	0.130–2.500	NR	[39]

(continued)

Table 2
(continued)

Synthetic Cathinones	Instrumental method	Collection Device	Sample preparation (including derivatization)	Column	Mobile phase	Run time (min)	LOD (ng/mL)	LOQ (ng/mL)	Authentic samples' concentrations (ng/mL)	References
bk-MBDB, PMMC, 4-MMC, 4-MEC, α-PVP, MABP, MDPV		(Salimetrics, Carlsbad, CA, USA)	ultrapure H_2O (total volume 200 μL) c. Vortexed + mixed 1 minute d. Sonicated 25 ° C for 6 minutes e. Centrifuged 15000 rpm/5 min Room temperature f. Supernatant Clean-up - MEPS syringe (C18 sorbent inside BIN)	(100 mm × 2.1 mm, 1.7 μm)						
Methylone, Pentedrone, Pyrovalerone, ETH-CAT, bk-MDEA, bk-MBDB, 4-MMC, MDPV	UPLC-MS/ MS	Salivette	a. 500 μL OF + deionized H_2O 1:4, v:v. + 1.4 mL MeOH centrifuged 3500 rpm/5 min b. Supernatant + 1.5 mL H_2O (pH 8.0) + 0.1 g NaCl + 200 μL $CHCl_3$ placed in ultrasound bath 5 min c. Centrifuged 3500 rpm/5 min d. Evaporated to dryness + reconstituted 50 μL mobile phase e. 5 μL injected into UPLC-MS/MS	Acquity UPLC™ BEH shield RP 18 (100 mm × 2.1 mm, 1.7 μm)	A: 0.1% FA in H_2O B: 0.1% FA in ACN	3 min	0.1–1.0	0.5–2.50	NR	[53]
Methylone, ETH-CAT, 3-MMC, 3-FMC, 4-MEC, 4-MeMABP, 4-CMC, 4-MMC-MeO, bk-MBDB, α-PVP, α-PHP, MDPPP, 4-CEC, NEP, MPHP, 4Cl-PVP, MDPV, TH-PVP, PV9, 3,4-MDPHP	IMS – UHPLC-MS/MS	NR	a. OF centrifuged 5000 rpm/2 min b. 1 mL sample loaded onto MIP column washed +1 mL 10% $CH_3CH(OH)CH_3$ + NaH2PO4•H2O (0.1 M, pH 9) + 1 mL deionized H_2O c. Cartridges dried + 0.5 mL CH_3COOH 1% (v/v) + MeOH for elution	1759 V total drift voltage, 7 cm drift tube length – BEH C18 (1.7 μm, 2.1 × 50 mm)	A: 0.1% FA in H_2O B: MeOH	NR	IMS 12–80 UHPLC-MS/MS 0.06–1.1	NR	NR	[55]

Pyrovalerone, Naphyrone, Cathinone, Methcatinone, Pentedrone, Benzedrone, Methylone, bk-MBDB, bk-MBDP, 4-MMC, PMMC, 4-CEC, DMBDP, bk-MDDMA, MDPV, bk-MDEA, α-PVP, 4-FMC	UHPLC-MS/MS	Quantisal	a. 500-μL OF Quantisal elution buffer + 25 μL IS + 500 μL saturated $Na_2B_4O_7$ aqueous solution + 2 mL MTBE b. Vortexed 2 min c. Centrifuged 4500 rpm/5 min + 1.7-mL organic phase + nitrogen (40 °C) d. Reconstituted +100 μL MeOH	Raptor™ biphenyl (100 mm × 2.1 mm, 2.7 μm)	A: H_2O + 0.1% FA (v/v) + 2 mmol/L NH_4HCO_2 B: MeOH + 0.1% FA (v/v) + 2 mmol/L NH_4HCO_2	NR	0.05–1.00	NR	NR	[30]
Methcatinone, Methylone, Diethylcathinone, Pentedrone, Naphyrone, 4-FMC, 4-MMC, MABP, bk-MDEA, bk-MBDB, N,N-DMC, ETH-CAT, 3,4-DMMC, 4-MEC, bk-MBDP, MDPV	UPLC-MS/MS	NR	a. 100 μL OF +1 μL IS +500 μL M3® buffer solution (pH 3.5) b. 1 μL injected into UPLC-MS/MS system	Waters oasis HLB (5 μm 4.6 × 20 mm × 5 μm)	A: NH_4HCO_2 pH 9.5 B: ACN	10	0.15	0.45	0.1–120 (bk-MBDB)	[56]

temperature, 4 °C, and − 20 °C for up to 1 month was investigated by Miller et al. using a new sensitive UHPLC-MS/MS method. Authentic OF from SCs consumer could not be tested, but the study demonstrated that basic buffers and room storage temperature negatively influence SCath stability in OF [50]. In the same year, to assess the prevalence of NPS positivity in French and Belgian drivers' OF samples, 492 OF screening tests collected during checks on road were analyzed using liquid chromatography high resolution mass spectrometry (LC-HRMS) and LC-MS/MS. 3-Methylmethcathinone (3-MMC) or mephedrone (4-MMC), 3,4-methylenedioxypyrovalerone (MDPV), and 4-methylethcathinone (4-MEC) were found in 10, 3, 4 drivers, respectively. Etylone (bk-MDEA) and methylone were detected in other two OF samples [51].

The following year, Ares et al. applied an UHPLC-MS/MS method with sample MEPS pre-treatment for the determination of SCath and classic drugs (opiates, cocaine) in OF from 12 patients on a methadone or buprenorphine/naloxone substitution therapy program. None of them resulted positive to SCaths [52]. Similarly, Fernandez et al. applied their UPLC-MS/MS method for the quantification of drugs and SCath to 15 OF samples from patients on a detoxification program and SCaths were not detected [53].

In the same year, Williams et al. presented a new UHPLC-MS/MS method for the determination of "bath salts" in OF. Since the Australian Standard Procedures were not comprehensive of NPS and SCath may cross-react in drug screen test kits giving false-positive results for amphetamine, the authors developed and validated a specific method for NPS including cathinones. The method was designed for workplace drug testing and applied to pre-screened OF samples. About 10 of 12 samples tested negative including 4 that previously resulted positive for amphetamines. In one OF sample, 4 ng/mL of MPBP (4-′-Methyl-α-pyrrolidinobutiophenone) and of 1-[3-(trifluoromethyl)phenyl]-piperazine [54] were found. In 2018, Rocchi et al. developed a method for the quantification of NPS, including 10 SCath, in OF. No authentic OF samples were tested but the method was developed to be used in a high-throughput laboratory for confirmatory analyses [39].

A screening method based on molecularly imprinted polymer (MIP) using methamphetamine template was created by Sorribes-Soriano et al. for the detection of SCaths in OF. The assay was developed for police checks on the road using a portable IMS (ion mobility spectrometry) instrument. A methamphetamine-based molecularly imprinted polymer was synthetized and placed in empty polypropylene cartridges used for SCaths extraction from OF. Then the extracts were introduced into the IMS instrument. Moreover, to prove the validity of IMS technique, UHPLC-MS/MS was used as confirmation. Synthetic saliva and OF from volunteers with interferences substances were used for the method development but no real samples were tested [55].

da Cunha et al. presented a screening method for the detection in oral fluid of 104 NPS including 19 SCaths. Seven OF samples from volunteers of an electronic music festival were analyzed to prove the applicability of the method and in one specimen methylone, N-ethylpentylone (NEP) along with methylphenidate and THC were detected [30].

Finally, with a comprehensive UHPLC-MS/MS method for the detection and quantification in OF and other matrices of 77 NPS, 24 classic drugs and 18 related metabolites, Di Trana et al. (2020) included 16 different SCaths. Real cases were tested, and in three authentic OF samples butylone was found [56].

4 Phenethylamines

The use of liquid chromatographic separations coupled to tandem mass spectrometry detection has been the predominant methodology applied in the determination of phenethylamines in OF (Table 3). One of the most recent methods, by da Cunha et al. in 2020, consisted of an LC-MS/MS-based targeted OF screening technique that covered a broad range of phenethylamines from NBOMes derivatives and from 2C family. The Quantisal OF device was used to collect OF samples [30]. LOD, recovery, matrix effects, interference studies, carryover, and stability were the parameters considered. Unfortunately, this methodology was developed for research purposes without any clinical or forensic application to any real sample.

No application was present neither in the LC-MS/MS assay proposed by Accioni et al. in the same 2020 for the quantification of phenethylamines of different families (e.g., 2C family, amphetamine derivatives). Sample treatment and extraction of analytes were simultaneously achieved by applying supramolecular solvents (SUPRAS) tool and the efficacy of this approach was confirmed by recovery and matrix effect results. The international guidelines were used to validate the entire analytical procedure. Calibration curves for all analytes were linear in the concentration range 10–500 ng/mL [57].

LC-MS/MS was also applied in 2019 to measure 4 fluoroamphetamine (4-FA) in OF, collected over 12 h after administration of 100 mg 4-FA to 12 subjects and 150 mg 4-FA to 5 subjects [58]. 4-FA peaked at 2 h post administration and it could be detected for 12 h with concentrations higher than in serum, most probably due to oral contamination. In 2018, Rocchi et al. developed and validated a UHPLC–MS/MS assay with no application to authentic samples in OF for 31 NPS belonging to different chemical classes, including 2-CB. OF sample preparation was based on protein precipitation followed by MEPS. The use of the MEPS

Table 3
Analytical methods for the screening and determination of phenethylamines in oral fluid

Phenetilamines	Instrumental method	Collection device	Sample preparation (including derivatization)	Column	Mobile phase	Run time (min)	LOD (ng/mL)	LOQ (ng/mL)	Authentic samples' concentrations (ng/mL)	References
25B-NBOH, 25B-NBOMe, 25C-NBOH, 25C-NBOMe, 25D-NBOMe, 25E-NBOH, 25E-NBOMe, 25G-NBOMe, 25H-NBOMe, 25I-NBOH, 25I-NBOMe, 25 N-NBOMe, 25 T2-NBOMe, 2C-B, 2C-C, 2C-D, 2C-E, 2C-G, 2C-I, 2C-T, 2C-T-2, 2C-T-4	LC-MS-MS	Quantisal	a. 500-μL OF Quantisal elution buffer + 25 μL IS + 500 μL saturated $Na_2B_4O_7$ aqueous solution + 2 mL MTBE b. Vortexed 2 min c. Centrifuged 4500 rpm/ 5 min + 1.7-mL organic phase + nitrogen (40 °C) d. Reconstituted +100-μL MeOH	Raptor™ biphenyl (100 × 2.1 mm, 2.7 μm)	A: H_2O + 0.1% FA (v/v) + 2 mmol/L NH_4HCO_2 B: MeOH + 0.1% FA (v/v) + 2 mmol/ L NH_4HCO_2	13.5	0.05–1	NR	NR	[30]
PMA, PMMA, TMA, 2C-T-7, 2C-T-2, DOET, DOM, 2C-I, DOB, 2C-B, DMA	LC–MS/MS	Glass test tube	a. 0.675 mL OF basified + NH_3 25% (OF 0.1 M) b. 1.5 mL Eppendorf + 0.150 mL $C_6H_{14}O$ + 0.675 mL THF d. Centrifuged 21125 g/5 min e. 2 mL extracted Injected into LC–MS/MS system	Security guard cartridge (4 × 2.1 mm i.d., 2.6 μm) PFP (75 × 2.1 mm i.d., 2.6 μm)	A: 0.1% FA in H_2O B: 0.1% FA in ACN	12.5	1.32–2.94	4.39–9.80	NR	[57]

4-FA	LC-MS/MS	Quantisal	a. 100 µl OF + 50 µl IF + 0.02 ng/µl 4-FA-d5 + amphetamine-d8 in ACN + 50 µl MeOH b. Mixed two minutes + centrifuged 13000 g c. 100 µl in autosampler vials for analysis	Kinetex (2.6 µm XB-C18 100 Å, LC column 100 × 2.1 mm)	A: 5 mM NH_4HCO_2 + 0.01% FA B: ACN + 0.01% FA	a. 95% A at a flow rate of 0.6 ml/ min b. Decreased during 4 min to 5%	0.25	0.95	NR	[58]
2C-B	UHPLC -MS/MS	Plastic disposable containers (Salimetrics, Carlsbad, CA, USA)	a. 90 µL OF + 20 µL MeOH + IS (100 ng/mL) b. Diluted 30 µL MeOH + 60 µL ultrapure H_2O (total volume 200 µL) c. Vortexed + mixed 1 minute d. Sonicated 25 ° C 6 minutes e. Centrifuged 15000 rpm/ 5 min room temperature f. Supernatant clean-up - MEPS syringe (C18 sorbent inside BIN)	Reverse-phase Kinetex C_{18} (1.7 µm, 2.1 × 100 mm)	A: 5 mM FA + H_2O B: MeOH:ACN (80: 20 v/v) + 5 mM FA	8.3	0.050	0.150	NR	[39]
MDA, PMA, TMA, MDEA, DMA, MBDB, MTA, 2C-B, DOB, DOET, 2C-T-7, 25C-NBOMe, 25B-NBOMe, 25T4-NBOMe	LC–MS-MS	Sterile tube	a. 100 µL of + 300 µL H_2O + 200 µL ACN + IS Centrifuged 5000 rpm/5 min b. 1 µL supernatant injected into LC–MS-MS system	Kinetex F5 (50 mm × 3 mm × 2.6 µm)	A: 0.1% FA B: ACN	8.5	1	2.5	4 (2C-B, DOB)	[54]

(continued)

Table 3
(continued)

Phenetilamines	Instrumental method	Collection device	Sample preparation (including derivatization)	Column	Mobile phase	Run time (min)	LOD (ng/mL)	LOQ (ng/mL)	Authentic samples' concentrations (ng/mL)	References
2C-B, 2C-I, 25C-NBOMe, 25I-NBOMe, Ethylphenidate	UHPLC–MS/MS	Intercept	a. 2 ml MeOH + OF mixed 20 min b. 0.50 mL sample + 50 mL IS + LLE c. 10 mL triton-X100 (0.1%) + 250 mL $(NH_4)_2CO_3$ buffer pH 9.3 + 1.3 mL EtOAc:C_7H_{16} (4:1, v/v) mixed 10 min d. Centrifuged 3900 g/5 min 4 °C e. Supernatant + 30 mL of 0.1% HNO_3 + MeOH f. Evaporated to dryness at 40 °C + reconstituted + 60 mL ACN/H_2O (10:90, v/v) g. Autosampler vials and analysed	UPLC HSS T3 (100 × 2.1 mm, 1.8 μm)	A: 10 mM NH_4HCO_2 pH 3.1 B: MeOH	9	NR	0.006-0.22	NR	[59]

MDA	LC-MS/MS	Collection tube	a. 150 μL OF + 20 μL IS (0.1 ng/mL) + MeOH vortexed 30 s b. Sonicated 6 min + centrifuged 10000 rpm/ 5 min c. Supernatant passed through a 0.45 μm PTFE filter	Reversed-phase C18 Alltima (250 mm, 4.6 mm,5 μm)	A: ACN B: H_2O + 5 mM FA	38	0.7	1.5	NR	[60]

clean-up was very important because it provided an effective removal of interfering compounds [39].

In 2017, Williams et al. provided another validated method for the rapid detection of 14 phenethylamines, among other NPS in OF. The phenethylamines included MDA, PMA, TMA, MDEA, MBDB, MTA, 2C-B, MDPV, DOB, DOET, 2C-T-7, 25C-NBOMe, 25B-NBOMe, and 25 T4-NBOMe. Sample preparation was undertaken using a simple protein precipitation in acetonitrile. Chromatographic separation was achieved in 7.5 min on a Kinetex F5 column using 0.1% formic acid in water and acetonitrile as the mobile phases. The method was validated and limit of detection, limit of quantitation, selectivity, linearity, accuracy, and precision were the parameters considered. This method was applied to 12 samples previously submitted for routine testing and two were found to contain 2-CB and DOB (5 and 4 ng/mL), and MPBP and TFMPP (both at 4 ng/mL) [54].

A UHPLC–MS/MS method was developed by Gjerde et al. to analyze phenethylamines in OF. Sample collection was conducted on Friday and Saturday nights between March and May 2014 outside nine popular nightclubs with different profiles and patron-characteristics in downtown Oslo. Participation was voluntary and anonymous. OF was collected using the Intercept ® OF sampling device (OraSure Technologies Inc., Bethlehem, PA, USA). Sample preparation was performed using liquid–liquid extraction with ethyl acetate/heptane (4:1) after adding internal standards, ammonium carbonate buffer pH 9.3 and Triton X100 [59].

The last analytical procedure for the determination of methylenedioxyamphetamine (MDA) in human OF used protein precipitation in methanol. Sergi et al. used liquid chromatography–tandem mass spectrometry to perform the quantitative analysis. All the analytes were detected in positive ionization mode using a TurboIonSpray source. Sensitivity (limits of detection and LOQs), linearity and carryover, precision, accuracy, specificity, recovery, stability, and matrix effect were the parameters used to validate the method [60]. This methodology was developed only for research purposes without any clinical or forensic application.

5 Designer Piperazines

Up to now, only three different analytical methods have been set up and validated to determine the presence of designer piperazines in OF (Table 4). All reported methodologies were developed for research purposes without any clinical or forensic application, and LC-MS/MS was the most used technique for the purpose.

In the most recent method, described by da Cunha et al., LC-MS/MS was used to develop a sensitive screening method for

Table 4
Analytical methods for the screening and determination of designer piperazines in oral fluid

Designer Piperazines	Instrumental method	Collection device	Sample preparation (including derivatization)	Column	Mobile phase	Run time (min)	LOD (ng/mL)	LOQ (ng/mL)	Authentic samples' concentrations (ng/mL)	References
BZP, TFMPP, mCPP	LC-MS-MS	Quantisal	a. 500 μL OF Quantisal elution buffer + 25 μL IS + 500 μL saturated $Na_2B_4O_7$ aqueous solution + 2 mL MTBE b. Vortexed 2 min c. Centrifuged 4500 rpm/ 5 min + 1.7-mL organic phase + nitrogen (40 °C) d. Reconstituted +100-μL MeOH	Raptor™ biphenyl (100 × 2.1 mm, 2.7 μm)	A: H_2O + 0.1% FA (v/v) + 2 mmol/ L NH_4HCO_2 B: MeOH + 0.1% FA (v/v) + 2 mmol/ L NH_4HCO_2	13.5	0.1–1	NR	NR	[30]
Piperonyl piperazine, 4-MeOPP	UHPLC -MS/MS	Plastic disposable containers (Salimetrics, Carlsbad, CA, USA)	a. 90 μL OF + 20 μL MeOH + IS (100 ng/mL) b. Diluted 30 μL MeOH + 60 μL ultrapure H_2O (total volume 200 μL) c. Vortexed + mixed 1 minute d. Sonicated 25 ° C for 6 minutes e. Centrifuged 15000 rpm/5 min room temperature f. Supernatant clean-up - MEPS syringe (C18 sorbent inside BIN)	Reverse-phase Kinetex C_{18} column (1.7 μm, 2.1 × 100 mm)	A: H_2O + 5 mM FA B: MeOH/ACN (80:20 v/ v) + 5 mM FA	8.3	0.830–0.850	2.5–2.6	NR	[39]

(continued)

Table 4 (continued)

Designer Piperazines	Instrumental method	Collection device	Sample preparation (including derivatization)	Column	Mobile phase	Run time (min)	LOD (ng/mL)	LOQ (ng/mL)	Authentic samples' concentrations (ng/mL)	References
mCPP, TFMPP	LC MS-MS	Salivette Quantisal	a. 25 μL IStd + 2 mL BBS pH 9 + 0.5 mL OF b. Loaded + 2 mL MeOH + 2 mL H_2O d. Washing steps + 2 mL H_2O:MeOH (95:5, v/v) + 2 mL H_2O:MeOH: NH_4OH (70:29.5:0.5, v/v/v) e. Eluted with 3 mL CH_2Cl_2:$CH_3CH(OH)CH_3$ (75:25) f. Added 100 μL HCl 0.1% g. Evaporated nitrogen 35 °C h. Reconstituted 75 μL 0.1% FA:ACN (90:10, v/v) i. Injected into LC–MS/MS system	Atlantis® T3 (2.1 mm × 50 mm, 3 μm) reversed-phase	A: 0.1% FA B: ACN	10	0.05	0.2	89.8 (mCPP)	[48]

piperazines in OF. The Quantisal OF device was used to collect OF samples [30]. Previously, Rocchi et al. developed an analytical method for screening and quantification piperazines in OF by using UHPLC–MS/MS. The procedure involved a rapid sample preparation based on protein precipitation followed by clean-up utilizing MEPS. The MEPS clean-up, regardless of non-quantitative recoveries, provided an effective removal of interfering compounds. SWGTOX guidelines were used for the validation protocol [61]; linearity range ranged over 3 or 4 orders of magnitude; precision and accuracy tests gave RSD% values below 18%; and accuracy ranged from 88.4% to 101.2%. LODs ranged between 0.83 ng/mL and 0.85 ng/mL and LOQs from 2.5 to 2.6 ng/mL [39].

The first analytical method published for the determination of piperazines in OF was based on LC–MS/MS technique. The method was developed and validated by de Castro et al. in 2014 for the determination of mCPP and 3-trifluoromethylphenylpiperazine (TFMPP) in OF. Sample extraction was performed using Strata X cartridges and the chromatographic separation was achieved using an Atlantis® T3 column. Formic acid 0.1% and acetonitrile was used as mobile phase. The method was validated in neat OF and in OF mixed with Quantisal buffer. Selectivity, linearity, LOD, LOQ, imprecision, accuracy, matrix effect, Quantisal and extraction recovery, dilution integrity, and stability under different conditions were studied for method validation [48].

6 Tryptamines

Currently, only two analytical methods have been developed and validated for the determination of psychotropic tryptamines in OF (Table 5). Both assays included UHPLC-MS/MS. The first method was developed by Gjerde et al. to investigate drug use profile among Norwegian nightclub patrons, as described in the previous section (5. Designer piperazines). OF was collected using the Intercept OF collection device and screened using UHPLC–MS/MS [59]. Dimethyltryptamine (DMT), 5-methoxy-α-methyltryptamine (5-MeO-DMT) and α-methyltryptamine (AMT) were under investigation, but only one participant resulted positive for DMT with a cut-off concentration of 0.014 ng/mL.

Acetyl-O-dimethyltryptamine (AcO-DMT), 5-methoxy-α-methyltryptamine (5-MeO-AMT), 4-hydroxy-diethyltryptamine (4-OH-DET), 5-methoxy-N-methyl-N-isopropyltryptamine (5-MeO-MIPT), 4-acetoxy-diisopropyltryptamine (4-AcO-DIPT), 5-methoxy-dipropyltryptamine (5-MeO-DPT), and 5-methoxy-diallyltryptamine (5-MeO-DALT) were comprised in a second method for NPS detection in OF samples of unknown

Table 5
Analytical methods for the screening and determination of tryptamines in oral fluid

Tryptamines	Instrumental method	Collection device	Sample preparation (including derivatization)	Column	Mobile phase	Run time (min)	LOD (ng/mL)	LOQ (ng/mL)	Authentic samples' concentrations (ng/mL)	References
DMT, 5-MeO-DMT, AMT	UHPLC-MS/MS	Intercept	a. 2 ml MeOH + OF mixing 20 min b. 0.50 mL sample + 50 mL IS + LLE c. 10 mL triton-X100 (0.1%) + 250 mL $(NH_4)_2CO_3$ buffer pH 9.3 + 1.3 mL $EtOAc/C_7H_{16}$ (4:1, v/v) mixed 10 min d. Centrifuged 3900 g/5 min 4 °C e. Supernatant + 30 mL of 0.1% HNO_3 + MeOH f. Evaporated to dryness at 40 °C + reconstituted + 60 mL ACN/H_2O (10:90, v/v) g. Autosampler vials and analysed	UPLC HSS T3 (100 × 2.1 mm, 1.8 μm)	A: 10 mM NH_4HCO_2 pH 3.1 B: MeOH	9	NR	NR	NR	[59]

AcO-DMT, 5-MeO-AMT, 4-OH-DET, 5-MeO-MIPT, 4-AcO-DIPT, 5-MeO-DPT, 5-MeO-DALT	UHPLC-MS/MS	NR	a. 100 μL OF+ 5 μL of ISs solution + 500 μL of M3® buffer solution (pH 3.5) b. Injected 1.0 μL	Waters Oasis HLB (5 μm 4.6 × 20 mm)	A: NH_4HCO_2 pH 9.5 B: ACN	10	0.03–0.25	0.06–0.5	0.6–120 (AcO-DMT)	[56]

consumers, presented in 2020 by Di Trana et al. using UHPLC separation by reversed phase column and LC-MS/MS. AcO-DMT was detected in two OF samples at an average concentration of 60 ng/mL [56].

7 Non-fentanyl-Derived Synthetic Opioids

Analytical techniques for detecting non-fentanyl-derived novel synthetic opioids in alternative matrices such as OF are minimal with only two published methods (Table 6). A first study of Griswold et al. in 2017 investigated adult Emergency Department patients who presented after a reported heroin overdose requiring naloxone administration [62]. Thirty participants provided paired OF and urine specimens. At the time of testing, only the non-fentanyl-related opioids U-47700, U-50488, AH-7921, and MT-45 were available. OF was collected using the Quantisal OF collection device and analyzed by two separate LC-QTOF-MS analytical methods, with distinctive LC separation and mass acquisition for screening and confirmation of transitional and emerging opioids [63]. U-47700 was the only new synthetic opioid detected in paired OF and urine specimens from two participants.

Quantisal device was also used to collect OF from 18 anonymous detainees in a Texas adult detention center by Truver et al. in 2018 to test a developed and validated analytical method for the detection and quantification of U-47700, U-49900, U-50488, AH-7921, MT-45, W-18, and W-15 [64]. Linearity was observed between 10 and 500 ng/mL ($r^2 \geq 0.9959$). Bias and imprecision were $<\pm 11.1\%$ and matrix effects ranged from -21.1 to 13.7%. No carryover was detected following injection of the highest calibrator. All analytes were stable (within $\pm 15\%$ change from baseline) under all tested conditions (24 h at room temperature, 72 h at 4 °C, and in the autosampler for 60 h at 4 °C). Unfortunately, no OF samples gave any positivity for new synthetic opioids under investigation.

8 Fentanyl and Analogues

It is apparent that fentanyl analogues' metabolites disposition in OF is limited due to its acidic pH at physiological condition. Therefore, the analytical methods mainly focused on the parent drugs detection, although some N-dealkylated metabolites were included in certain assays [65]. To date, seven analytical methods for fentanyl and its analogues in OF have been published in the literature, mostly based on chromatography coupled to mass spectrometry (Table 7).

Already in 2005, Gunnar et al. [66] developed a GC-MS assay to quantify 30 drugs of abuse in OF, including fentanyl. The

Table 6

Analytical methods for the screening and determination of non-fentanyl-derived synthetic opioids in oral fluid

Non-fentanyl-derived synthetic opioids	Instrumental method	Collection device	Sample preparation (including derivatization)	Column	Mobile phase	Run time (min)	LOD (ng/mL)	LOQ (ng/mL)	Authentic samples' concentrations (ng/mL)	References
U-47700, U-50488, AH-7921, MT-45	LC-QTOF-MS	Quantisal	a. 1 mL OF +3 mL buffer b. Single step (pH 10.4) LLE	NR	NR	NR	1	NR	NR	[62]
U-47700, AH-7921, U-49900, U-50488, MT-45, W-18, W-15	LC-MS/MS	Quantisal	a. 1 mL OF + buffer + 25 μL ISTD + 25 μL calibrator or Qc solution + 2 mL 0.1 M NaH2PO4•H2O buffer (pH 6) b. Vortexed, load onto SPE columns, flow under gravity c. Rinsed with 1 mL deionized H_2O + 1 mL CH_3COOH 1 M d. Dried under 20 psi nitrogen 5 minutes e. Washed + 1 mL C_6H_{14} f. Eluted + 1 mL EtOAc g. Washed + 1 mL CH_3OH h. Eluted using	Agilent Poroshell 120 EC-C18 column (100 mm × 3.0 mm, 2.7 μm) + Poroshell 120 EC-C18 guard column (2.1 × 5 mm, 2.7 μm)	A: 0.05% FA + 5 mM NH_4HCO_2 + H_2O B: 0.1% FA + ACN	13	5	10–500	NR	[64]

(continued)

Table 6
(continued)

Non-fentanyl-derived synthetic opioids	Instrumental method	Collection device	Sample preparation (including derivatization)	Column	Mobile phase	Run time (min)	LOD (ng/mL)	LOQ (ng/mL)	Authentic samples' concentrations (ng/mL)	References
			1 mL of CH_2Cl_2: $CH_3CH(OH)CH_3$: NH_4OH (80:20:5) i. Evaporated under nitrogen at 50 °C l. Reconstituted + 1 mL of mobile phase A: Mobile phase B (95:5) m. Injected 5 μL							

Table 7
Analytical methods for the screening and determination of fentanyl and analogues in oral fluid

Fentanyl and analogs	Instrumental method	Collection device	Sample preparation (including derivatization)	Column	Mobile phase	Run time (min)	LOD (ng/mL)	LOQ (ng/mL)	Authentic samples' concentrations (ng/mL)	Reference
Fentanyl	GC-EI-MS	NR	a. 200 μL OF + 2 ml NaH2PO4•H2O b. SPE (elution with ACN and 4% NH_4OH) c. Dried at 65 °C under air stream d. Derivatization with MSTFA in CAN 85 °C 30 min e. Inject 1 μL	J&W Scientific DB-35 ms (30 m × 0.32 mm × 0.25 μm)	Helium	9.30	NR	2	NR	[66]
Fentanyl, Carfentanl	UHPLC-MS/MS	Intercept	a. 2 ml MeOH + OF mixing 20 min b. 0.50 mL sample + 50 mL IS + LLE c. 10 mL triton-X100 (0.1%) + 250 mL $(NH_4)_2CO_3$ buffer pH 9.3 + 1.3 mL EtOAc/ C_7H_{16} (4: 1, v/v) mixed 10 min d. Centrifugation 3900 g/5 min 4 °C e. Supernatant + 30 mL of 0.1% HNO_3 + MeOH f. Evaporated to dryness at 40 °C + reconstituted + 60 mL ACN/H_2O (10:90, v/v) g. Autosampler vials and analysed	UPLC HSS T3 (100 × 2.1 mm, 1.8 μm)	A: 10 mM NH_4HCO_2 pH 3.1 B: MeOH	9	NR	NR	NR	[59]

(continued)

Table 7
(continued)

Fentanyl and analogs	Instrumental method	Collection device	Sample preparation (including derivatization)	Column	Mobile phase	Run time (min)	LOD (ng/mL)	LOQ (ng/mL)	Authentic samples' concentrations (ng/mL)	Reference
Fentanyl, Carfentanil	LC-QTOF-MS	Quantisal	a. 1 mL OF + 3 mL buffer b. Single step (pH 10.4) liquid–liquid extraction	NR	NR	NR	1	NR	NR	[62]
Norfentanyl, Norcarfentanyl, Remifentanil Acetyl Fentanyl, Alfentanyl, 4-ANPP Fentanyl, Sufentanil, Valeryl fentanyl	LC-Q-TOF-MS	NR	a. 500 µL OF + 500 µL Quantisal buffer + 2 ml NaH2PO4•H2O b. SPE extraction with EtOAc + 4% $NHOH_4$ c. Dried at 50 °C under nitrogen d. Reconstitute in 100 µl mobile phase	Matching precolumn (2.1 × 5 × 2.7 µ) + Poroshell 120 EC-C18 (2.1 x 100mm x 2.7 µm)	A: 0.1% HCOOH in H_2O B: 0.1% HCOOH in ACN	7	0.25–2.5	NR	NR	[67]
Fentanyl, Furanyl fentanyl, Remifentanil, Sufentanil	TS-MS	NR	a. Direct immersion of VAMS in OF b. pH equilibration with NaH2PO4•H2O buffer	NR	NR	NR	0.08–0.45	0.25–1.49	NR	[68]
Fentanyl, Norfentanyl	UPLC-MS/MS	NR	a. 100 µl OF+ 1 ml H_2O+ phosphate buffer until pH 6.0 b. SPE with CH_2Cl_2: $CH_3CH(OH)CH_3$: NH_4OH (78:20:2, v/v/v) as eluent c. Dry under nitrogen reconstitute 100 µl aqueous 0.1% FA and 10 mM NH_4HCO_2	Allure bisphenyl column (2.5 × 100 mm × 5 µm)	A: 0.1% CH_2O_2 B: MeOH	4	NR	NR	28–79	[69]

Methoxyacetyl Norfentanyl, Acetyl Norfentanyl, Carboxy butyryl fentanyl, Carboxy valeryl fentanyl, Methoxyacetyl fentanyl, Furanyl norfentanyl, Cis-3-methyl norfentanyl, Trans-3-methyl norfentanyl, Butyryl norfentanyl, Cyclopropyl norfentanyl, β-hydroxyfentanyl, Alfentanyl, Furanyl fentanyl, Butyryl fentanyl, Despropionil-p-F-fentanyl, 4-ANPP, Phenyl acetyl fentanyl	UPLC-MS/MS	NR	a. 100-μL OF + 1 μL IS + 500 μL M3® buffer solution (pH 3.5) b. 1 μL injected into UPLC-MS/MS system	Waters oasis HLB (4.6 mm × 20 mm × 5 μm)	A: HCO_2NH_4 pH 9.5 B: ACN	10	0.03	0.07	12 (Carfentanil)	[56]

quantitative method was fully validated on a calibration range between 2 and 200 ng/mL, achieving good linearity ($r^2 \geq 0.995$), intra-assay (average 8.33%) and inter-assay precision (average 6.91%), and recovery (75%). The developed method was applied to subjects suspected of intoxication in road traffic by police, but out of the 140 authentic analyzed OF samples, none was found positive to fentanyl.

A UHPLC-MS/MS assay was developed in 2014 by Gjerde et al. to quantify 122 psychotropic drugs in 500 OF samples from nightclub patrons and criminal offenders' OF samples, as previously described. Among all the drugs, the method allowed the detection of fentanyl and carfentanil, even if the authors did not report any positive sample for these two compounds among the 500 samples screened in the study [59]. A UHPLC-QTOF-MS analytical method for 9 fentanyl analogues, 1 metabolite, and 4 novel opiates in OF was developed by Griswold et al. to analyze 30 heroin users' real samples in 2017 [62]. The method was applied to compare OF and urine fentanyl analogues disposition to evaluate OF as alternative to urine in opioid overdose cases. Only fentanyl and acetylfentanyl were detected in the OF samples and compared to paired urine samples, reporting an overall agreement between the two matrices of 93.3% for fentanyl and 80.0% for acetylfentanyl.

In 2019, Palmquist and Swortwood developed and validated a LC-QTOF-MS-based method to screen 14 fentanyl analogues in 30 post-mortem blood and 20 oral fluid samples. Six fentanyl analogues were detected in the blood samples with furanyl fentanyl and 4-ANPP being the most prevalent, whereas no fentanyl analogues were detected in OF samples [67]. In the same year, a simple and rapid quantitative touch spray-MS assay to detect 30 drugs of abuse including fentanyl, furanyl fentanyl, remifentanil, and sufentanil in OF was developed by Morato et al. [68]. The analytical technique allowed the direct analysis of oral fluid samples absorbed on a volumetric absorptive microsampling (VAMS) swab, previously immersed in OF sample. The method showed good linearity ($r^2 > 0.97$ for all analytes) on a five-point calibration range between 25 and 1000 ng/mL, with precision, accuracy, matrix effect and recovery better than the established acceptance criteria at all the assessed concentration (100 ng/mL and 600 ng/mL).

In 2020, OF was used as alternative matrix to monitor fentanyl neonatal exposure by Gesseck et al. through an UHPLC-MS/MS assay. Out of the three authentic OF samples collected from infants exposed to fentanyl, parent drug and norfentanyl were detected only in one sample at a concentration of 28 ng/ml and 79 ng/ml, respectively [69].

In the same year, several fentanyl analogues, including methoxyacetylnorfentanyl, acetyl-norfentanyl, carboxy butyryl fentanyl, carboxy valeryl fentanyl, methoxyacetyl fentanyl, furanyl norfentanyl, cis-3-methyl norfentanyl, trans-3-methyl norfentanyl, butyryl norfentanyl, cycloproyl norfentanyl, β-hydroxyfentanyl, alfentanyl, furanyl fentanyl, butyryl fentanyldespropionil-p-F-fentanyl, 4-ANPP, and phenyl acetyl fentanyl, were included in a panel of several NPS to investigate authentic OF samples by Di Trana using UHPLC-MS/MS [56]. Only 1 out of 14 authentic OF samples resulted positive for carfentanil (12 ng/mL).

9 Designer Benzodiazepines

Currently, three different analytical methods have been set up and validated to determine the presence of designer benzodiazepines in OF (Table 8). The first method was the same already described for tryptamines and published by Gjerde et al. in 2016 concerning the investigation of 500 Norwegian partygoers' OF samples [59]. A number of designer benzodiazepines were included in the method, including bentazepam, diclazepam, etizolam, flubromazepam, phenazepam, and pyrazolam. Two participants resulted positive for phenazepam and one for etizolam.

Using a similar methodology, metizolam was detected in OF, hair, sweat, and exhaled breath of a 54-year-old healthy man orally administered with 2 mg tablet of metizolam. OF was collected over 8 h using the NeoSal® (Neogen, Lexington, KY, USA) device and analyzed using UHPLC-MS/MS. Metizolam was detectable in oral fluid during 8 h, with concentrations lower than 1 ng/mL [70].

The third method was used to assess the detectability of two designer benzodiazepines in OF in two different self-administration studies. Two of the authors ingested 0.25 mg and 6 mg of flunitrazolam and deschloroetizolam, respectively. OF was collected over 8 h using the NeoSal device and analyzed using UHPLC-MS/MS. T_{max} was obtained at 3 h for flunitrazolam and at 30 min for deschloroetizolam [71, 72].

Table 8
Analytical methods for the screening and determination of designer benzodiazepines in oral fluid

Designer benzodiazepines	Instrumental method	Collection device	Sample preparation (including derivatization)	Column	Mobile phase	Run time (min)	LOD (ng/mL)	LOQ (ng/mL)	Authentic samples' concentrations (ng/mL)	References
Bentazepam, Diclazepam, Etizolam, Flubromazepam, Phenazepam, Pyrazolam	UHPLC-MS/MS	Intercept	a. 2 ml MeOH + OF mixing 20 min b. 0.50 mL sample + 50 mL IS + LLE c. 10 mL triton-X100 (0.1%) + 250 mL $(NH_4)_2CO_3$ buffer pH 9.3 + 1.3 mL EtOAc/ C_7H_{16} (4:1, v/v) mixed 10 min d. Centrifugation 3900 g/5 min 4 °C e. Supernatant + 30 mL of 0.1% HNO_3 + MeOH f. Evaporated to dryness at 40 °C + reconstituted + 60 mL ACN/H_2O (10:90, v/v) g. Autosampler vials and analyzed	UPLC HSS T3 (100 × 2.1 mm, 1.8 μm)	A: 10 mM NH_4HCO_2 pH 3.1 B: MeOH	9	NR	NR	NR	[59]
Metizolam	UHPLC-MS/MS	NeoSal	a. 1 mL mix (OF + buffer) + 1 ng diazepam-d5 + 1 mL saturated NH_4Cl buffer pH 9.5 b. Extract with 5 mL of CH_2Cl_2/ C_7H_{16}/ $CH_3CH(OH)CH_3$ (25: 65:10, v/v) c. Agitate d. Centrifuge e. Evaporate to dryness f. Reconstitute with 50 μL ACN/FA 0.1% (50/50, v/v) g. Inject 10 μL	Acquity BEH C18 column (100 × 2.1 mm, 1.7 μm)	A: ACN B: 0.1% FA	5.5	NR	0.002	0.77	[70]
Flunitrazolam	UPLC-MS/MS	NeoSal	a. 500 μL mix (OF + buffer) + 1 ng of diazepam-d5 + 500 μL saturated BBS pH 9.5 b. Extract with 2.5 mL $(C_2H_5)_2O$/ CH_2Cl_2/ C_6H_{14}/ $(CH_3)_2CHCH_2CH_2OH$ (50:30:20:0.5, v/v) c. Agitate 15 min d. Centrifuge 3000 rpm/15 min e. Evaporate to dryness 45 °C f. Reconstitute with 50 μL 5 mM NH_4HCO_2 buffer pH 3	Waters Acquity HSS C18 column (150 × 2.1 mm × 1.8 μm)	A: 5 mM NH_4HCO_2 pH 3 B: 0.1% FA in ACN	15	0.005	0.010	0.007–0.178	[71]

Deschloroetizolam	UHPLC-MS/MS	NeoSal	a. 500 µL mix (OF + buffer) + 1 ng of diazepam-d5 + 500 µL saturated borate buffer pH 9.5 b. Extract with 2.5 mL $(C_2H_5)_2O$/ CH_2Cl_2/ C_6H_{14}/ $(CH_3)_2CHCH_2CH_2OH$ (50:30:20: 0.5, v/v) c. Agitate 15 min d. Centrifuge 3000 rpm/15 min e. Evaporate to dryness 45 °C f. Reconstitute with 50 µL 5 mM NH_4HCO_2 buffer pH 3	Waters Acquity HSS C18 column (150 × 2.1 mm × 1.8 µm)	A: NH_4HCO_2 pH 3 B: 0.1% FA in ACN	15	0.001	0.001	0.003–6.491	[72]

References

1. Bosker WM, Huestis MA (2009) Oral fluid testing for drugs of abuse. Clin Chem 55: 1910–1931
2. Allen KR (2011) Screening for drugs of abuse: which matrix, oral fluid or urine? Ann Clin Biochem 48:531–541
3. De La Torre R, Farré M, Navarro M, Pacifici R, Zuccaro P, Pichini S (2004) Clinical pharmacokinetics of Amfetamine and related substances: monitoring in conventional and non-conventional matrices (2004). Clin Pharmacokinet 43:157–185
4. Pichini S, Altieri I, Zuccaro P, Pacifici R (1996) Drug monitoring in nonconventional biological fluids and matrices. Clin Pharmacokinet 30:211–228
5. Pujadas M, Pichini S, Civit E, Santamariña E, Perez K, de la Torre R (2007) A simple and reliable procedure for the determination of psychoactive drugs in oral fluid by gas chromatography-mass spectrometry. J Pharm Biomed Anal 44:594–601
6. Busardò FP, Pichini S, Pellegrini M, Montana A, Lo Faro AF, Zaami S, Graziano S (2018) Curr Neuropharmacol 16:84–96
7. Blandino V, Wetzel J, Kim J, Haxhi P, Curtis R, Concheiro M (2017) Oral fluid vs. urine analysis to monitor synthetic cannabinoids and classic drugs recent exposure Vincent. Curr Pharm Des 18:796–805
8. Navarro M, Pichini S, Farré M, Ortuño J, Roset PN, Segura J, de la Torre R (2001) Usefulness of saliva for measurement of 3,4-methylenedioxymethamphetamine and its metabolites: correlation with plasma drug concentrations and effect of salivary pH. Clin Chem 47:1788–1795
9. Desrosiers NA, Huestis MA (2019) Oral fluid drug testing: analytical approaches, issues and interpretation of results. J Anal Toxicol 43: 415–443
10. Graziano S, Anzillotti L, Mannocchi G, Pichini S, Busardò FP (2019) Screening methods for rapid determination of New Psychoactive Substances (NPS) in conventional and non-conventional biological matrices. J Pharm Biomed Anal 163:170–179
11. Drummer OH (2006) Drug testing in oral fluid. Clin Biochem Rev 27:147–159
12. Vindenes V, Yttredal B, Øiestad EL, Waal H, Bernard JP, Mørland JG, Christophersen AS (2011) Oral fluid is a viable alternative for monitoring drug abuse: detection of drugs in oral fluid by liquid chromatography-tandem mass spectrometry and comparison to the results from urine samples from patients treated with Methadone or Buprenorphine. J Anal Toxicol 35:32–39
13. Jufer R, Walsh SL, Cone EJ, Sampson-Cone A (2006) Effect of repeated cocaine administration on detection times in oral fluid and urine. J Anal Toxicol 30:458–462
14. Crouch DJ (2005) Oral fluid collection: the neglected variable in oral fluid testing. Forensic Sci Int 150:165–173
15. Choo RE, Huestis MA (2004) Oral fluid as a diagnostic tool. Clin Chem Lab Med 42: 1273–1287
16. Concheiro M, Jones H, Johnson RE, Choo R, Huestis MA (2011) Preliminary buprenorphine sublingual tablet pharmacokinetic data in plasma, oral fluid and sweat during treatment of opioid- dependent pregnant women. Ther Drug Monit 23:1–7
17. Smink BE, Hofman BJA, Dijkhuizen A, Lusthof KJ, De Gier JJ, Egberts ACG, Uges DRA (2008) The concentration of oxazepam and oxazepam glucuronide in oral fluid, blood and serum after controlled administration of 15 and 30 mg oxazepam. Br J Clin Pharmacol 66:556–560
18. Jenkins AJ, Oyler JM, Cone EJ (1995) Comparison of heroin and cocaine concentration in saliva with concentrations in blood and plasma. J Anal Toxicol 19:359–374
19. Newmeyer MN, Swortwood MJ, Barnes AJ, Abulseoud OA, Scheidweiler KB, Huestis MA (2016) Free and glucuronide whole blood cannabinoids' pharmacokinetics after controlled smoked, vaporized, and oral Cannabis administration in frequent and occasional Cannabis users: Identification of recent Cannabis intake. Clin Chem 62:1579–1592
20. Cone EJ, Oyler J, Darwin WD (1997) Cocaine disposition in saliva following intravenous, intranasal, and smoked administration. J Anal Toxicol 21:465–475
21. Schepers RJF, Oyler JM, Joseph RE, Cone EJ, Moolchan ET, Huestis MA (2003) Methamphetamine and amphetamine pharmacokinetics in oral fluid and plasma after controlled oral methamphetamine administration to human volunteers. Clin Chem 49:121–132
22. European Drug Report 2021: Trends and Developments at https://www.emcdda.europa.eu/publications/edr/trends-developments/2021_en. Last accessed 14 June 2021
23. Langel K, Engblom C, Pehrsson A, Gunnar T, Ariniemi K, Lillsunde P (2008) Drug testing in

oral fluid – evaluation of sample collection devices. J Anal Toxicol 32:393–401
24. Lee D (2018) Oral fluid drug testing in pain management practice: factors to consider before choosing the alternative biological matrix. J Appl Lab Med 2:598–609
25. Marchei E, Malaca S, Graziano S, Gottardi M, Pichini S, Busardò FP (2020) Stability and degradation pathways of different psychoactive drugs in neat and in buffered oral fluid. J Anal Toxicol 44:570–579
26. Kerrigan S (2013) Sampling, storage and stability. Sample chapter from Clarke's Analytical Forensic Toxicology, second edition 335–356
27. Cohier C, Mégarbane B, Roussel O (2017) Illicit drugs in oral fluid: evaluation of two collection devices. J Anal Toxicol 41:71–76
28. La Maida N, Pellegrini M, Papaseit E, Pérez-Mañá C, Poyatos L, Ventura M, Galindo L, Busardò FP, Pichini S, Farré M, Marchei E (2020) Determination of the synthetic cannabinoids JWH-122, JWH-210, UR-144 in oral fluid of consumers by GC-MS and quantification of parent compounds and metabolites by UHPLC-MS/MS. Int J Mol Sci 21:9414
29. Mulet CT, Tarifa A, DeCaprio AP (2020) Comprehensive analysis of synthetic cannabinoids and metabolites in oral fluid by online solid-phase extraction coupled to liquid chromatography-triple quadrupole-mass spectrometry. Anal Bioanal Chem 412:7937–7953
30. da Cunha KF, Oliveira KD, Huestis MA, Costa JL (2020) Screening of 104 new psychoactive substances (NPS) and other drugs of abuse in oral fluid by LC-MS-MS. J Anal Toxicol 44: 697–707
31. Sorribes-Soriano A, Verdeguer J, Pastor A, Armenta S, Esteve-Turrillas FA (2021) Determination of third-generation synthetic cannabinoids in Oral fluids. J Anal Toxicol 45: 331–336
32. Cannaert A, Ramírez Fernández MDM, Theunissen EL, Ramaekers JG, Wille SMR, Stove CP (2020) Semiquantitative activity-based detection of JWH-018, a synthetic cannabinoid receptor agonist, in oral fluid after vaping. Anal Chem 92:6065–6071
33. May B, Naqi HA, Tipping M, Scott J, Husbands SM, Blagbrough IS, Pudney CR (2019) Synthetic cannabinoid receptor agonists detection using fluorescence spectral fingerprinting. Anal Chem 91:12971–12979
34. Anzillotti L, Marezza F, Calò L, Andreoli R, Agazzi S, Bianchi F, Careri M, Cecchi R (2019) Determination of synthetic and natural cannabinoids in oral fluid by solid-phase microextraction coupled to gas chromatography/mass spectrometry: a pilot study. Talanta 201: 335–341
35. Malaca S, Busardò FP, Gottardi M, Pichini S, Marchei E (2019) Dilute and shoot ultra-high performance liquid chromatography tandem mass spectrometry (UHPLC-MS/MS) analysis of psychoactive drugs in oral fluid. J Pharm Biomed Anal 170:63–67
36. Deriu C, Conticello I, Mebel AM, McCord B (2019) Micro solid phase extraction surface-enhanced Raman spectroscopy (mu-SPE/SERS) screening test for the detection of the synthetic cannabinoid JWH-018 in Oral fluid. Anal Chem 91:4780–4789
37. Williams M, Martin J, Galettis P (2019) A validated method for the detection of synthetic cannabinoids in Oral fluid. J Anal Toxicol 43: 10–17
38. Fojtíková L, Šuláková A, Blažková M, Holubová B, Kuchař M, Mikšátková P, Lapčík O, Fukal L (2017) Lateral flow immunoassay and enzyme linked immunosorbent assay as effective immunomethods for the detection of synthetic cannabinoid JWH-200 based on the newly synthesized hapten. Toxicol Rep 5:65–75
39. Rocchi R, Simeoni MC, Montesano C, Vannutelli G, Curini R, Sergi M, Compagnone D (2018) Analysis of new psychoactive substances in oral fluids by means of microextraction by packed sorbent followed by ultra-high-performance liquid chromatography-tandem mass spectrometry. Drug Test Anal 10: 865–873
40. Amaratunga P, Thomas C, Lemberg BL, Lemberg D (2014) Quantitative measurement of XLR11 and UR-144 in oral fluid by LC-MS-MS. J Anal Toxicol 38:315–321
41. Øiestad EL, Johansen U, Christophersen AS, Karinen R (2013) Screening of synthetic cannabinoids in preserved oral fluid by UPLC-MS/MS. Bioanalysis 5:2257–2268
42. Rodrigues WC, Catbagan P, Rana S, Wang G, Moore C (2013) Detection of synthetic cannabinoids in oral fluid using ELISA and LC-MS-MS. J Anal Toxicol 37:526–533
43. Kneisel S, Speck M, Moosmann B, Corneillie TM, Butlin NG, Auwärter V (2013) LC/ESI-MS/MS method for quantification of 28 synthetic cannabinoids in neat oral fluid and its application to preliminary studies on their detection windows. Anal Bioanal Chem 405: 4691–4706
44. Strano-Rossi S, Anzillotti L, Castrignanò E, Romolo FS, Chiarotti M (2012) Ultra-high performance liquid chromatography-electrospray ionization-tandem mass

spectrometry screening method for direct analysis of designer drugs, "spice" and stimulants in oral fluid. J Chromatogr A 1258:37–42

45. Peters FT, Wissenbach DK, Busardò FP, Marchei E, Pichini S (2017) Method development in forensic toxicology. Curr Pharm Des 23:5455–5467
46. Coulter C, Garnier M, Moore C (2011) Synthetic cannabinoids in oral fluid. J Anal Toxicol 35:424–430
47. Amaratunga P, Lorenz Lemberg B, Lemberg D (2013) Quantitative measurement of synthetic cathinones in oral fluid. J Anal Toxicol 37: 622–628
48. de Castro A, Lendoiro E, Fernández-Vega H, Steinmeyer S, López-Rivadulla M, Cruz A (2014) Liquid chromatography tandem mass spectrometry determination of selected synthetic cathinones and two piperazines in oral fluid. Cross reactivity study with an on-site immunoassay device. J Chromatogr A 1374: 93–101
49. Mercolini L, Protti M, Catapano MC, Rudge J, Sberna AE (2016) LC–MS/MS and volumetric absorptive microsampling for quantitative bioanalysis of cathinone analogues in dried urine, plasma and oral fluid samples. J Pharm Biomed Anal 123:186–194
50. Miller B, Kim J, Concheiro M (2017) Stability of synthetic cathinones in oral fluid samples. Forensic Sci Int 274:13–21
51. Richeval C, Wille SMR, Nachon-Phanithavong M, Samyn N, Allorge D, Gaulier J (2018) New psychoactive substances in oral fluid of French and Belgian drivers in 2016. Int J Drug Policy 57:1–3
52. Ares AM, Fernández P, Regenjo M, Fernández AM, Carro AM, Lorenzo RA (2017) A fast bioanalytical method based on microextraction by packed sorbent and UPLC–MS/MS for determining new psychoactive substances in oral fluid. Talanta 174:454–461
53. Fernández P, Regenjo M, Ares A, Fernández AM, Lorenzo RA, Carro AM (2019) Simultaneous determination of 20 drugs of abuse in oral fluid using ultrasound-assisted dispersive liquid–liquid microextraction. Anal Bioanal Chem 411:193–203
54. Williams M, Martin J, Galettis P (2017) A validated method for the detection of 32 Bath salts in oral fluid. J Anal Toxicol 41:659–669
55. Sorribes-Soriano A, Esteve-Turrillas FA, Armenta S, Amorós P, Herrero-Martínez JM (2019) Amphetamine-type stimulants analysis in oral fluid based on molecularly imprinting extraction. Anal Chim Acta 1052:73–83
56. Di Trana A, Mannocchi G, Pirani F, Maida NL, Gottardi M, Pichini S, Busardò FP (2020) A comprehensive HPLC-MS-MS screening method for 77 new psychoactive substances, 24 classic drugs and 18 related metabolites in blood, urine and oral fluid. J Anal Toxicol 44: 769–783
57. Accioni F, Nieddu M, Corona P, Boatto G (2020) Hexanol-based supramolecular solvents tool for the determination of 11 illicit phenethylamines in oral fluid by LC-MS/MS. J Anal Toxicol 44:15–21
58. Toennes SW, Schneider D, Pogoda W, Paulke A, Wunder C, Theunissen EL, Kuypers KPC, de Sousa Fernandes Perna E, Ramaekers JG (2019). Pharmacokinetic properties of 4-fluoroamphetamine in serum and oral fluid after oral ingestion. Drug Test Anal 11: 1028–1034
59. Gjerde H, Nordfjærn T, Bretteville-Jensen AL, Edland-Gryt M, Furuhaugen H, Karinen R, Øiestad EL (2016) Comparison of drugs used by nightclub patrons and criminal offenders in Oslo, Norway. Forensic Sci Int 265:1–5
60. Sergi M, Bafile E, Compagnone D, Curini R, D'Ascenzo G, Romolo FS (2009) Multiclass analysis of illicit drugs in plasma and oral fluids by LC-MS/MS. Anal Bioanal Chem 393: 709–718
61. Scientific Working Group for Forensic Toxicology (2013) Scientific Working Group for Forensic Toxicology (SWGTOX) standard practices for method validation in forensic toxicology. J Anal Toxicol 37:452–474
62. Griswold MK, Chai PR, Krotulski AJ, Friscia M, Chapman BP, Varma N, Boyer EW, Logan BK, Babu KM (2017) A novel Oral fluid assay (LC-QTOF-MS) for the detection of fentanyl and clandestine opioids in Oral fluid after reported heroin overdose. J Med Toxicol 13: 287–292
63. Mohr AL, Friscia M, Logan BK (2016) Identification and prevalence determination of novel recreational drugs and discovery of their metabolites in blood, urine and oral fluid. US Department of Justice, Washington, DC
64. Truver MT, Swortwood MJ (2018) Quantitative analysis of novel synthetic opioids, morphine and buprenorphine in Oral fluid by LC-MS-MS. J Anal Toxicol 42:554–561
65. Peters FT (2014) Recent developments in urinalysis of metabolites of new psychoactive substances using LC-MS. Bioanalysis 6: 2083–2107
66. Gunnar T, Ariniemi K, Lillsunde P (2005) Validated toxicological determination of 30 drugs

of abuse as optimized derivatives in oral fluid by long column fast gas chromatography/electron impact mass spectrometry. J Mass Spectrom 40:739–753
67. Palmquist KB, Swortwood MJ (2019) Data-independent screening method for 14 fentanyl analogs in whole blood and oral fluid using LC-QTOF-MS. Forensic Sci Int 297:189–197
68. Morato NM, Pirro V, Fedick PW, Cooks RG (2019) Quantitative swab touch spray mass spectrometry for oral fluid drug testing. Anal Chem 91:7450–7457
69. Gesseck AM, Poklis JL, Wolf CE, Xu J, Bashir A, Hendricks-Muñoz KD, Peace MR (2020) A case study evaluating the efficacy of an Ad Hoc hospital collection device for fentanyl in infant oral fluid. J Anal Toxicol 44: 741–746
70. Kintz P, Jamey C, Ameline A, Richeval C, Raul JS (2017) Characterization of metizolam, a designer benzodiazepine, in alternative biological specimens. Toxicol Anal Clin 29: 57–63
71. Ameline A, Richeval C, Gaulier JM, Raul JS, Kintz P (2018) Characterization of flunitrazolam, a new designer benzodiazepine, in oral fluid after a controlled single administration. J Anal Toxicol 42:e58–e60
72. Ameline A, Arbouche N, Raul JS, Kintz P (2018) Documentation of a little-studied designer benzodiazepine after a controlled single administration: II. Concentration profile of deschloroetizolam in saliva. Ther Drug Monit 40:759–761

Chapter 8

Target Analysis of NPS in Other Alternative Matrices: Dried Blood Spots and Meconium

Luca Morini, Claudia Carelli, and Matteo Moretti

Abstract

Dried Blood Spots (DBS) and meconium represent valid alternative matrices for detection of New Psychoactive Substances (NPS), both having the advantages of noninvasive collection procedures. DBS offers an easy storage as well, and meconium is also characterized by a wide window of detection. Most existing studies about the detection of NPS in these two matrices relied on liquid chromatography coupled with tandem mass spectrometry systems (LC-MS/MS) or high-resolution methods, and concentrated on target analyses rather than on screening tests. Up to date, methods have been validated only for some categories of NPS. While literature about synthetic cathinones and phenethylamines, synthetic opioids, and cannabinoids in DBS is present, there are no current studies on new designer benzodiazepines. In meconium, only methods for synthetic cathinones have been published.

Key words DBS, Meconium, NPS, LC-MS/MS, GC-MS/MS, Synthetic Cathinones, Phenethylamines, Synthetic opioids, Synthetic cannabinoids, New designer benzodiazepines

1 Dried Blood Spot

DBS technique is a form of microsampling method for blood collection and storage, first developed in 1963 to be used in the neonatal screening for phenylketonuria and for both qualitative and quantitative detection of metabolic disorders [1]. Due to the emergence of more sensitive analytical techniques over the last two decades, the DBS methodology has been increasingly used and adapted in the development and validation of new methods, with an exponential growth of publications for applications in several areas, such as pharmacokinetics [2–4], toxicokinetics [5], and therapeutic drug monitoring [6].

In recent years, several target methods for the analysis, identification, and quantification of drugs of abuse [7–10] and psychoactive substances [11, 12] have been developed and validated, even in postmortem samples. The use of DBS for a fast straightforward

Marta Concheiro and Karl B. Scheidweiler (eds.), *Methods for Novel Psychoactive Substance Analysis*,
Methods in Pharmacology and Toxicology, https://doi.org/10.1007/978-1-0716-2605-4_8,

approach helps to simplify and shorten sample preparation significantly.

1.1 Advantages and Disadvantages

DBS offers several advantages compared to whole blood collection. DBS is simple, minimally invasive, involves collecting a small sample volume, is easy to transport and store, and cost saving [13]. It also may allow minimizing the need for pre-analytical sample processing. Furthermore, DBS reduces the risk of infection associated with sample handling, since it is a dry matrix, thereby promoting pathogens inactivation. The stability of analytes on DBS seems to be superior to plasma, serum, or whole blood. This is primarily due to inactivation of degradation processes such as hydrolysis, photolytic processes, esterase, and RNAase action.

Besides these benefits, there are also drawbacks in DBS analysis. The small volume of blood in DBS requires highly sensitive analytical instrumentation, thus explaining the need for target analysis. Yet, the limited sample amount potentially affects DBS utility for repeated testing. Moreover, the sampling and drying procedures, however simple, can introduce several biases. DBS measurements can be impacted by chromatographic or matrix effects within the filter paper itself, which can lead to uneven spreading of blood or distribution of analytes within a spot, depending on their physicochemical properties.

At present, the Food and Drug Administration does not accept DBS as a stand-alone sample and requires bridging studies for comparing DBS with traditional models, which adds cost and labor [14].

1.2 Types of Cards

The two types of cards mostly available are cotton paper filters of different qualities and glass microfiber filter papers. The main difference between the two supports is that the glass fiber does not soak up reagents, which diminishes nonspecific analyte adsorption on the membrane.

Some cards are soaked in chemicals that cause denaturation of proteins, preserving DNA (such as Flinders Technology Associates -FTA®- cards). Other cards do not include cellulose material, which would reduce nonspecific binding, improving mass spectrometer analyte response and increasing signal to noise ratios.

Volumetric absorptive microsampling (VAMS™), a second-generation dried miniaturized sampling technology, has been developed just in order to provide a more accurate sampling but still granting feasible sample processing.

1.3 Sampling

Originally, the DBS sampling technique involved the collection of a small volume of blood, by puncturing the fingertip or heel, using an automatic, sterile, disposable lancet, which was then directly deposited onto filter paper.

The sampled blood can also be collected using a precision micro capillary technique. It must be noted that most of available publications describe the analysis of DBS that have been obtained by pipetting existing venous blood samples, rather than setting up a true capillary DBS approach. In general, drying the samples for at least 2–3 h in an open space at room temperature (15–22 °C) is necessary. The DBS cards are then moved into a waterproof plastic bag, possibly along with a desiccant and a humidity indicator.

1.4 Sample Preparation

The volume of blood generally used to generate DBS for detecting drugs and metabolites ranges from 5 μL (a quantity that could create problems with analytical sensitivity) to 100 μL (which are not easily obtained via finger puncture).

For volumetric application, the complete DBS analysis is often preferred. When DBS are obtained by direct application from the fingertip onto filter paper, typically a smaller disk (3–6 mm) is punched from the card. Analyzing a complete spot largely avoids the hematocrit effect and facilitates validation, though may increase the paper amount to be extracted and, consequently, the potential matrix effects.

1.5 Extraction

Extraction of the analytes from DBS specimens needs to be achieved using a standard procedure. DBS retains the major components of blood (proteins and salts) on the paper, so the target analytes can be extracted from the DBS and put into a solution for subsequent analysis. The desorption of analytes from DBS is usually performed using organic solvents, water, or a mixture of both. However, the use of water and aqueous solutions leads to an increase in interferences, specifically the presence of endogenous compounds in the sample [15]. Methanol seemed to increase the extraction performance whenever FTA® cards are used [16].

Punched or whole spots are placed in an elution buffer for variable time spans, according to the procedure; the DBS extracted is then treated as a hemolyzed whole blood sample and tested with methods often intended for this type of sample.

To limit interferences, additional treatments are often needed to separate the analyte from the matrix, including adding an organic solvent and/or using liquid/liquid extraction (LLE) or solid phase extraction (SPE) [17–20]. The use of organic solvents, such as methanol and acetonitrile or mixtures of organic solvents, in the sample preparation stages, provides the transition of the desired analytes to solvent by holding the protein and salts in the blood on paper [9, 12, 21, 22].

The desorption process can be performed by passive infusion of the filter paper, but is most often facilitated by sonication, agitation, or heating. However, the increase of temperature may affect substance stability; therefore, agitation and immersion in ultrasonic baths are the preferred procedures.

A derivatization reagent could also be directly added onto the filter paper, without the use of an extraction solvent, followed by drying and redissolution in a suitable solvent, ensuring a minimal and less time-consuming sample pretreatment.

Automated sample preparation procedures have also been described [23]. Automation of extraction processes minimizes the number of manual steps, reducing potential errors. Recent technologies for automating the extraction and/or mass spectrometric detection of DBS include paper spray, thin-layer chromatography, liquid-surface analysis, and online DBS sample extraction with or without additional clean-up. Flow-through extraction of DBS coupled to online SPE has also been developed; this system performs an automated desorption of DBS, complete with the addition of internal standard, followed by online SPE with subsequent analysis by a mass spectrometer.

1.6 Factors Influencing the Analysis

The analytical result could be greatly influenced by sample hematocrit, hemoglobin concentration, blood volume spotted, and site of punching (peripheral versus central). Blood with low hematocrit values results on regular cellulose-based DBS cards in larger spot sizes compared to blood with high values, influencing the accuracy of quantitative results obtained from DBS [24, 25]. Even for a qualitative method, this problem may affect the reproducibility of results, especially around the LOD [12].

The VAMS™ sampler can help overcome the hematocrit issue, absorbing a fixed volume of blood (~10 μL) in 2–4 s with less than 5% volume variation across the hematocrit range of 20–70% with low tip-to-tip variability.

1.7 Analytical Method

Such small blood volumes require a highly sensitive technique for identification and quantification of the compounds. The evaluation of target analytes from DBS is performed by using several analytical techniques, according to the application: for low molecular weight molecules, the most frequently used analytical techniques are immunoassays, liquid chromatography (LC) coupled to ultraviolet (UV) and fluorescence detectors, gas chromatography (GC), and LC coupled to mass-spectrometry (MS) or tandem mass spectrometry (MS/MS). Rapid technological advances over the last decade allowed that currently the LC instruments coupled to tandem MS achieve high levels of sensitivity and specificity. In particular, the triple quadrupole detectors (triple quad) operating in the selected reaction monitoring mode (SRM), and the new high-resolution analyzers time of flight (TOF) and Orbitrap, that allow the determination of exact masses, enable the qualitative and quantitative analysis with high precision of substances with special relevance in forensic context, using DBS samples [5, 26–30].

Nowadays, several hundred monitored compounds are included in the group of NPS; hence, it is difficult for laboratories to purchase all the standards and to develop and optimize analytical procedures for the detection of these molecules in biological specimens.

A target screening LC-MS/MS approach, based on parameters provided by previously published studies, is useful for a tentative detection of NPS in DBS. Although devoid of forensic validity, this procedure might help in preliminary evaluation of the potential presence of substances in the sample, eventually to be confirmed by a validated method.

1.8 Synthetic Cathinones, Phenethylamines, Piperazines, and Tryptamine Derivatives

1.8.1 Extraction

Methanol is the preferred extraction solvent, though buffered aqueous solutions can achieve adequate recovery percentages as well. Butylone, buphedrone, and N-ethylcathinone yielded better extraction efficiencies in diethyl ether, while methanol guarantees better results for metabolites. However, ethyl acetate seems to provide the best compromise for most synthetic cathinones [15].

The optimal volume of blood was considered 30 μL, because it demonstrated greater extraction efficiency [31]. A sonication step could be included because of its improved extraction efficiency, particularly in the case of 4-fluoromethcathinone (4-FMC), 5,6-methylenedioxy-2-aminoindane (MDAI), butylone, and eutylone.

Compared to the analysis of liquid samples, the sensitivity of the DBS method is expected to be lower because of the small spotting volume. An evaporation and reconstitution procedure is commonly implemented to counteract the loss of sensitivity [31]. Four different extraction procedures are summarized in Table 1.

1.8.2 Instrumental Analysis

Target LC-MS/MS analysis provided the highest sensitivity and was generally applied for identification and quantification of synthetic cathinones, phenetylamines, piperazines, and tryptamine derivatives in DBS. Reversed phase is the most frequently used chromatographic separation mode. Different C18 columns, with different lengths, internal diameters, and particle sizes, can reach adequate results, in terms of separation and peak resolution; the broader spectrum of monitored analytes, the longer the column and/or the smaller the particle sizes. The addition of formic acid (0.1%) in mobile phases is always advisable, while buffer solutions must be carefully evaluated, and the salts concentrations kept as low as possible, since ESI sources are used for ionization. The instrumental parameters used for the previous extraction procedures are listed in Table 2.

1.8.3 Stability

Considering that cathinones have been shown to be susceptible to degradation in blood samples under different storage conditions, sample drying might offer increased stability. Actually, this fact has already been demonstrated in dried *Catha Edulis* plant material

Table 1
Extraction procedures for synthetic cathinones analysis in DBS

Procedure	Cards	Amount of blood spotted (μL)	Extraction solution	Extraction procedure
1. [31]	Whatman™ FTA®	30	Buffer solution at pH 9.2	LLE (3 mL ethylacetate)
2. [32]	VAMS™	15 (plasma)	Methanol	Direct injection
3. [33]	Whatman™ 903	15	Methanol	Direct injection
4. [34]	Whatman™ 903	85	Methanol	SPE: mixed-mode

[35]. Wang et al. observed a good stability for most of the 37 monitored synthetic cathinones when stored at low temperature for 1 month, while 7 out of 37 substances remained stable for only 24 h at room temperature [31].

Da Cunha et al. compared the stability of butylone, benzedrone, mephedrone, and pentylone in whole blood and DBS during storage at −20 °C, 4 °C, and room temperature for 90 days [33]. At room temperature, a marked instability was found in all analytes when stored for 90 days in both samples. However, analytes remained detectable in DBS over the 90 days but undetectable in whole blood. At −20 °C for 90 days, all analytes had no or only minor effects on the concentrations in both samples. Likewise, 4 °C for 90 days led to no significant losses for butylone and pentylone in both samples but to a significant to complete degradation for mephedrone and benzedrone under the same storage temperature.

1.8.4 Real Samples

Synthetic cathinones are generally measured in blood at levels of dozens or higher ng/mL. Hence, LOQs of 10–20 ng/mL appear as adequate to monitor a potential intoxication. On the contrary, the evaluation of phenethylamines and other NPS requires a higher sensitivity. In Figs. 1 and 2 the chromatograms of α-PHP (synthetic cathinone) and 25C-NBOMe (phenetylamine) measured in DBS in concentrations lower than 10 ng/mL, are reported.

1.9 New Designer Benzodiazepines

The class of new designer benzodiazepines is not yet included in the list of substances detected in DBS. However, different analytical approaches for the identifications of classic benzodiazepines are developed and validated, and could be easily applied also to new designer ones, due to the similar chemical structures and properties.

1.9.1 Extraction

Buffer solution is generally preferred. However, organic solvents such as methanol or acetonitrile should provide sufficient extraction

Table 2
Instrumental parameters for synthetic cathinones analysis in DBS

Procedure	Column	Mobile phases	Flow rate (mL/min)	Mass Spectrometry (MS)	MS operating mode	Number of synthetic cathinones monitored	LOD (ng/mL)	LOQ (ng/mL)
1. [31]	C18 (100 mm × 2.1 mm, 1.8 μm)	20 mM ammonium acetate/ formic acid 0.1% (a) and acetonitrile (b): gradient elution	0.250	Triple quad with ESI^+ source	SRM (two transitions for analyte)	37	0.5–2.0	2.0–10.0
2. [32]	C18 (50 mm × 2.1 mm, 3.5 μm)	0.1% formic acid in acetonitrile (a) and 0.1% formic acid in water (b): gradient elution	0.300	Triple quad with ESI^+ source	SRM (two transitions for analyte)	7	3.0	10.0
3. [33]	C18 (50 mm × 4.6 mm, 2.6 μm)	Water (a) and methanol (b): gradient elution	0.500	Triple quad with ESI^+ source	SRM (two transitions for analyte)	4	10.0	25.0
4. [34]	C18 (100 mm × 2.1 mm, 2.6 μm)	Water (a) and methanol (b): gradient elution	0.350	Triple quad with ESI^+ source	SRM (two transitions for analyte)	16	0.1	0.5

LOD limit of detection, *LOQ* limit of quantification

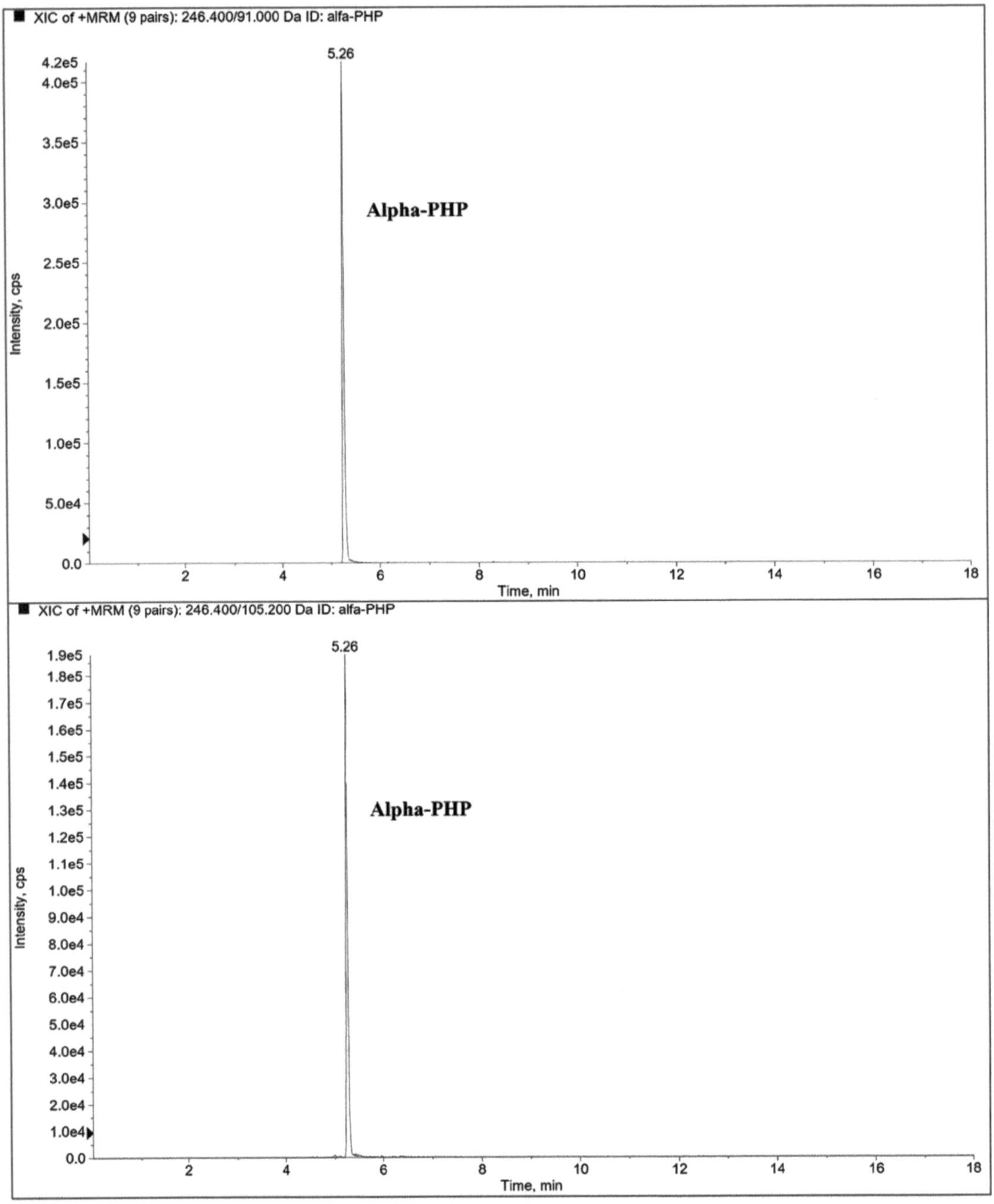

Fig. 1 SRM transitions of α-PHP in an authentic DBS sample

efficiencies as well. The amount of spotted blood is relatively high, up to 100 μL [36]. A purification step is always advisable to increase sensitivity, and mixed-mode SPE is the most frequently used technique. Table 3 reports an example of extraction procedure.

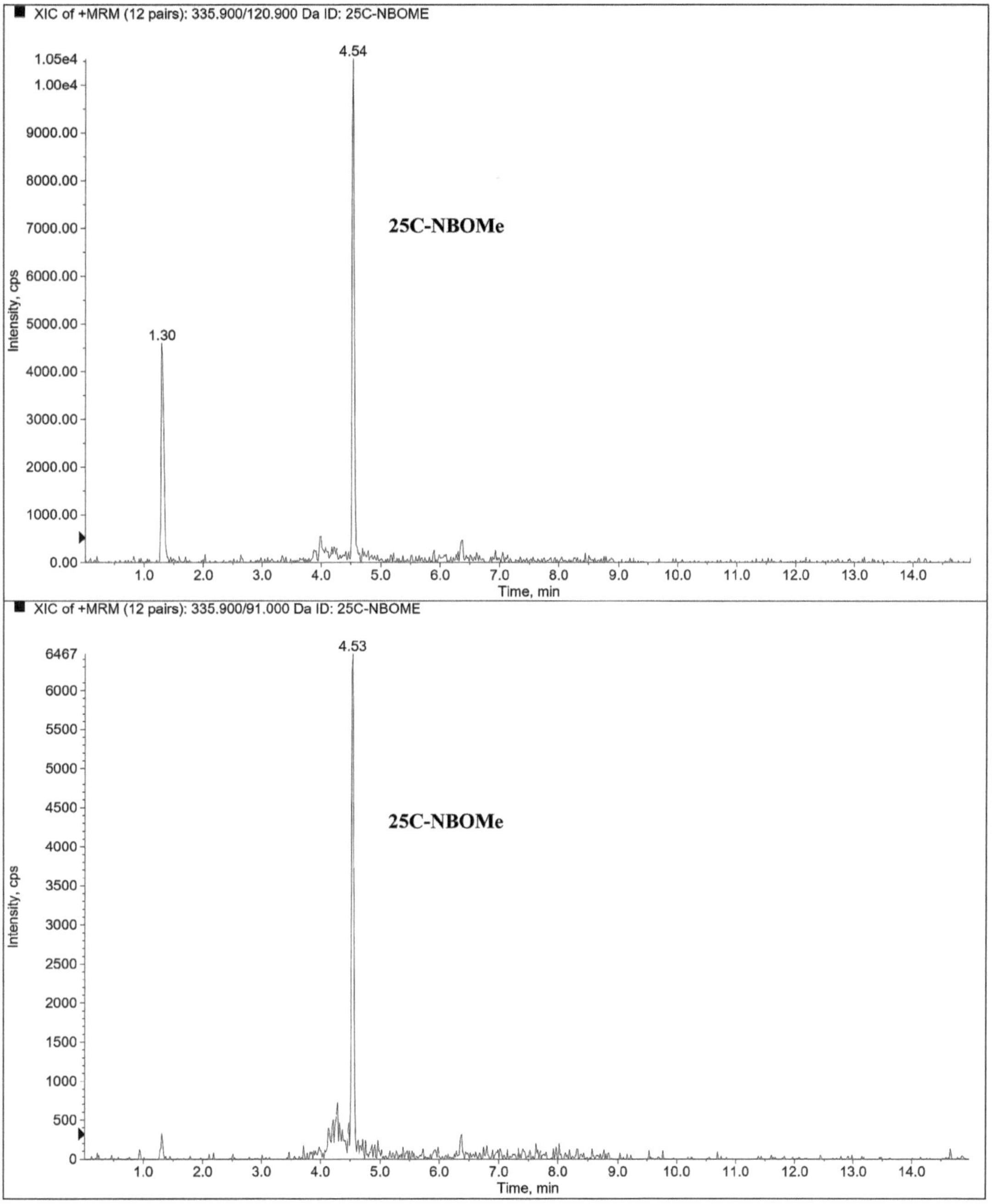

Fig. 2 SRM transitions of 25C-NBOMe in an authentic DBS sample

1.9.2 Instrumental Analysis

LC-MS/MS systems are the most frequently used for identification of benzodiazepines in DBS. The instrumental parameters, such as chromatographic conditions, do not differ from the ones used for the identification of other NPS. This aspect represents an important advantage when a sample must be checked for different classes of

Table 3
Extraction procedure for benzodiazepines analysis in DBS

Procedure	Cards	Amount of blood spotted (μL)	Extraction solution	Extraction procedure
1. [37]	Whatman™ 903	85	Buffer solution at pH 6.0	SPE: mixed-mode: strong cation exchange, reversed phase

compounds. The LODs and LOQs are within a wide range (Table 4), probably due to the different moieties of the benzodiazepines that can affect the ionization.

1.9.3 Stability

The stability of traditional benzodiazepines in DBS appears also acceptable when storage is carried out at room temperature. In particular, diazepam, nordiazepam, alprazolam, chlordesmethyldiazepam, and flurazepam are stable for up to three months. By contrast, midazolam and flurazepam metabolites may undergo a significant loss, independently of the storage conditions and temperatures [37].

1.9.4 Real Samples

Moretti and coauthors detected nine different traditional benzodiazepines and metabolites in DBS collected from 14 postmortem blood samples, at therapeutic and toxic levels [37]. A good qualitative and quantitative correlation was observed between DBS and blood analyses, thus confirming that target methods, for benzodiazepines monitoring in DBS, are adequately sensitive, specific, and reliable.

1.10 Synthetic Cannabinoids

Compared to blood and plasma, DBS and VAMS™ can provide the same information on cannabinoid levels. Different hematocrit levels demonstrated only a minimal effect on measured cannabinoid levels with whole-spot analysis; a further improvement could be obtained using a VAMS™ procedure [34].

1.10.1 Extraction

The best results in terms of analyte extraction efficiency and negligible ion suppression have been described using methanol to desorb the analytes from the dried sample, under vortex mixing and agitation in an ultrasonic bath. Another promising cleanup procedure of synthetic cannabinoids from biological matrices is the supported liquid extraction technique, consisting of liquid immobilized in an inert medium packed, and immiscible with the solution containing the analytes, that percolates through the immobilized one [38]; however, this technique has not been yet applied in DBS analysis. Two extraction procedures are reported in Table 5.

Table 4
Recommended instrumental parameters for benzodiazepines analysis in DBS

Procedure	Column	Mobile phases	Flow rate	Mass Spectrometry	MS operating mode	# analytes	LOD	LOQ
1. [37]	C18 (100 mm × 2.1 mm, 5 μm)	Water with 0.1% formic acid (a) and acetonitrile with 0.1% formic acid (b): gradient elution	0.200 mL/min	Triple quad with ESI^+ source	SRM (two transitions for analyte)	18	0.2–50.0 ng/mL	1.0–100.0 ng/mL

LOD limit of detection, *LOQ* limit of quantification.

Table 5
Extraction procedures for synthetic cannabinoids analysis in DBS

Procedure	Cards	Amount of blood spotted (μL)	Extraction solution	Extraction procedure
1. [34]	Whatman™ FTA® VAMS™	10	Methanol	Direct injection
2. [39]	Whatman™ FTA® Sartorius™ DBS cards	20	Methanol/acetonitrile/2% acetic acid (1:1:1, v/v/v).	Direct injection

1.10.2 Instrumental Analysis

Again, target analysis by LC-MS/MS systems should be the preferred technique. Though a chromatographic separation in reversed phase mode is the most frequently used, a particular attention to mobile phases and gradient elution parameters is needed. Indeed, several synthetic cannabinoids and their main metabolites have different polarities and acidic or basic properties. The instrumental parameters suggested are listed in Table 6.

1.10.3 Stability

Stability experiments have been performed only in fortified DBS and VAMS ™ samples, and not in authentic samples [34]. Ten different synthetic cannabinoids have been tested within 30 days of storage at room temperature in bags containing desiccant packages. Satisfactory results were observed, proving that these NPS are adequately stable without the need of cold temperatures.

1.10.4 Real Samples

Three different synthetic cannabinoids have been detected in both VAMS™ and DBS, collected from two drug users. The subjects self-declared an herbal mixture consumption, through smoking, about four hours prior to sample collection. JWH-81 was measured at a concentration of about 7.7 ng/mL, while JWH-122 and JWH-250 at about 4.1 and 1.3 ng/mL [34], confirming that DBS are promising alternative matrices for monitoring NPS, even at low levels.

1.11 Synthetic Opioids

1.11.1 Extraction

Due to the alkaline nature of synthetic opioids, a basic extraction could increase the efficiency. Given the high solubility of fentanyl in methanol, the addition of this organic solvent to the desorption solvent could improve recovery. The sonicated assisted extraction increases recovery percentages and limits the loss of time during sample treatment. A mixture of 85% aqueous, 1% formic acid, and 15% methanol seems to maximize desorption of fentanyl analogs while maintaining sufficient water content; this allows to maximize retention on the SPE cartridge for most of the analytes avoiding an intermediate evaporation step [40]. A higher solvent temperature

Table 6
Instrumental parameters for synthetic cannabinoids analysis in DBS

Procedure	Column	Mobile phases	Flow rate (mL/min)	Mass Spectrometry	MS operating mode	# analytes	LOD (ng/mL)	LOQ (ng/mL)
1. [34]	C18 (50 mm × 2.1 mm, 2.5 μm)	0.1% FA in acetonitrile (a) and 0.1% FA in water (b): gradient elution	0.350	Triple quad with ESI^+ source	SRM (two transitions for analyte)	10	0.03–0.1	0.1–0.5
2. [39]	C6, (10 mm × 2.1, 3.0 μm)	0.2% formic acid (a) and acetonitrile (b): gradient elution	0.250	Orbitrap mass spectrometer with ESI^+ source	HRMS (single ion monitoring)	1	2.0	–

LOD limit of detection, *LOQ* limit of quantification

Table 7
Extraction procedures for synthetic opioids analysis on DBS

Procedure	Cards	Amount of blood spotted/analyzed (μL)	Extraction solution	Extraction procedure
1. [40]	Whatman™ FTA®	5	15 : 85 methanol : aqueous 1% formic acid	Online SPE
2. [8]	Bond Elut Dried Matrix Spotting cards	30	Methanol/0.1% formic acid	Direct injection

seems to improve extraction, without affecting molecules' stability. Different extraction methods are reported in Table 7.

1.11.2 Instrumental Analysis

Though common chromatographic conditions are normally applied for separation of synthetic opioids (two examples of instrumental analysis are reported in Table 8), an online SPE procedure followed by a direct injection into the mass spectrometer system has been recently proposed [40]. An automated preparation with online SPE amenable to DBS samples could provide a streamlined laboratory workflow and maximize sample throughput.

1.11.3 Stability

To date, no specific studies have been performed on the stability of synthetic opioids on DBS.

It has been demonstrated that, in DBS, fentanyl, norfentanyl, and despropionyl fentanyl were stable at room temperature for three days and for a longer time at −20 °C and −80 °C [41].

1.11.4 Real Samples

Analysis of clinical samples with the described methods confirmed the ability to positively identify fentanyl analogs in individual samples with such a good sensitivity to easily detect overdose cases. However, evaluation of clinical samples could require a lower limit of detection, as the potency for these analogs can vary greatly. A postmortem blood sample, collected from a subject who died by furanyl fentanyl overdose, was deposed on DBS and, eventually analyzed through a LC-MS/MS method, previously validated for blood. A concentration of furanyl fentanyl lower than 10 ng/ml was clearly detected in DBS, confirming the sensitivity of the target ESI-SRM developed. Figure 3 reports the chromatogram of a real DBS sample.

2 Meconium

Meconium is the first neonatal feces; however, contrarily to stools, these odorless, green-to-black feces, with a viscous texture

Table 8
Instrumental parameters for synthetic opioids analysis in DBS

Procedure	Column	Mobile phases	Flow rate (mL/min)	Mass Spectrometry	MS operating mode	# analytes	LOD (ng/mL)	LOQ (ng/mL)
1. [40]	Online SPE cartridge	1% formic acid in water (a) and 100% acetonitrile (b): gradient elution	0.500	Triple quad with ESI^+ source	SRM (two transitions for analyte)	6	0.15–0.66	–
2. [8]	C18 (100 mm × 2.1 mm, 2.6 μm)	5 mM ammonium formate with 0.1% formic acid (a) and methanol/ acetonitrile, ratio of 1:1 with 0.1% of formic acid (b): gradient elution	0.300	Triple quad with ESI+ source	SRM (two transitions for analyte)	2	0.05	0.2

LOD limit of detection, *LOQ* limit of quantification

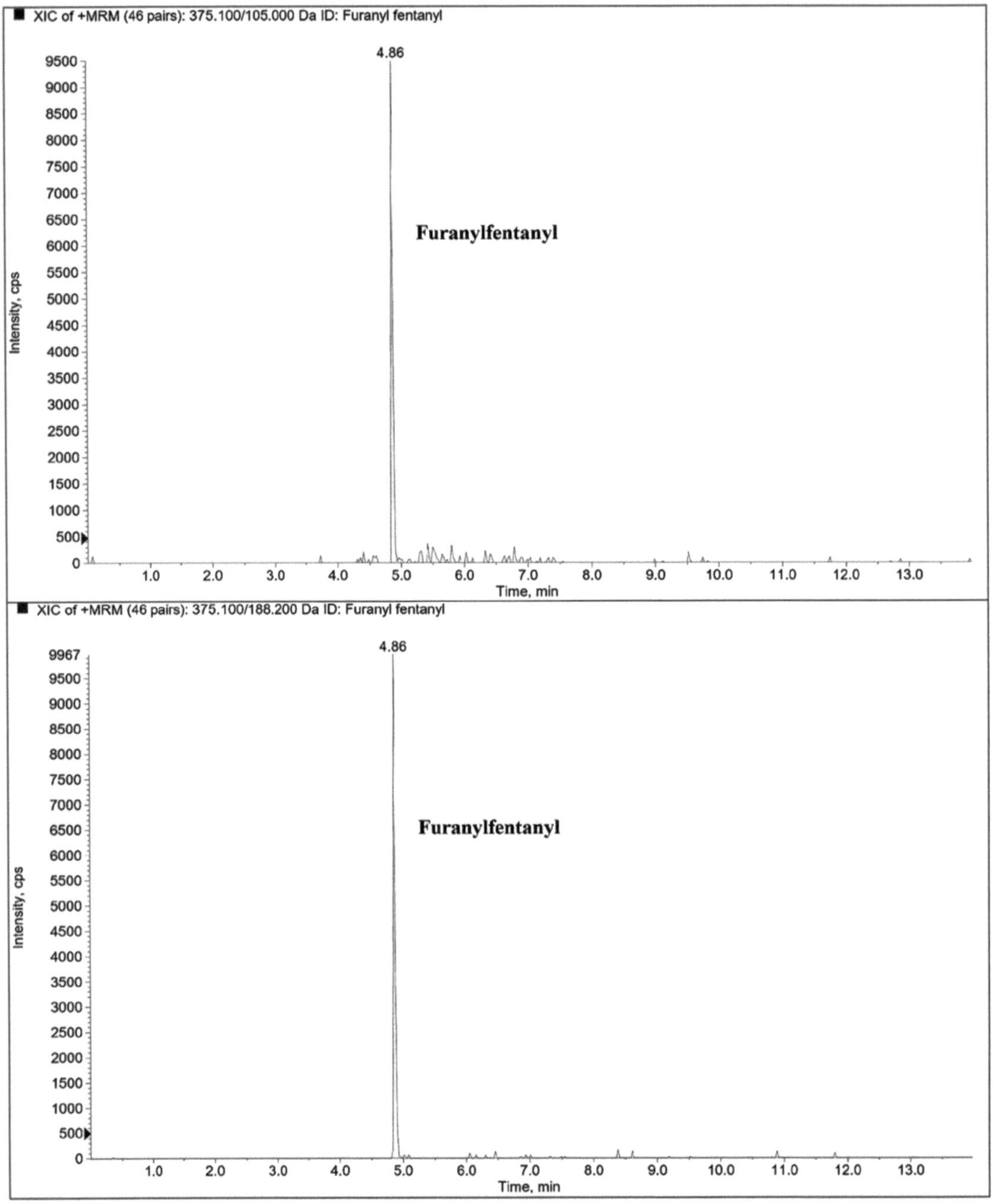

Fig. 3 SRM transitions of furanyl fentanyl in a real DBS sample

biological matrix, are primarily made of water, desquamated epithelial cells from gastrointestinal tract and skin, lipids deriving from vernix caseosa, biliary acids and salts, proteins, sugars, various pancreatic and gastrointestinal secretions, ingested amniotic fluid and, in minority, of other different substances. Other exogenous molecules, such as drugs of abuse and metabolites, have been found over time in meconium, assessing it as an optimal biological matrix to evaluate fetal exposure to different compounds [42]. Meconium formation in fetal bowel begins approximately around the 11th week of gestation, and generally corresponds to the beginning of amniotic fluid ingestion. For this reason, meconium can be helpful in monitoring fetal exposure to exogenous compounds at least during the third trimester, though a previous study identified cocaine exposure at the 17th week, so in the second trimester [43]. Moreover, it is thought that meconium positivity to substances could also result from direct ingestion of liquids (such as urine) by the fetus itself through amniotic fluid; this means that the parent drug could be found together with its main metabolites, taking into account the fact that hepatic fetal metabolism will play a minor role [44]. However, other studies indicated that some drugs and their metabolites could directly pass the placental barrier. The high complexity and heterogeneity of meconium has always represented a big problem for development of specific and sensitive methods for determination of exogenous substances [45]. The window for sample collection, though being concentrated within the first 72 h, can last for up to 125 h after birth [42].

2.1 Advantages and Disadvantages

Meconium offers several advantages in comparison to the other biological matrices utilized for the evaluation of perinatal exposure to xenobiotics, such as breast milk, neonatal nails and hair, umbilical cord, and placenta. It is a waste material and it is relatively easy to collect and store; the window of detection is wide and reflects, at least, the third trimester of pregnancy [46]; though the topic is still under debate [47], meconium should be a sterile biological matrix, thus guaranteeing a safer handling without risk of infection with pathogens. The stability of common xenobiotics and metabolites in meconium seems acceptable [48], in particular when stored at −20 °C.

The main disadvantage is represented by the difficulty to collect a sufficient volume of sample. Indeed, several times the total amount of meconium available for the analysis is less than 1 g, and, because of its viscosity, another significant portion may be lost during sample transfer and sampling. Moreover, urinary contamination is not rare, and it could lead to unreliable analytical results.

2.2 State of the Art on Drugs of Abuse Analysis in Meconium

Most traditional drugs of abuse have already been detected in real samples, through different analytical methods, based both on immunoassay screening tests and confirmatory analyses. In particular, for what concerns screening methods, immunochemical tests had been initially validated for urine, and then modified to be applied on meconium [42].

Extraction procedures depend both on physicochemical characteristics of substances (i.e., polarity, pKa, etc.) and on the kind of metabolites that are to be searched. Several drugs of abuse are subjected to an extensive phase II metabolism [49]; hence, different methods avoid strong extraction procedures that could lead to hydrolysis phenomena, by preferring buffer aqueous and/or methanolic solutions.

Confirmatory analyses are based on chromatographic techniques coupled with mass spectrometry detectors; recently, LC-MS/MS is rapidly replacing the old gas chromatographic techniques in meconium analysis. Whenever it becomes necessary to identify a broad spectrum of substances, target high-resolution mass spectrometric systems are preferred to triple quad detectors, though a lower sensitivity may represent an important drawback. Unfortunately, up to date, very few methods focusing on identification and quantitative determination of NPS in meconium have been developed, validated, and/or applied to real positive samples.

2.3 Synthetic Cathinones

Among the NPS, synthetic cathinones represent the only investigated class in meconium. The first identification was in 2014, when Pichini and coauthors detected 4-methylethcathinone (4-MEC) in meconium of a term infant born from a drug abuser mother [50]. The woman was in withdrawal syndrome with methadone during pregnancy and reported daily smoking of 4-MEC.

More recently, two different methods focusing on the identification and quantification of several synthetic cathinones in meconium have been developed and validated [51, 52]. All the methods are based on similar instrumental analysis: LC-MS/MS has been used in Selected Reaction Monitoring mode, operating in positive polarization.

2.3.1 Extraction Procedures

The developed pre-analytical procedures are applied on a relatively low amount of meconium (~0.2 g) (Table 9). The material used for extraction seems not to interfere with the procedure, since both pyrex® and plastic tubes and vials have been employed. Extraction can be performed with organic solvents, such as methanol, or through the addition of a mixture of organic and aqueous solutions (acetonitrile and water). An important aspect is represented by the pH of all the used solvents, that must be kept as low as possible, since synthetic cathinones are not stable in basic solutions; hence, the addition of a small percentage of inorganic (e.g., HCl) or organic (e.g., formic acid) acid can avoid any potential degradation of the investigated compounds.

Apparently, a direct injection of the extraction solution into the LC-MS/MS system does not appear to yield an adequate sensitivity, so a purification procedure should be mandatory. In case of extraction with an organic solvent, the use of SPE cartridges requires a previous solvent drying step, followed by reconstitution

Table 9
Extraction procedures for synthetic cathinones in meconium

Procedure	Amount of meconium (g)	Tube for extraction	Extraction solution	Extraction procedure	Final solution
1. [51]	0.25	Pyrex®	Methanol	Mixed-mode SPE: strong cation exchange, reversed phase	100 μL (0.1% formic acid: acetonitrile (90: 10 *v/v*)
2. [52]	0.20	Plastic	Acetonitrile: water with formic acid 1% (1:1)	Consecutive purification steps by aqueous and organic solutions	250 μL (water: methanol (90:10 *v/v*)

of the sample in an aqueous solution. The suggested mixed-mode polymeric sorbent includes a strong cation exchange and a reversed phase.

2.3.2 Instrumental Analysis

The LC-MS/MS system appears to be the best instrumental technique. Though a high-resolution mass spectrometric system (HRMS) could guarantee the monitoring of a broader number of molecules, the need for high sensitivity still forces to choose triple quad systems. As for natural alkaloids, such as cathine and cathinone, also synthetic cathinones are chromatographically separated through C18 columns, operating in reversed phase conditions. Due to matrix complexity of the meconium, the use of a guard column is advisable whenever a SPE procedure is avoided. Common chromatographic conditions can be applied to meconium samples analysis (Table 10). ESI source is preferred, and positive polarization guarantees a more intense signal. Though two SRM transitions are generally sufficient for analyte identification, increasing the number to three or four different transitions may help in discriminating also some isobaric compounds (i.e., MBDB, MDEA, methylone) with similar retention times.

2.3.3 Real Samples

To date, only 4-MEC and methylone have been detected in real meconium samples [50, 52]. Though synthetic cathinones represent one of the most frequently detected classes of NPS in biological samples, their use remains limited in comparison to classic drugs of abuse, such as opiates, cocaine, and cannabis. The lack of positive results may be explained by different reasons: first, very few meconium samples have been tested for synthetic cathinones; secondly, there is limited data about the sensitivity that is needed to identify those molecules in meconium. Moreover, it is not clear if the parent drug would be accumulated in a higher amount in respect to metabolites. A study on pharmacokinetics of

Table 10
Instrumental parameters and sensitivity for synthetic cathinones analysis in meconium

Procedure	Column	Mobile phases	Flow rate (mL/min)	Mass Spectrometry	MS operating mode	# analytes	LOD (ng/g)	LOQ (ng/g)
1. [51]	C18 (50 mm × 2.1 mm, 3 μm)	Water (a) and acetonitrile (b)	0.250	Triple quad with ESI^+ source	SRM (two transitions for analyte)	6	2–10	10.0–25.0
2. [52]	C18 (50 mm × 2.1 mm, 1.8 μm)	Water (a) and methanol (b)	N/S	Triple quad with ESI^+ source	SRM (two or three transition for analyte	7	0.5–1	1.0–2.0

three synthetic cathinones, namely mephedrone, methylone, and MDPV, carried out on pregnant mice, demonstrated that relatively high concentrations can be detected in the fetal brain [53]. Hence, theoretically, it should be advisable to monitor the exposure to synthetic cathinones through the analysis of meconium. A sensitivity lower than 10 ng/g appears to affect the reliability of the results. Indeed, the two compounds detected in real samples were measured at a lower concentration. In particular, a self-declared consumption of 4-MEC, on a daily basis, by a pregnant woman, yielded to a concentration of the parent drug in meconium lower than 1 ng/g [50].

2.4 New Designer Benzodiazepines

Though studies and methods about the detection and quantification of new designer benzodiazepines in meconium are not yet published, several analytical procedures for monitoring the classic ones in the same matrix have already been developed, validated, and applied to real samples [54–57]. Since new and classic benzodiazepines share similar structures, chemical characteristics, and, mostly, therapeutic levels, the already developed methods should be successfully applied for monitoring the whole family of these anxiolytic/hypnotic compounds.

2.4.1 Extraction

Though recent methods suggest that the amount of meconium for analysis should be kept as low as possible, other methods used up to 2 g of material, thus ensuring the chance to monitor a broader range of molecules within the same analytical procedure. Benzodiazepines detection is not affected by the material used for extraction. Methanol is the preferred extraction solvent, though it generally requires a further pretreatment before proceeding with the purification step.

The use of purification procedures is advisable to achieve a cleaner chromatogram. A mixed-mode SPE extraction, combining strong cation exchange with reversed phase (i.e., C8, C18, etc.), is the most applied technique. This choice is probably due to increasing the number of monitored substances through the same method, rather than as the best extraction procedure for the benzodiazepine family. Several extraction procedures for traditional benzodiazepines are reported in Table 11.

2.4.2 Instrumental Analysis

GC-MS methods for detection of benzodiazepines in biological matrices were widely used in the past years. More recently, LC-MS/MS ones have been preferred, because they achieve a significantly higher sensitivity. Reversed-phase columns yield to an adequate separation of most of the benzodiazepines, though pentafluorophenyl ones guarantee good results in terms of retention time and clean chromatograms. The use of buffered mobile phases, with addition of ammonium acetate or formate, may help in peak resolution and reproducibility. However, since an ESI source is

Table 11
Extraction procedures for benzodiazepines analysis in meconium

Procedure	Amount of meconium (g)	Tube for extraction	Extraction solution	Extraction procedure	Final solution
1. [54]	0.25	Pyrex®	Methanol	Mixed-mode SPE: strong cation exchange, reversed phase	100 μL 0.1% formic acid mixture (2 mM ammonium formate: acetonitrile (70:30 *v/v*))
2. [56]	1.00	Polypropylene	Methanol	Mixed-mode SPE: strong cation exchange, reversed phase	150 μL (acetonitrile/water 1:1)
3. [55]	0.50	Plastic	Methanol	L/L: ethyl acetate:hexane (80:20 *v/v*)	100 μL (0.1% formic acid: acetonitrile (90:10 *v/v*))
4. [57]	2.00	Polypropylene	Methanol	Mixed-mode SPE: strong cation exchange, reversed phase	160 μL 0.1% formic acid mixture (10 μL acetonitrile and 150 μL 10 mM ammonium acetate)

generally chosen for ionization, concentrations of the salts should be kept low (about 2 mM). Table 12 reports some of the different analytical methods for traditional benzodiazepines.

2.4.3 Real Samples

Though new designer benzodiazepines have not been detected yet in meconium samples, several studies on real material demonstrated that benzodiazepines consumption during pregnancy may lead to an accumulation of these compounds in meconium. The developed LC-MS/MS methods reached a sensitivity considered sufficient to detect both parent drugs and metabolites. Nordiazepam and α-hydroxy-alprazolam, metabolites of diazepam and alprazolam, respectively, are measured at higher levels compared to the parent drugs [54]. In contrast to phase I metabolites, the concentration of the ones derived from conjugation, such as with glucuronic acid, seems to be negligible in meconium. Indeed, levels of parent drugs, measured after enzymatic or alkaline hydrolysis, do not significantly differ from the ones evaluated in nonhydrolysed meconium samples [54]. Stability of benzodiazepines represents another important aspect for analytical results. Storage at −20 °C is always advisable, whenever analyses are not carried out within a few days from sample collection. However, a good stability was observed for most of the benzodiazepines when stored at room temperature for a two-week period [48].

Table 12
Instrumental parameters and sensitivity for benzodiazepines analysis in meconium

Procedure	Column	Mobile phases	Flow rate (mL/min)	Mass Spectrometry	MS operating mode	# analytes	LOD (ng/g)	LOQ (ng/g)
1. [54]	C18 (100 mm × 2.1 mm, 3 μm)	2 mM ammonium formate (a) and acetonitrile (b): gradient elution	0.300	Triple quad with ESI^+ source	SRM (two transitions for analyte)	13	1.0–20.0	5.0–20.0
2. [56]	C18 (150 mm × 2.1 mm, 3.5 μm)	Acetonitrile, water and 100 mM ammonium formate buffer: isocratic solution	0.125	Triple quad with ESI^+ source	SRM (two transitions for analyte)	13	10.0	20.0
3. [55]	C18 (100 mm × 2.1 mm, 1.8 μm)	0.1% formic acid in acetonitrile (a) and 5 mM ammonium formate pH3 (b): gradient elution	0.350	Triple quad with ESI^+ source	SRM (two transitions for analyte)	8	1.5–7.0	5.0–20.0
4. [57]	Pentafluorophenyl (PFP) (100 mm × 2.0 mm, 3 μm)	0.1% formic acid with 2 mM ammonium acetate (a) and methanol (b): gradient elution	0.300	QTOF	High-resolution m/z in the range of 50–800	–	–	–

2.5 Synthetic Opioids and Synthetic Cannabinoids

None of the substances included in the class of fentanyl analogs and synthetic cannabinoids has been analyzed in meconium samples yet. There are several methods for the identification of traditional opioids (i.e. oxycodone, tramadol, etc.) [58, 59] and opiates (morphine, codeine, etc.) [57, 60–63]. Cannabinoids are largely studied in meconium; however, in contrast to benzodiazepines, the analytical approaches developed for monitoring the traditional drugs of abuse cannot be directly utilized for the detection of new opioids and cannabinoids. First, the sensitivity needed to confirm the exposure to most of the compounds is much higher: in particular, fentanyl analogs may be found in concentrations thousands of times lower than morphine. Also, several synthetic cannabinoids may exert a toxic effect at a very low dose. Yet, synthetic cannabinoids' chemical characteristics are not similar to those of phytocannabinoids, thus leading to the need for specific extraction procedures and instrumental analysis. Moreover, the mixed-mode cartridge based on strong cation exchange and nonpolar phase, that is widely used for pre-analytical treatments of biological matrices checked for the presence of NPS, may not provide adequate extraction efficiencies for synthetic cannabinoids. To conclude, it is not yet possible to describe an analytical protocol or provide any data about the potential determination of these classes of NPS.

References

1. Scriver CC (1998) A simple phenylalanine method for detecting phenylketonuria in large populations of newborn infants, by Robert Guthrie and Ada Susi, Pediatrics, 1963;32: 318–343. Pediatrics 102:236–237
2. Kulkarni P, Chaudhari GH, Sripuram V et al (2014) Oral dosing in adult zebrafish: proof-of-concept using pharmacokinetics and pharmacological evaluation of carbamazepine. Pharmacol Rep 66:179–183
3. Nijenhuis CM, Huitema AD, Marchetti S et al (2016) The Use of Dried Blood Spots for Pharmacokinetic Monitoring of Vemurafenib Treatment in Melanoma Patients. J Clin Pharmacol 56:1307–1312
4. Patel P, Mulla H, Tanna S, Pandya H (2010) Facilitating pharmacokinetic studies in children: a new use of dried blood spots. Arch Dis Child 95:484–487
5. Stove CP, Ingels AS, De Kesel PM, Lambert WE (2012) Dried blood spots in toxicology: from the cradle to the grave? Crit Rev Toxicol 42:230–243
6. Berm EJJ, Paardekooper J, Brummel-Mulder-E, Hak E, Wilffert B, Maring JG (2015) A simple dried blood spot method for therapeutic drug monitoring of the tricyclic antidepressants amitriptyline, nortriptyline, imipramine, clomipramine, and their active metabolites using LC-MS/MS. Talanta 134: 165–172
7. Moretti M, Visonà SD, Freni F et al (2018) A liquid chromatography-tandem mass spectrometry method for the determination of cocaine and metabolites in blood and in dried blood spots collected from postmortem samples and evaluation of the stability over a 3-month period. Drug Test Anal 10:1430–1437
8. Odoardi S, Anzillotti L, Strano-Rossi S (2014) Simplifying sample pretreatment: application of dried blood spot (DBS) method to blood samples, including postmortem, for UHPLC-MS/MS analysis of drugs of abuse. Forensic Sci Int 243:61–67
9. Chepyala D, Tsai IL, Liao HW, Chen GY, Chao HC, Kuo CH (2017) Sensitive screening of abused drugs in dried blood samples using ultra-high-performance liquid chromatography-ion booster-quadrupole time-of-flight mass spectrometry. J Chromatogr A 1491:57–66

10. Sadler Simões S, Castañera Ajenjo A, Dias MJ (2018) Dried blood spots combined to an UPLC-MS/MS method for the simultaneous determination of drugs of abuse in forensic toxicology. J Pharm Biomed Anal 147:634–644
11. Kyriakou C, Marchei E, Scaravelli G, García-Algar O, Supervía A, Graziano S (2016) Identification and quantification of psychoactive drugs in whole blood using dried blood spot (DBS) by ultra-performance liquid chromatography tandem mass spectrometry. J Pharm Biomed Anal 128:53–60
12. Ambach L, Hernández Redondo A, König S, Weinmann W (2014) Rapid and simple LC-MS/MS screening of 64 novel psychoactive substances using dried blood spots. Drug Test Anal 6:367–375
13. Wagner M, Tonoli D, Varesio E, Hopfgartner G (2016) The use of mass spectrometry to analyze dried blood spots. Mass Spectrom Rev 35:361–438
14. Spooner N (2013) A dried blood spot update: still an important bioanalytical technique? Bioanalysis 5:879–883
15. Sadones N, Capiau S, De Kesel PM, Lambert WE, Stove CP (2014) Spot them in the spot: analysis of abused substances using dried blood spots. Bioanalysis 6:2211–2227
16. Moretti M, Manfredi A, Freni F et al (2021) A comparison between two different dried blood substrates in determination of psychoactive substances in postmortem samples. Forensic Toxicol:1–9
17. Wilhelm AJ, den Burger JC, Swart EL (2014) Therapeutic drug monitoring by dried blood spot: progress to date and future directions. Clin Pharmacokinet 53:961–973
18. Jantos R, Veldstra JL, Mattern R, Brookhuis KA, Skopp G (2011) Analysis of 3,4-methylenedioxymetamphetamine: whole blood versus dried blood spots. J Anal Toxicol 35:269–273
19. Sosnoff CS, Ann Q, Bernert JT, Jr., Powell MK, Miller BB, Henderson LO, Hannon WH, Fernhoff P, Sampson EJ (1996) Analysis of benzoylecgonine in dried blood spots by liquid chromatography – atmospheric pressure chemical ionization tandem mass spectrometry. J Anal Toxicol 20:179-184
20. Garcia Boy R, Henseler J, Mattern R, Skopp G (2008) Determination of morphine and 6-acetylmorphine in blood with use of dried blood spots. Ther Drug Monit 30:733–739
21. Antelo-Domínguez Á, Cocho J, Tabernero MJ, Bermejo AM, Bermejo-Barrera P, Moreda-Piñeiro A (2013) Simultaneous determination of cocaine and opiates in dried blood spots by electrospray ionization tandem mass spectrometry. Talanta 117:235–241
22. Versace F, Déglon J, Lauer E, Mangin P, Staub C (2013) Automated DBS extraction prior to hilic/RP LC–MS/MS target screening of drugs. Chromatographia 76:1281–1293
23. Verplaetse R, Henion J (2016) Quantitative determination of opioids in whole blood using fully automated dried blood spot desorption coupled to on-line SPE-LC-MS/MS. Drug Test Anal 8:30–38
24. Holub M, Tuschl K, Ratschmann R et al (2006) Influence of hematocrit and localisation of punch in dried blood spots on levels of amino acids and acylcarnitines measured by tandem mass spectrometry. Clin Chim Acta 373:27–31
25. O'Broin SD, Kelleher BP, Gunter E (1995) Evaluation of factors influencing precision in the analysis of samples taken from blood spots on filter paper. Clin Lab Haematol 17:185–188
26. Meesters RJ, Hooff GP (2013) State-of-the-art dried blood spot analysis: an overview of recent advances and future trends. Bioanalysis 5: 2187–2208
27. Li W, Tse FL (2010) Dried blood spot sampling in combination with LC-MS/MS for quantitative analysis of small molecules. Biomed Chromatogr 24:49–65
28. Keevil BG (2011) The analysis of dried blood spot samples using liquid chromatography tandem mass spectrometry. Clin Biochem 44: 110–118
29. Thomas A, Geyer H, Schänzer W, Crone C, Kellmann M, Moehring T, Thevis M (2012) Sensitive determination of prohibited drugs in dried blood spots (DBS) for doping controls by means of a benchtop quadrupole/Orbitrap mass spectrometer. Anal Bioanal Chem 403: 1279–1289
30. Mercolini L, Protti M (2016) Biosampling strategies for emerging drugs of abuse: towards the future of toxicological and forensic analysis. J Pharm Biomed Anal 130:202–219
31. Wang Y, Shi Y, Yu Y et al (2020) Screening of synthetic cathinones and metabolites in dried blood spots by UPLC-MS-MS. J Anal Toxicol. https://doi.org/10.1093/jat/bkaa106
32. Mercolini L, Protti M, Catapano MC, Rudge J, Sberna AE (2016) LC-MS/MS and volumetric absorptive microsampling for quantitative bioanalysis of cathinone analogues in dried urine, plasma and oral fluid samples. J Pharm Biomed Anal 123:186–194

33. da Cunha KF, Eberlin MN, Costa JL (2018) Long-term stability of synthetic cathinones in dried blood spots and whole blood samples: a comparative study. Forensic Toxicol 36:424–434
34. Protti M, Rudge J, Sberna AE, Gerra G, Mercolini L (2017) Dried haematic microsamples and LC-MS/MS for the analysis of natural and synthetic cannabinoids. J Chromatogr B Anal Technol Biomed Life Sci 1044–1045:77–86
35. Chappell JS, Lee MM (2010) Cathinone preservation in khat evidence via drying. Forensic Sci Int 195:108–120
36. Alfazil AA, Anderson RA (2008) Stability of benzodiazepines and cocaine in blood spots stored on filter paper. J Anal Toxicol 32:511–515
37. Moretti M, Freni F, Tomaciello I et al (2019) Determination of benzodiazepines in blood and in dried blood spots collected from postmortem samples and evaluation of the stability over a three-month period. Drug Test Anal 11: 1403–1411
38. Ong RS, Kappatos DC, Russell SGG et al (2020) Simultaneous analysis of 29 synthetic cannabinoids and metabolites, amphetamines, and cannabinoids in human whole blood by liquid chromatography-tandem mass spectrometry – a New Zealand perspective of use in 2018. Drug Test Anal 12:195–214
39. Thomas A, Geyer H, Guddat S, Schänzer W, Thevis M (2011) Dried blood spots (DBS) for doping control analysis. Drug Test Analysis 3: 806–813
40. Shaner RL, Schulze ND, Seymour C, Hamelin EI, Thomas JD, Johnson RC (2017) Quantitation of fentanyl analogs in dried blood spots by flow-through desorption coupled to online solid phase extraction tandem mass spectrometry. Anal Methods 9:3876–3883
41. Clavijo CF, Thomas JJ, Cromie M et al (2011) A low blood volume LC-MS/MS assay for the quantification of fentanyl and its major metabolites norfentanyl and despropionyl fentanyl in children. J Sep Sci 34:3568–3577
42. Gareri J, Klein J, Koren G (2006) Drugs of abuse testing in meconium. Clin Chim Acta 366:101–111
43. Ostrea EM Jr, Romero A, Knapp DK, Ostrea AR, Lucena JE, Utarnachitt RB (1994) Postmortem drug analysis of meconium in early-gestation human fetuses exposed to cocaine: clinical implications. J Pediatr 124:477–479
44. Kwong TC, Ryan RM (1997) Detection of intrauterine illicit drug exposure by newborn drug testing. Nat Acad Clin Biochem Clin Chem 43:235–242
45. Bourget P, Roulot C, Fernandez H (1995) Models for placental transfer studies of drugs. Clin Pharmacokinet 28:161–180
46. Moore C, Negrusz A, Lewis D (1998) Determination of drugs of abuse in meconium. J Chromatogr B Biomed Sci Appl 713:137–146
47. Jiménez E, Marín ML, Martín R et al (2008) Is meconium from healthy newborns actually sterile? Res Microbiol 159:187–193
48. Wu F, Marin SJ, McMillin GA (2017) Stability of 21 cocaine, opioid and benzodiazepine drug analytes in spiked meconium at three temperatures. J Anal Toxicol 41:196–204
49. Christrup LL (1997) Morphine metabolites. Acta Anaesthesiol Scand 41:116–122
50. Pichini S, Rotolo MC, García J et al (2014) Neonatal withdrawal syndrome after chronic maternal consumption of 4-methylethcathinone. Forensic Sci Int 245: e33–e35
51. López-Rabuñal Á, Lendoiro E, Concheiro M, López-Rivadulla M, Cruz A, de-Castro-Ríos A (2019) A LC-MS/MS method for the determination of common synthetic cathinones in meconium. J Chromatogr B Anal Technol Biomed Life Sci 1124:349–355
52. Nemeškalová A, Bursová M, Sýkora D, Kuchař M, Čabala R, Hložek T (2019) Salting out assisted liquid-liquid extraction for liquid chromatography tandem-mass spectrometry determination of amphetamine-like stimulants in meconium. J Pharm Biomed Anal 172:42–49
53. Strange LG, Kochelek K, Keasling R, Brown SD, Pond BB (2017) The pharmacokinetic profile of synthetic cathinones in a pregnancy model. Neurotoxicol Teratol 63:9–13
54. López-Rabuñal A, Lendoiro E, Concheiro M, López-Rivadulla M, Cruz A, de-Castro-Ríos A (2020) LC-MS-MS method for the determination of antidepressants and benzodiazepines in meconium. J Anal Toxicol 44:580–588
55. Pichini S, Cortes L, Marchei E et al (2016) Ultra-high-pressure liquid chromatography tandem mass spectrometry determination of antidepressant and anxiolytic drugs in neonatal meconium and maternal hair. J Pharm Biomed Anal 118:9–16
56. Marin SJ, Coles R, Merrell M, McMillin GA (2008) Quantitation of benzodiazepines in urine, serum, plasma, and meconium by LC-MS-MS. J Anal Toxicol 32:491–498
57. Ristimaa J, Gergov M, Pelander A, Halmesmäki E, Ojanperä I (2010) Broad-spectrum drug screening of meconium by liquid chromatography with tandem mass

spectrometry and time-of-flight mass spectrometry. Anal Bioanal Chem 398:925–935

58. Le NL, Reiter A, Tomlinson K, Jones J, Moore C (2005) The detection of oxycodone in meconium specimens. J Anal Toxicol 29:54–57
59. Moore CM, Deitermann D, Lewis D, Leikin J (1995) The detection of hydrocodone in meconium: two case studies. J Anal Toxicol 19:514–518
60. Coles R, Kushnir MM, Nelson GJ, McMillin GA, Urry FM (2007) Simultaneous determination of codeine, morphine, hydrocodone, hydromorphone, oxycodone, and 6-acetylmorphine in urine, serum, plasma, whole blood, and meconium by LC-MS-MS. J Anal Toxicol 31:1–14
61. Ostrea EM Jr, Knapp DK, Tannenbaum L et al (2001) Estimates of illicit drug use during pregnancy by maternal interview, hair analysis, and meconium analysis. J Pediatr 138:344–348
62. Pichini S, Puig C, Zuccaro P et al (2005) Assessment of exposure to opiates and cocaine during pregnancy in a Mediterranean city: preliminary results of the "Meconium Project". Forensic Sci Int 153:59–65
63. Concheiro M, Lendoiro E, de Castro A et al (2017) Bioanalysis for cocaine, opiates, methadone, and amphetamines exposure detection during pregnancy. Drug Test Anal 9:898–904

Chapter 9

High-Resolution Mass Spectrometry in NPS Analysis

Sarah M. R. Wille, Jean-Michel Gaulier, Vincent Di Fazio, and Nele Samyn

Abstract

The gold standard for the new psychoactive substance (NPS) detection in toxicology is high-resolution mass spectrometry (HRMS), because of its flexibility concerning the detection of compounds that have a rapid turnover and a broad range of physicochemical properties. Therefore, this chapter discusses the choices toxicologists have to make during the analytical method development, the selection of the database and the data-acquisition parameters, as well as the development of workflows for data-interpretation (targeted and nontargeted). Moreover, issues that can occur during the validation process and quality assurance (QA) will be discussed and a validation procedure is proposed. The scientific community should improve the simplicity of the HRMS-procedures leading to less time-constraining methods and develop guidelines concerning identification criteria, as well as express requirements for adequate validation and QA (for the analytical method as well as the databases) to further improve its general routine use in a toxicological setting.

Key words HRMS, New Psychoactive Substances, Method development, Data acquisition, Targeted workflow, Untargeted workflow, Validation, Quality Assurance

1 Introduction

The detection of illicit and legal substances in biological samples is the basis of analytical toxicology performed in a clinical or forensic setting. One of the major challenges for toxicologists all over the globe is the continuous development of new psychoactive substances (NPS), making it difficult to obtain up-to-date techniques for monitoring these substances in bulk form or in biological samples. These NPS are defined by EU Council Decision 2005/387/JHA as psychotropic drugs in pure form or in a preparation that have not yet been scheduled under the 1961 or in the 1971 United Nations Single Convention on Narcotic Drugs and that may pose a threat to public health comparable to the substances listed in either schedule I, II, III, or IV [1]. In recent years, legislation has been introduced in many countries to address the NPS market, either via nominative or generic types of legislation

Marta Concheiro and Karl B. Scheidweiler (eds.), *Methods for Novel Psychoactive Substance Analysis*,
Methods in Pharmacology and Toxicology, https://doi.org/10.1007/978-1-0716-2605-4_9,

[2]. According to the EMCDDA 2020 drug report [3], new compounds are being introduced at a rate of more than 50 new drugs per year (since 2012), and include substances such as stimulants, synthetic cannabinoids, benzodiazepines, opioids, hallucinogens, and dissociatives. The most common NPS are synthetic cannabinoids and synthetic cathinones, together accounting for almost 80% of all NPS seizures in Europe and 40% in Australia [3, 4]. An increase in synthetic opioids has recently been observed. According to the UNODC, the emergence of synthetic opioids is worrisome as they account for most deaths attributed to drug use worldwide [5]. The very dynamic NPS market comprises a large number of substances which may have a short turnover time (enter and disappear from the drug market at a fast pace). By the end of 2019, the EMCDDA monitored around 790 NPS [3]. NPS enlarge the total drug market as it seems that they do not always replace classical drugs, but are often used by new groups of consumers. This is certainly the case for NPS with stimulant or hallucinogenic properties as their consumers are often seeking new drug effects. However, there are consumer groups that use these substances either as a substitute because they cannot obtain the classical drug easily (*e.g.* synthetic opioids *versus* heroin) or as a replacement to circumvent legislation (*e.g.* synthetic cannabinoids (SCRAs) *versus* cannabis). NPS consumption poses a serious problem not only for public order but also in terms of public health as their (side-) effects can have an enormous impact on the psychological and physical behaviour of its consumer.

The current challenges of a rapidly changing drug landscape make it clear that High-Resolution Mass Spectrometry (HRMS) has advantages over traditionally used techniques such as immunological analysis or even Liquid Chromatography (LC) coupled to Mass Spectrometry (MS). HRMS is defined by the Royal Society of Chemistry as "Any type of mass spectrometry where the 'exact' mass of the molecular ions in the sample is determined as opposed to the 'nominal' mass (the number of protons and neutrons)" [6]. HRMS has been identified as the method of choice for broad screenings of NPS in a wide range of analytical disciplines due to their main advantage: detection of analytes to the nearest 0.001 mass unit, thus via their "exact" mass, adding to the selectivity of the method. Detection of NPS is hampered via traditional MS techniques as certified reference materials (CRMs) for novel compounds are not easily at hand because the compounds have just appeared on the drug market and CRMs are thus not yet available. Moreover, it can be very costly for a laboratory to be up to date with the acquisition of CRMs and it can be too time-consuming to obtain them due to (inter)national regulations. As you need CRMs to develop targeted MS methods or to update existing ones, this can lead to a delay in the detectability of NPS for research or casework. In addition, the best target (*e.g.*, metabolites or degradation products) for the detection of a specific NPS may not always be known. HRMS can help handle the rapid proliferation of these (unknown) compounds, as an untargeted approach can be utilized looking into accurate mass. Today, a variety of HRMS techniques utilizing different technologies are commercially available with variable resolutions. These instruments include time-of-flight (TOF), Fourier transform ion cyclotron resonance (FT-ICR), and orbitrap (OT) mass analyzers [7, 8]. HRMS analyzers such as TOF are ideal to screen biological samples for (unknown) legal and illegal drugs because of their very high mass range (currently up to or higher than ± 500 kDa), their ability to collect a large number of mass spectra per second, and the fact that their resolution is independent of acquisition time. Moreover, HRMS techniques provide the possibility to thoroughly characterize each individual compound. The nature of the obtained data also makes a retrospective analysis possible. While low-resolution mass spectrometry is still the standard technique in clinical and forensic toxicology for quantitative analysis due to the ease of operation and cost, HRMS is becoming the gold standard for a nontargeted broad screening. The importance of HRMS will only increase in the future, certainly when the instrument prices decrease and the data interpretation becomes more user-friendly [9, 10].

HRMS for the detection of NPS is used in several ways: (1) (un)targeted screening and/or quantification of NPS in bulk samples or biological matrices or (2) elucidation of its metabolites which are then used for qualitative or quantitative determination of the compounds during initial analysis or in a retrospective way. Elucidation of NPS metabolization pathways is of utmost importance for an adequate detection of these compounds and the final interpetation of the obtained toxicological data. Therefore, a different chapter will be attributed to this topic, while the current chapter will focus on the development, the workflow, and validation of HRMS NPS detection methods.

2 HRMS Method Development and Acquisition Method

HRMS is a versatile analytical technique that is applied in different configurations with various ionization sources and data acquisition capabilities. Gas chromatography (GC), as well as LC, capillary electrophoresis (CE) [11], and direct sampling techniques *i.e.*, flow injection analysis (FIA), desorption electrospray ionization (DESI), matrix-assisted laser desorption ionization (MALDI) [12], and direct analysis in real-time (DART) [13, 14], can be combined with HRMS such as TOF, FT-ICR, and OT [8, 15]. As several NPS detection and quantification methods in specific biological matrices (*e.g.*, oral fluid, hair) will be described in detail in other chapters of this book, this chapter will rather focus on the workflow of the NPS detection with HRMS.

2.1 Sample Preparation

Sample preparation, meaning the optimization of the extraction of the compounds-of-interest from the biological matrix, is a critical step in the development of any analytical method, as it will affect the overall sensitivity of the technique. Sample preparation, however, will not only lead to highly concentrated extracts but also can remove potentially interfering matrix compounds, resulting in enhanced selectivity and a more reproducible method independent of variations in the sample matrix. In recent years, sample preparation techniques often look into reducing sample volume, analysis time, and cost, and to automation in order to encourage a reduced workload.

NPS are determined with HRMS in different matrices, such as seized samples [12] (*e.g.*, blotter paper [16]), urine [17–19], (*post-mortem* (PM)) blood [11, 20–22], hair [11, 20, 23–25], oral fluid [26, 27], and wastewater [28–30]. While these matrices have different properties resulting in a different approach to extract the compounds of interest, often sample preparation is very simple to ensure the detection of a multitude of compounds. Of course, if the focus is a certain group of compounds, with the same physiochemical properties, then a more selective type of sample preparation (*e.g.*, solid phase extraction (SPE) with a sorbent focusing on ion-exchange [19]) can be applied.

Sample preparations consist often of a protein precipitation [22], a dilution [17] or in the case of hair extraction, the extraction solvent itself [23]. However, liquid/liquid extraction (LLE) is also still popular either in a classic setup [11, 20, 25] or via techniques such as pressurized liquid extraction combined with dispersive liquid–liquid microextraction (PLE-dLLME) [24]; acetonitrile with a salting-out assisted LLE (SALLE) [21] or supported liquid extraction (SLE) [19]. Diamanti et al. used a generic four-sorbent SPE protocol to be able to retain compounds with wide physicochemical properties [28]. SPE is also often used in waste-water analysis as larger volumes of sample are applied here [28–30]. Another observation in the literature is the miniaturization of the sample preparation step due to the sensitivity of HRMS. The reduction of the necessary sample volume leads to a great advantage in the field of forensic toxicology, where the available sample volume for a full systematic toxicological analysis (STA) can be limited. When using HRMS, even a volume of only 100 μL blood or vitreous humor can result in an adequate screening of a wide range of compounds (Fig. 1). A step further is the development of mass spectrometric detection techniques without the need for sample preparation. DART, for example, allows the analyte to be ionized via desorption from the sample surface through interaction with metastable species generated from the ion source. This technique, however, is not applied to biological matrices but rather to seized samples [31]. Gwak et al. used a DART ionization source

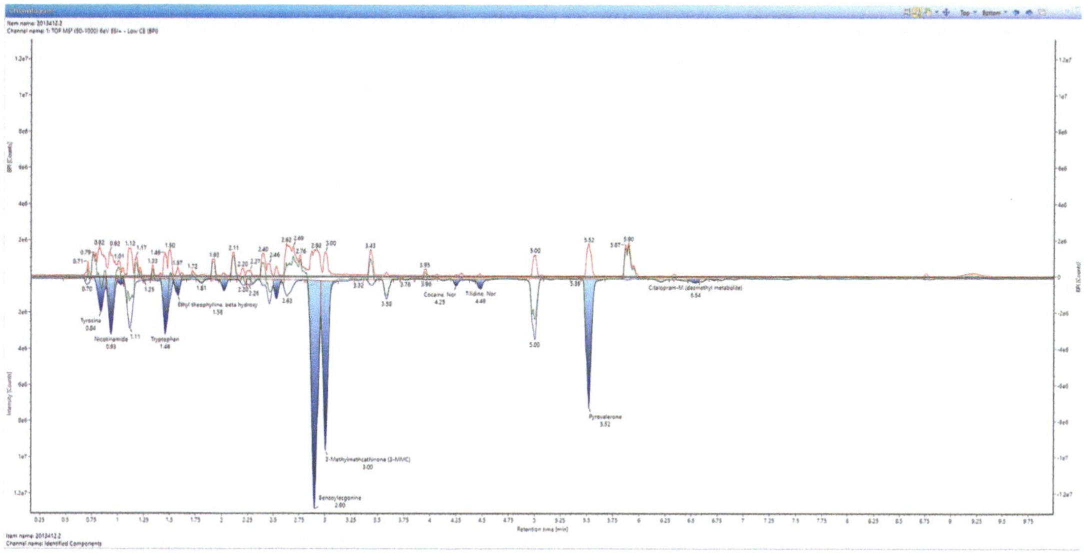

Fig. 1 CASE 1: Minimal volume necessary for sensitive and broad-scale screening. In forensic casework, the available amount of blood sample can be limited. While the analysis of other autopsy samples (*e.g.,* liver, kidney, bile) will give an overview of relatively recent substance use, they will not necessarily link to the cause of death. The use of a sensitive HRMS technique can give you a good indication of the compounds present in the sample and can lead you towards the next steps: which compounds need to be quantified (*e.g.,* via HRMS or via a separate LC-MS/MS target method) and which candidates need to be elucidated. In this case example, a post-mortem blood sample (100 μL) after a protein precipitation with acetonitrile, was found positive for several compounds. Suspect compounds were identified via the in-house library: tyrosine (RT 0.84 min), nicotinamide (RT 0.93), tryptophan (RT 1.46), beta hydroxyl ethyl theophylline (RT 1.58; Internal Standard), benzoylecgonine (RT 2.9), 3-methylmethcathinone (RT 3.00), pyrovalerona (RT 5.52). Via a mirror plot candidate peaks for possible elucidation can be found. This approach is also known as the top-down approach, which is best described by selecting the abundant peaks from the visual inspection of a Total Ion Current (TIC) or Base Peak Intensity (BPI) chromatogram with a plot indicating which peaks are already identified via the applied database. Several candidates for possible elucidation can be observed at RT 1.11, 3.58, and 5.00

and coupled ion-mobility spectrometry (IMS) with a Q-TOF-MS for the analysis of 35 NPS (mostly synthetic cathinones and cannabinoids) [14].

2.2 Separation Method

Conventional chromatographic methods such as reversed-phase LC and GC are commonly used in toxicological analysis as this leads to an adequate chromatographic separation for complex matrices. Other methods that are applied are CE [11], direct sampling techniques (mainly for seized sample analysis) [12–14], and IMS [14, 31]. However, in the field of toxicology, LC has been the most applied separation technique, because it is well-known in and very accessible by forensic and clinical laboratories. Moreover, NPS have the physicochemical properties for an adequate LC analysis (*e.g.*, not-volatile). In general, a lot of methods have been published that determine a wide range of compounds including NPS. Pasin et al. have reviewed the use of the above-mentioned techniques for the detection of NPS in seized as well as biological samples [15]. Therefore, in Table 1, only recently published methods for the detection of NPS are presented. The development of the chromatographic separation for HRMS methods is not different from classical GC or LC methods with other types of detectors. It is important to have knowledge concerning the general physicochemical properties of the compounds of interest and to adapt the method to these properties. Moreover, knowledge of the metabolism of NPS is a prerequisite for the development of reliable analytical methods. If for example SCRAs should be monitored in urine, the method developer should be aware that these compounds are severely metabolized with minimal to no unchanged parent compound presence in blood or urine; so, metabolites identification is crucial to detect SCRAs exposure [18].

Another important issue to take into consideration is the necessity of adequate separation for some compounds. While analysts may have assumed that accurate mass could make analytical separation unnecessary, this idea has proven to be wrong. Precise identification of a specific NPS is of utmost importance not only because they differ in potency and thus potential toxicity but also because their legal status can differ. However, this identification can sometimes be challenging because of the existence of (structural) isomers having the same molecular formula (*e.g.*, morphine/hydromorphone, temazepam/clobazam, 3,4-methylenedioxymethamphetamine (MDMA)/methedrone, methamphetamine/phentermine). In order to solve this issue, retention time can be an essential identification criterion. Structural isomers that produce very similar mass spectra should be identified as individual isomers; otherwise structural isomers should be reported as "isomer of X." Of course, during method development one should look for unique MS/MS ions to monitor if possible. In that way, Scheidweiler et al. [18] have separated the two isobaric and co-eluting SCRAs, JWH-019 hydroxyhexyl, and JWH-122 hydroxypentyl, in their Sequential Window Acquisition of All Theoretical Mass Spectra (SWATH) QTOF method. Another idea for correct identification, in wide-range screening methods for which separation of all compounds cannot be guaranteed, is the use of related compounds such as metabolites or degradation products. Skov-Skov Bergh et al. looked into the separation of cyclopropylfentanyl and crotonylfentanyl, two structural fentanyl analogue isomers which have almost identical mass spectra when analyzed with an MS-technique [20]. Using a BEH-column and an adequate gradient, both compounds could be separated via UHPLC-MS/MS, moreover, via UHPLC-QTOF-MS after human liver microsomes (HLM) incubation, they were able to detect unique metabolites for each isomer. Additionally, recent implementation of ultra-high performance super- or subcritical fluid chromatography (UHPSFC) with HRMS in the field of metabolomics could be an elegant answer to NPS isomers identification

Table 1
Recently published NPS HRMS detection methods

NPS	Matrix	Sample prep	Separation	Workflow	System	Data analysis	References
Fentanyl-analogues	Hair (50 mg)	Dilution	UHPLC Phenomenex Kinetex C18 (100 × 2.1 mm, 1.7 μm)	Data acquisition in nontargeted; data processing targeted m/z range 230-450 SWATH	SCIEX X500R QTOF with Turbo V™	SCIEX 0S 1,5	[23]
DOA+ NPS	Hair	PLE with dLLME	UHPLC Phenomenex Kinetex PFP (100 × 2.1 mm, 2.6 μm)	Targeted PMR acquisition mode Resolution 35,000FWHM	THERMO QExactive mass spectrometer Orbitrap	/	[24]
DOA + NPS	Hair	Extraction with phosphate buffer and LLE	UHPLC Thermo Accucore™ phenyl-hexyl (PH, 100 x2.1, 2.6 μm) A ccucore™ pentafluoro-phenyl (PFP, 100 × 2.1 mm, 2.6 μm)	Untargeted Mass range 125–650 and 125–1000 m/z DDA	THERMO Q-Exactive with HESI operating in PIM	Compound Discoverer 2.0 software (Thermo). Identification by comparison with mzCloud™ data	[25]
Cyclopropylfentanyl and Crotonyl-fentanyl	PM blood	LLE	UHPLC Waters Acquity HSS T3 (100 × 2.1 mm, 1.8 μm)	Targeted; Untargeted for metabolites m/z range	AGILENT 6500 iFunnel QTOF (PIM)	Masshunter Qualitative Analysis (version B.07.00, Agilent)	[20]

(continued)

Table 1 (continued)

NPS	Matrix	Sample prep	Separation	Workflow	System	Data analysis	References
				50–1000 DDA			
DOA + NPS	PM blood	Protein Precipitation with acetonitrile	UHPLC Waters Acquity HSS T3 column (100 mm × 2.1 mm, 1.8 μm)	Targeted m/z range 50–1000 DDA	AGILENT 6540 QTOF	MassHunter Acquisition software B.05.01	[47]
DOA + NPS	(PM) blood	LLE	UHPLC Waters Acquity BEH C 18 (50 × 3.0 mm, 1.7 μm)	Untargeted and targeted (quantitative) m/z range 100–1000 m/z range 40–700	AGILENT 6545 QTOF	MassHunter Qualitative and Quantitative Analysis version B07	[48]
Ethylphenidate and Mephedrone	Urine	Dilution	UHPLC Thermo Accucore™ P henyl-Hexyl column (100 × 2.1 mm, 2.6 μm)	Targeted m/z range 70–750	THERMO Q-Exactive	TraceFinderTM software version 3.3	[17]
DOA + NPS	Dried saliva spots	Methanol soak	LC	Targeted	WATERS Xevo-G2-QTOF-XS	/	[27]
NPS	Waste-water	SPE	UHPLC Thermo Acclaim RSLC 120 C18 (100 × 2.1 mm, 2.2 μm)	Untargeted and targeted (278 NPS) m/z range 50-1000 DIA and DDA	BRUKER Maxis Impact QTOF	TASQ Client 1.0 Bruker and DataAnalysis 4.3	[28]

NPS	Waste-water	SPE	UHPLC Phenomenex Kinetex PFP column (100 × 2.1 mm, 1.7 μm) Shimadzu LC-30CE	Targeted (187 NPS) m/z range 50-600 SWATH	SCIEX Triple TOF 5600	MultiQuant 2.1.1 PiekView 2.2.	[29]
DOA + NPS	Waste-water	SPE	HPLC XBridge®C18 (100 × 2.1 mm, 3.5 μm)	Targeted and nontargeted m/z range 100-500 DIA	THERMO Q-Exactive hybrid quadrupole-orbitrap	Thermo Xcalibur™ 4.0 software Trace Finder™ 3.1	[30]
DOA + NPS	Seizures	MALDI	–	Targeted; non targeted; quantitative m/z range 100-1000 DDA	THERMO MALDI-LTQ-Orbitrap XL	Tracefinder (Thermo Scientific), Xcalibur, MSD Enhanced ChemStation (Agilent)	[12]
NPS	Seizures	Methanol desolvation	UHPLC Thermo Accucore™ C 18 (30 × 2.1 mm, 2.6 μm)	Untargeted m/z range 50-850 DDA	THERMO QExactive O rbitrap MS With HESI	MSConvert, GNPS network, Cytoscape 3.8	[35]

DDA data-dependent acquisition, *DIA* data-independent acquisition, *dLLME* dispersive liquid microextraction, *HESI* Heated electrospray ionization, *HSS* high strength silica, *m/z* mass to charge ratio, *PIM* positive ionization mode, *PLE* pressurized liquid extraction, *PM* post-mortem, *PMR* parallel-reaction monitoring, *SWATH* sequential window acquisition of all theoretical mass spectra

[33]. According to Polettini et al., the strong correlation between log *P* and the LC retention, as well as the correlation between the electrophoretic mobility and the calculated pK_a will be useful to discriminate different isobaric compounds in the future [11]. Moreover, IMS combined with Q-TOF-MS can prove to be of exceptional value in toxicological analysis as it can separate isomers and isobars that may exhibit the same chromatographic retention times [14]. IMS is an analytical technique based on the gas phase separation of ionized compounds under an electric field at ambient pressure and is one of the UNODC recommended analytical methodologies to be used for testing of illicit drugs in seized materials [32].

2.3 Data Acquisition

HRMS combines the full scan accurate mass MS data acquisition method with MS/MS data either via tandem-in-space such as the Quadrupole TOF (QTOF) or tandem-in-time such as the Orbitrap instruments. Thus, combining all the different scan modes of a "classical" quadrupole with the fast analysis speed and the high mass resolution of the HRMS instruments, these instruments are of interest for a successful screening and quantification in toxicology because of the capacity to perform both targeted and nontargeted data acquisition with a high selectivity. However, other HRMS equipment such as the Hybrid Quadrupole-Orbitrap (Q exactive) [22, 24] are also applied.

The high mass accuracy and resolution enable measurement of the elemental composition of molecules and fragments, which can be used to elucidate possible structures of unknowns. HRMS is thus a useful tool in cases of intoxication related to a compound that is yet unknown, even more so in the case of NPS as this drug market has a very fast turnover. In the nontargeted data acquisition mode, data can be collected without any knowledge of the identity of the compounds in the sample. This acquisition mode can be divided into two groups: data-dependent acquisition (DDA) and data-independent acquisition (DIA).

In DDA mode, the intact precursor ion is first measured in the quadrupole, and then fragmented in the collision cell before subsequent analysis produces high-resolution product ion spectra. There are predefined criteria such as signal intensity in DDA, which determines which ions are selected for fragmentation. One disadvantage of DDA is that low intensity ions can be missed or lost in the signal of highly abundant peaks, and thus that potentially relevant analytes of low abundance may not be selected for further fragmentation.

In DIA, all ions of a specified mass to charge (*m/z*) range are fragmented with no other predefined criteria, which allows collection of comprehensive spectral data. When comparing the MS^2 spectra in DIA and DDA, the spectra quality seems to be higher in DDA, as there is less "background" noise [34]. Nevertheless, DIA techniques such as MS^E or SWATH allow a comprehensive collection of both precursor and product ion spectra. MS^E works by alternating low and high collision cell energy with precursor ion spectra being obtained at the low energy mode and precursor ion fragment spectra being acquired at the high energy mode. In SWATH acquisition mode, data are collected by an initial precursor ion survey scan, followed by product ion generation using many consecutive isolation windows, ensuring that the defined mass range is stepped through continuously during each scan cycle. SWATH leads to an increased selectivity and provides clearer MS/MS spectra, enabling better sensitivity. DIA generates significantly larger amounts of data compared with DDA. As a result, data interpretation is a bottleneck and therefore an adequate data platform is necessary. Laboratories must therefore have computer systems capable of managing this high amount of data. Indeed, several bio-informatics issues remain to be addressed in processing DIA data, and the development of more effective computational programs, such as deconvolution algorithms, is highly demanded to exploit the full potential of DIA data.

The major issue is that current software cannot associate product ions with their correct precursor ions. In this way, molecular networking can be a useful complement to conventional approaches for untargeted screening interpretation, for example, for NPS identification or NPS metabolism elucidation [35, 36]. By performing nontargeted screening using DIA with MS^E/SWATH, spectra on all parent ions and their fragments can be collected and a possible drug candidate can be found by searching databases with spectra and retention times of various drugs of abuse, including NPS. These large data files can be used in a retrospective way. The most abundant fragments of the compound can also be useful for identification if they are characteristic for the compound or a group of compounds [25, 37, 38]. Moreover, if the compound is not found in the database, possible chemical formulas can be elucidated from the mass of the compound, and online searches for its identity can be performed.

3 HRMS Data-Processing Workflow and Acceptance Criteria

The workflow concerning the analysis of the HRMS data and the acceptance criteria for identification are crucial and should be defined in detail. Data-processing workflows and acceptance criteria for compound identification can vary depending on the instrument type and vendor. Moreover, data-processing can occur in a targeted or nontargeted fashion. Most workflows, even with untargeted data acquisitions, start with a targeted data processing approach, in the sense that the acquired data are compared to an in-house library built up with CRMs (target screening) or a commercially available one with a CRMs confirmation afterwards

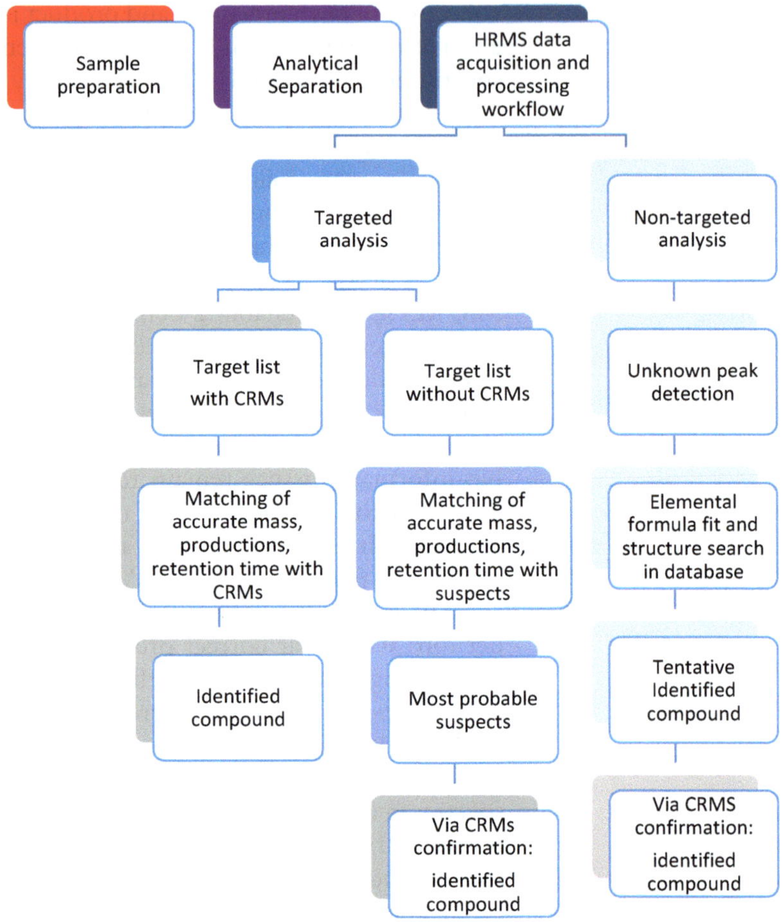

Fig. 2 Schematic representation of the HRMS workflows

(suspect screening). In a final phase, unknown peaks are evaluated via software processing to determine potential candidate formulas and structures using the experimentally determined high resolution and accurate mass of the detected unknown peaks (Fig. 2).

3.1 Targeted Analysis

For the first step, the targeted data processing approach compound identification is done via library matching involving the comparison of accurate mass, isotope pattern, and unique product ion spectra generated from the unknown sample with a database, containing this information of known compounds, using sophisticated computer algorithms and a scoring system.

If an in-house library is created, available CRMs are thus injected into the system to determine the retention time and adequate fragments to monitor. A major benefit of the use of spectral library matching in screening applications is the potentially unlimited number of compounds that can be included in a library, providing large comprehensive screening profiles. However, sometimes the NPS that are previously described in the literature and are available as CRMs are already outdated. Moreover, as NPS can be intensively metabolized it is crucial to determine the correct targets for an adequate detection in order to increase the compounds detection window in biological matrices (Fig. 3) [39].

	Name	Urine
Parent	U-47700	100
M01	*N*-Desmethyl	**660**
M02	*N,N*-Didesmethyl	**530**
M03	*N*-Desmethyl-hydroxy	220
M04-A	*N,N*-Didesmethyl-hydroxy	326
M04-B	*N,N*-Didesmethyl-hydroxy	260

Urinary chromatographic peak area of molecular ion ([M+H]+) of substances of interest normalized to chromatographic peak area of U-47700 in the fatal case

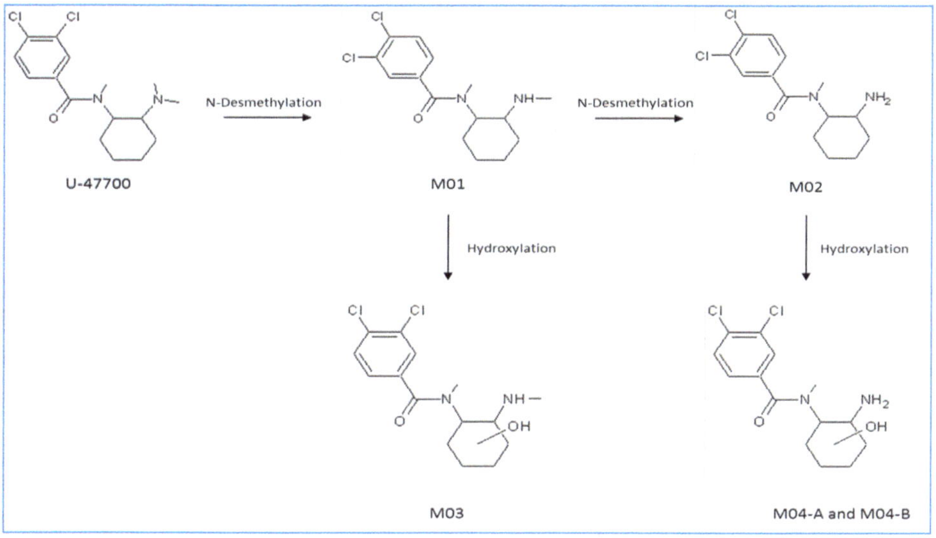

Fig. 3 CASE 2: Relevance of NPS metabolites identification in intoxication cases [39]. U-47700, an emerging new synthetic opioid, was identified in a fatal case. U-47700 metabolites were produced in vitro using human liver microsomes and their mass spectra were added in the MS/MS and HRMS libraries. U-47700 was identified in the powder obtained at the crime scene and in the peripheral PM blood (3.0 μg/L). In addition, high amounts of several U-47700 metabolites, especially *N*-desmethyl-U-47700, were observed in urine. Even if metabolite formation may largely depend on the enzymatic activity as well as on the length of the survival time, confrontation of these results to data found in the scientific literature strongly suggests that this metabolite is regularly a better blood and (mainly) urine biomarker of U-47700 intake than U-47700 itself. Indeed, in this fatality, *N*-desmethyl-U-47700 produced the main observed chromatographic signal (i) systematically in vitro and (ii) commonly in vivo, especially in urine samples. *N,N*-didesmethyl-U-47700 also seems to be a better biomarker of U-47700 intake than U-47700 itself

Another point to make is that maintaining a reference standard library for compounds that are only encountered either briefly or infrequently is not feasible for most laboratories due to constraints on costs and resources. Therefore, worldwide there are efforts to gather and share spectral data via online and commercially available and free libraries such as mzCloud^{TM}, GNPS, Metlin, SWGDRUG, Merck Index, NIST Chemistry Webbook, PubChem Compound [11, 25, 40, 41]. An interesting database is HighResNPS that is built up in such a way that it can be easily converted to an NPS screening library compatible with many vendor screening software, because it is based on diagnostic fragment ions [42, 43]. Each data entry of the HighResNPS must have the protonated molecular ion and possibly one to three diagnostic fragment ions. Moreover, each user inserting a compound in the database adds information concerning their analytical setup; in that way, retention times together with MS data can provide an increased confidence in detection (of course for the users with the same analytical setup) [40]. As analyzed samples may have varying degrees of complexity because of differences in matrix, increased identification confidence and/or the decrease of potentially false positive (FP) identifications can be established by implementation of predicted retention times as described by Diamanti et al. [44]. Some vendors even provide a complete methodology and compound database with their equipment (*e.g.*, Waters Forensic Toxicology Screening Application with UNIFI; 1700 compounds). However, when applying such external libraries or databases without the confirmation of CRM, one should be careful with reporting the data as being identified as demonstrated by Allen et al. [45]. They observed that the identification of 4′-Methyl-α-pyrrolidinopropiophenone (4-MePPP), a synthetic cathinone, using the Waters UNIFI library – with acceptance criteria including mass error, retention time error, and the number of expected fragments detected following collision-induced dissociation of the precursor ion – was not confirmed by subsequent CRM analysis. The compound resulting in this library match was actually a metabolite of tramadol: *N,N*-bisdesmethyl-tramadol. Addition of ion ratio tolerances for 4-MePPP to the library was the most effective approach to eliminate this possible false positive (FP) result. While FP hits can thus be observed when applying external made databases, application of such databases is the only way a laboratory can move forward at a non-time- and cost-consuming way. Polettini et al. concluded that the probability that a search retrieves a different compound having identical chemical formula is higher in larger databases (such as *e.g.*, PubChem Compound) in comparison to smaller databases [11]. The average number of hits with identical chemical formula in the database they evaluated was 1.82 ± 2.27 (range 1–39), typically decreasing at increased mass. As a result, the applied database should be evaluated.

The workflow and identification acceptance criteria are dependent on the manufacturer of the applied instrument and data interpretation software. A common software approach for compound identification includes a decision tree algorithm, where individual thresholds for each analyte characteristic must be met to provide a positive identification. The selection of appropriate identification criteria -especially for which no CRM is available- is a controversial topic. Identification criteria can use library matching scores such as fit and reversed fit or defining a minimum number or structural-specific that must be present to obtain an objective identification. Colby et al. compared four data analysis workflows (one targeted and three suspect screen workflows) for a panel of 170 drugs and metabolites in 100 urine samples [40]. They observed that retention time was not necessary for drug identification; when using accurate mass, isotope pattern, and product ion spectrum, a library matching was able to detect more than 80% of drugs present in the sample [40]. The factor that was of utmost importance was the inclusion of product ion spectral matching to decrease FP as well as false negative (FN) results; a weighting of 70% for product ion spectrum, 20% for accurate mass and 10% for isotope pattern was deemed to be best. Accurate mass and isotope pattern are intrinsic to a compound and can be predicted by their chemical formula. Product ion spectra can also be predicted via the molecular structure using a stand-alone in silico prediction program such as Mass Frontier or platform-specific options even though the applied algorithms are not perfect and can be time-consuming [40]. It is clear from the work of Colby et al. that looking into accurate mass alone or in combination with isotope patterns are not efficient options as they generate many FP matches. FP results occur because some compounds practically have the same exact mass (*e.g.*, scopolamine and cocaine) and are structural isomers (*e.g.*, phentermine and methamphetamine). FP results can also occur for compounds with simple fragmentation patterns (*e.g.*, norbuprenorphine). Some of these false matches maybe discovered when looking into the data manually; however, this will be very time-consuming. The attempts to confirm the found matches will often lead to higher costs as "wrong" CRMs will be purchased for confirmation. Additionally, one should be aware when generating product ion scans to not only collect spectra obtained with lower collision energy as this can lead to more FP results as fewer fragment ions are produced and as a result it is harder for the library search algorithm to accurately determine the match.

In the workflow suggested in the Waters UNIFI TOF systems, compound identification is based on mass accuracy (less than 5 ppm error) of the precursor and product ion, an average root-mean square (RMS) isotopic match intensity within 20% compared with the theoretical ratio and retention time within 0.3 min. These criteria are often combined with a criterion considering an arbitrary signal intensity threshold (*e.g.*, higher than 10,000). Data processing performed using SCIEX Software for positive analyte identification are based on four main confidence criteria: precursor and fragment mass error should be within ± 5 ppm, a library hit score, and a combined score calculated on the third confidence categories (MRIL) with custom weightings [23, 40, 41]. In the publication by Scheidweiler et al. [18] using the ABSciex software, the following identification criteria were applied: mass error for the molecular ion <15 ppm, retention time error <2%, isotopic pattern <80% difference from the theoretical, and library fit >60%. An additional weighting of respectively, 25%, 25%, 10%, and 40% was applied to the above-mentioned identification criteria. When applying the above-mentioned criteria, one should always be aware that overloaded peaks could result in false-negative results as specific criteria such as retention time and isotopic match intensity could be out of range. Therefore, it is of interest to run samples with two different injection volumes: one to eliminate possible "overloaded" peaks of compounds that are (ab)used in higher doses and one to be sure low-dosed compounds can be detected with adequate sensitivity.

3.2 Nontargeted Analysis

While this first step in the workflow can be very time-efficient for forensic and clinical casework, the second step in the workflow – untargeted data processing – is necessary to identify the "unknowns." HRMS techniques have the advantage to monitor accurate mass and can thus be developed in such a way that unknowns can be elucidated and confirmed with a CRM afterwards. When elucidating unknown peaks, chemical formulas are proposed by the applied elucidating software using accurate mass, double bond equivalents (or degree of unsaturation), and spectra obtained via collision-induced dissociation (CID) and comparing these results with available databases such as FDAUNII and Drugbank. Often extensive databases such as ChemSpider are used to verify whether the predicted in silico fragmentation pattern of the candidate structures corresponds to the experimental obtained MS/MS spectra. ChemSpider can generate candidate structures for each formula with a matching of HRMS and MS/MS spectra to predicted fragment ions [46].

As a lot of data are generated when using untargeted screening methods, the first problem to tackle is which compounds to elucidate. This is really a crucial step and *top-down* as well as *bottom-up* approaches are observed. In a top-down approach, the most abundant nonidentified peaks in the Total Ion Chromatogram (TIC) are selected for further elucidation (*e.g.*, mirror plot Fig. 1). In the bottom-up approach, characteristic class-specific mass spectral information such as common product ion and/or neutral losses are used to select specific targets. Klingberg et al. have demonstrated the utility of monitoring common fragments via their study looking into the CID of synthetic opioids to obtain diagnostic product ions which were used to identify opioid classes in a non-targeted workflow [38]. Allard et al. [36] use molecular networking to visually display the link between different compounds in the generated screening data: each ion in their fragmentation spectra is indicated, whereas structural relationships are also indicated. This technique could be able to help out when looking into the identification of new metabolites or structural analogues. Both approaches, top-down or bottom-up, have their pros and cons. In the top-down approach, less abundant peaks or peaks that are co-eluting with highly concentrated compounds or endogenous substances could be missed. While the bottom-up approach requires DIA and needs data processing software capable of following up numerous characteristic product ions in screening methods. Moreover, before even attempting the elucidation of unknowns, the elimination of endogenous or background peaks can be of interest by using data processing algorithms to subtract the TIC data from a blank or representative pooled matrix from the TIC of the authentic case file of interest or via a technique used in metabolomics which involves the binary comparison of *m/z* data from representative matrix blanks and authentic samples to determine exogenous compounds [15].

Mollerup et al. have described a workflow for unidentified candidate masses in five steps [22] when using the Forensic Screening Solution provided by UNIFI Waters. First, (1) candidates were filtered via their observed *m/z* (no common contaminant masses; masses no more than five times present in a batch), retention time (between 1.14 and 11.00 min.), response (higher than 1200; higher than 1/10 of a co-eluting compound), and identification status (response threshold higher than 10 times blank samples). While the set criteria should maybe be altered depending on your instrument type, as this step is based on low collision energy acquired data, this step should be applicable to any full-scan HRMS acquisition, independent of the product ion acquisition mode. Secondly, (2) the molecular formula was elucidated via the UNIFI "chemical composition" tool which uses the *i*-FIT^{TM} algorithm (based on mass and isotope intensity error). Only compounds with an *i*-FIT higher than 10% could go to the next step in the process. This next step (3) was the comparison of the candidates with external structure databases imported into the UNIFI software. Common fragmentation and common neutral losses were evaluated and the observed mass differences between precursor ions in low collision energy are used to elucidate possible structural similarities. Then (4), the candidate structures are in silico fragment matched via the UNIFI software, using a combinatorial fragmentation approach attempting to match observed product ions to candidate structures by disconnecting covalent bonds in the candidate structure. This gives a ranking of possible structures with the most probable one, the one with the highest predicted intensity (%-

matched intensity of total intensity) and the highest number of matched product ions. Finally, a tentative identification (5) is achieved by searching for the selected candidate structure in the literature. It is clear that the overall process time for the (tentative) identification of unknowns is time-consuming. According to Mollerup et al. it takes 5–30 min per sample for an experienced user [22].

3.3 Retrospective Data Analysis

Another possibility in HRMS, in contrast to analytical methods based on single ion monitoring or multiple reaction monitoring, is the use of retrospective or post-target data analysis if the data acquisition is done in a nontargeted way. New library entries or diagnostic fragment ions from emerging NPS or drug classes can be used to screen for these substances in "older" case work, via reprocessing of these data files using updated spectral databases without the need for reextraction or reanalysis [47, 48]. This is of major interest for epidemiological analysis or in specific clinical or forensic cases for which the observed effects or actions could not be related to the compounds detected in the sample (Fig. 4). However, clear criteria for case work should be set on how to handle the always enhancing knowledge of NPS compounds or their targets (*e.g.* metabolites, degradation products) as possible ethical dilemmas can occur in a judicial or even clinical process.

3.4 Screening vs. Quantitative Analysis

The above-mentioned workflows are qualitative workflows, looking into the identification of the compounds. HRMS instruments are seen as less sensitive and less exact for quantification purposes as compared to "classical" LC-MS/MS techniques. Thoren et al. [49] have compared matrix effects and sensitivity of a tandem quadrupole TOF versus a triple quadrupole linear ion trap (QqLIT), as well as their ability to identify drugs in 100 routine urine samples. While the QqLIT was more sensitive, it did not significantly affect compound identification. Elmiger et al. [50] have validated a quantitative HRMS method with SWATH acquisition in which they applied protein precipitation to post-mortem blood (200 μL). Twenty-four substances met the required validation criteria and could be quantified with acceptable accuracy, while in addition 52 other substances were monitored. In the study, different calibration models (full, one-point external and one-point internal) to achieve a reliable and practical method for triage of post-mortem cases (clearance versus full autopsy/toxicology) were also evaluated. The developed method was successfully applied to more than 100 blood samples and the one-point calibration resulted in a very time-saving technique. As a result, neither sensitive screening nor adequate quantification is thus problematic, but will be compound dependent and should be evaluated and validated thoroughly.

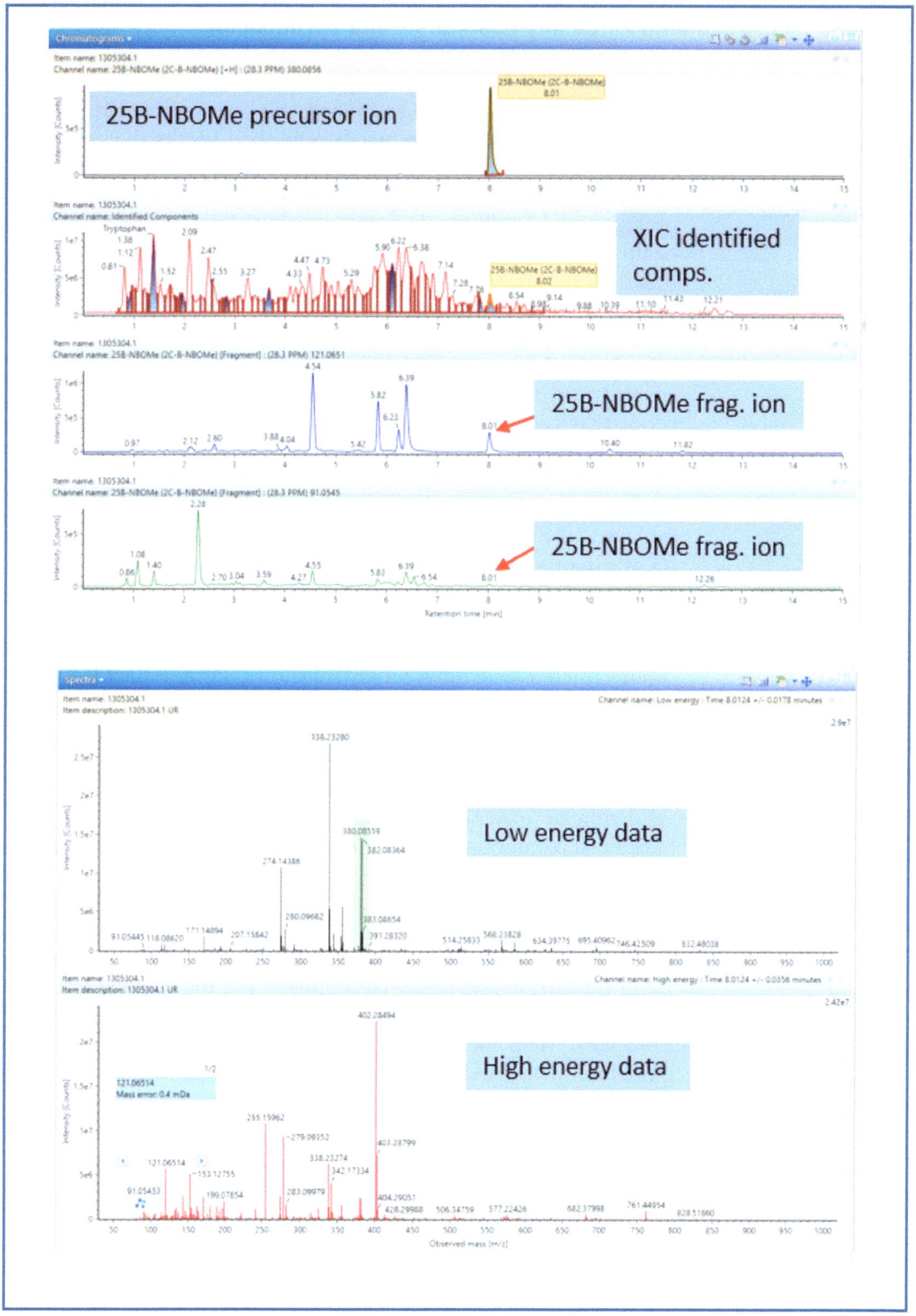

Fig. 4 CASE 3: Retrospective analysis: advantage of HRMS versus LC-MS/MS. When a man was found dead at a music festival in 2013, an overdose was suspected, but the drug screening and targeted methods available

4 HRMS Method Validation, Quality Assurance, and Certification

4.1 Validation

Reporting reliable analytical data is a prerequisite for the development of analytical toxicology. Indeed, it is obvious that the international scientific community needs published results that are valid, reproducible, and comparable. In addition, casework in a forensic or clinical context must be of high quality to ensure correct interpretation and finally result in proper judicial measures or adequate healthcare. Proper method development, a standardization or description of the applied identification criteria, and the validation of the developed method or workflow are thus of utmost importance. Validation is not the starting point of reliable analytical data; it is seen as the process which demonstrates the inherent quality of an analytical method through investigation, generating objective proof that certain demands are achieved ensuring a specific use. Thus, first the aim of the analytical method should be established, then the method development should be performed according to scientifically accepted quality standards, and finally validation will evaluate "the fitness for use" via predefined performance requirements. In other words, the confirmation criteria should be specified in advance. It is important to realize that inherent method quality is achieved during method development and based on the competence of the analysts. In NPS detecting methods applying HRMS, often targeted and untargeted workflows are combined. Several aspects that need to be developed and validated are observed: the analytical method, the workflow/data-processing criteria and the databases applied.

While high standards concerning method development and validation of analytical methods -certainly quantitative methods- have been paramount during the last decades and are set by international organizations [51–53], validation guidelines for HRMS methods and workflows are not separately specified or described. The general guidelines most referred to are the ones published by the US Food and Drug Administration (FDA), the Center for Drug Evaluation and Research (CDER), and the Center for Veterinary Medicine (CVM) [51], the guideline published by the European Medicines Agency (EMA) [52] and the toxicology guidelines from AAFS standards board (ASB) [53]. The goal of all of these guidelines was to create a consistency in the validation procedures and terminology for bioanalytical method validation in particular in the pharmaceutical field for full method comparison. Despite the impact the published guidelines have had on

Fig. 4 (continued) at the time did not offer any evidence. Further investigations were made in 2018 in light of analytical developments and new toxicological expertise. Blood and urine samples were re-analyzed by GC-MS and Q-TOF. Even with updated MS libraries, the results with GC-MS remained negative. Q-TOF results, however, confirmed the presence of 25B-NBOMe in the urine sample. This molecule is an NPS that belongs to the phenylethylamine chemical class and is often sold for its hallucinogenic effects

bioanalytical method validation in the last decades, it is important to emphasize that they remain nonbinding protocols and are intended to reflect a minimum standard of practice. Depending on the area of application and the particular purpose of analytical methods validation parameters and acceptance criteria may need to be adapted according to the particular analytical requirements. However, in some countries, specific legislations (*e.g.*, Belgium's Driving Under the Influence legislation) will refer to these guidelines and make the proposed validation criteria binding.

The field of forensic bioanalysis is continuously advancing and therefore method validation guidelines including its experimental setup and acceptance criteria are evolving considering the developments in different areas such as for example the field of statistics [54–60]. The experimental setup as well as the acceptance criteria will depend on the nature of the method: qualitative or quantitative. For qualitative or identification methods, parameters such as selectivity, limits of detection, and carryover will be most important [57]. For quantitative methods additional parameters such as calibration model, accuracy (bias and precision), and limits of quantification will be required. The evaluation of matrix effects is necessary –both for qualitative as well as quantitative methods- in case an LC-MS method is applied. Other parameters such as dilution integrity and stability studies depend on the use of the method. Robustness, i.e., the ability of a given method to remain unaffected by variations in the analytical conditions to be expected under routine conditions, and extraction efficiency are not considered an essential validation parameter but should be evaluated preferably in the development phase.

Figure 5 proposes the required validation parameters for both qualitative and quantitative guidelines, based on the publications by Wille et al. [58, 59], looking into a practical approach for analytical method validation in forensic toxicology. While the validation experiments and criteria for quantitative MS-methods are often well addressed in literature and guidelines as discussed above, validation and acceptance criteria for HRMS (un)targeted qualtitative screening methods are less specified. Different approaches are proposed in the literature and often, for the validation of wide-scope screening methods, a subset of representative compounds with a wide range of physicochemical properties can be chosen [23] to check for stability, limit of detection (LOD; signal to noise (S/N) is higher than 3) or limit of identification (LOI; all identification criteria are met), selectivity, carry-over, and matrix effects (ME) (Fig. 5). The subset should be chosen so that the most frequently encountered compounds in the routine forensic analysis are covered. Meanwhile, if a new NPS is discovered, retrospectively the necessary validation parameters can be assessed. However, if a subset of representative compounds is chosen, one is never sure of the validity of the method for a specific compound.

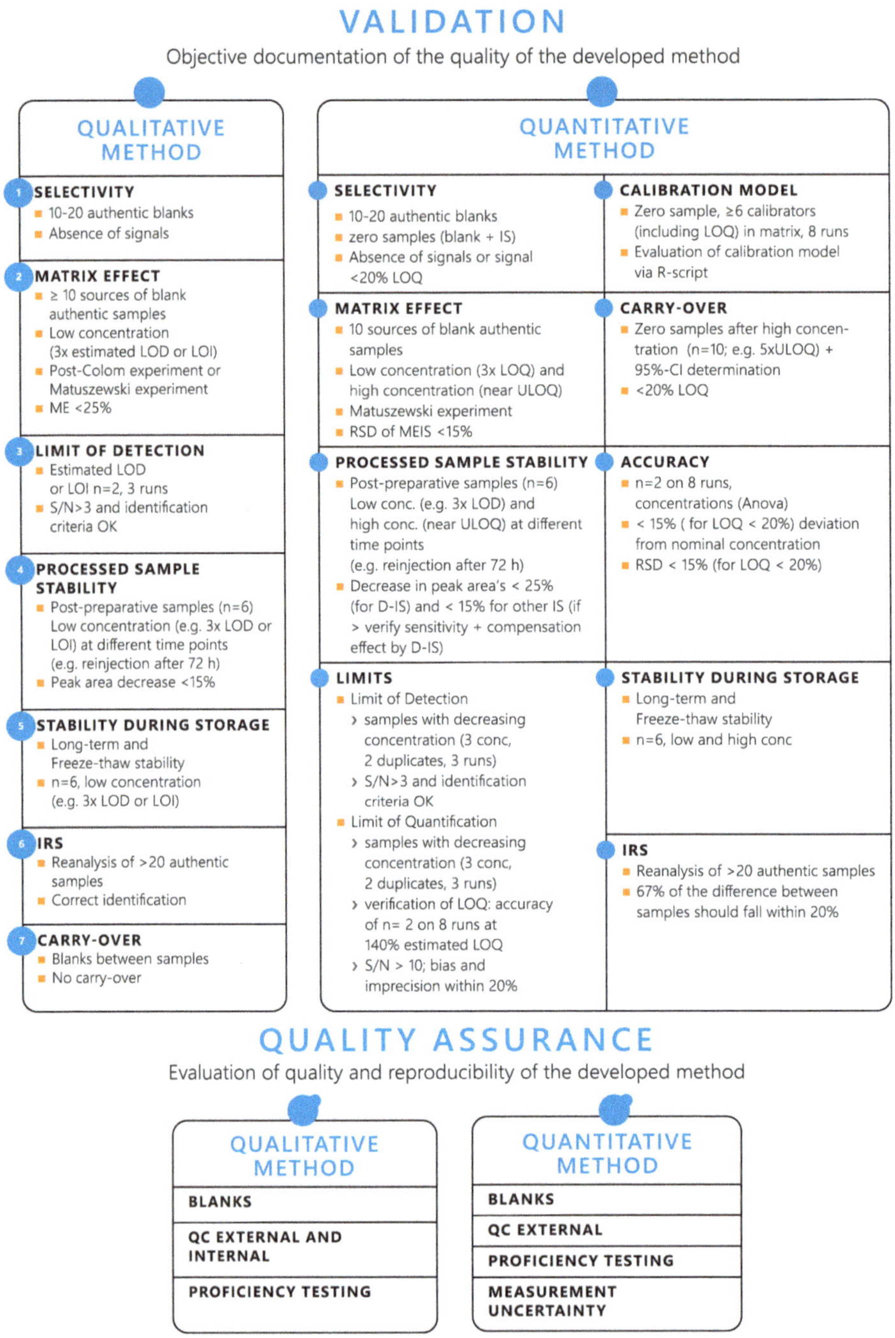

Fig. 5 Validation parameters for qualitative and quantitative validation based on the publications of Wille et al. [58, 59]. The parameters are indicated in the timely way that they should be evaluated. Basic information concerning choice of samples and criteria are given *CI* confidence interval, *Conc.* concentration, *D-IS* deuterated internal standard, *IRS* incurred reanalysis of samples, *IS* internal standard, *LOI* limit of identification, *LOQ* limit of quantification, *ME* matrix effect, *MEIS* relative matrix effect, *RSD* relative standard deviation, *S/N* signal to noise, *ULOQ* upper limit of quantification, *Zero sample* blank fortified with internal standard

Another way to tackle a qualitative validation of a HRMS screening approach is by looking into the sensitivity, selectivity, and reliability via calculation of the amount of FP and FN linked to a set decision cut-off or the LOI through analysis of spiked samples [61–63]. Desharnais et al. [61, 63] evaluated the effectiveness of a screening procedure with a cut-off by spiking samples with 50% and 150% of the cut-off concentration and calculating the false negative rate (FNR [FN/(FN + TP) × 100]), false positive rate (FPR [FP/(FP+FN)x100), selectivity (SLR [TN/(TN + FP) × 100]), sensitivity (SNR [TP/(TP + FN) × 100]), and reliability (RLR [(TP + TN)/(TP + TN + FP + FN) × 100]). These method performance estimators take into account all processes of the method, such as recovery variability and matrix effects. The acceptance criteria – via evaluation of two measured threshold samples, and the performance estimated over 30 samples – were as follows: FNR <7%, FPR = 0%, RLR >93%, SLR = 100%, SNR >93% [61, 63]. This is a very practical approach; however, again not all compounds will be covered. In addition, even if compounds are spiked in authentic blank samples, spiked samples may not always mimic the actual samples due to lack of *e.g.*, protein-binding or incorporation of analytes into the matrix, metabolite conversion to the precursor drug, and sample inhomogeneity. Therefore, Incurred Sample Reanalysis (ISR), a repeated measurement of analyte concentration from a portion of incurred samples to demonstrate reproducibility of the analytical results [64], has been introduced in bioanalytical method validation. These samples are preferably not pooled, but actual individual authentic samples [65]. This is rather a quantitative validation parameter, looking into the accuracy of the data as observed by the acceptance criteria as established by the EMA guideline: 5–10% of the number of samples in the study should be reanalyzed (depending on the number of samples being lower or higher than 1000) [52] and in order to pass the acceptance criteria, 67% of the reanalyzed samples should agree within 20% with the original measurement. However, reanalysis of ISR is a good way to validate qualitative methods as well as it can be seen as a cross-validation with an already established screening method on a large batch of samples. Here again the validation criteria should be set with regard to the number of samples to reevaluate, the quality/diversity of the samples, and the acceptance criteria. Moreover, the gold standard (*e.g.*, GC, LC screening or target methods or a combination) should be chosen wisely and according to the aim of the cross-validated HRMS method. As a result, the most ideal and elegant way to validate the whole HRMS workflow is an inter-laboratory study comparing samples via the same type of HRMS instrument with pre-defined acceptance criteria and adequate authentic samples (*e.g.*, containing NPS). The interest of such a procedure was demonstrated by Mardal et al. with their evaluation of the online-crowd-sourced

HRMS database HighResNPS.com in laboratories using different screening workflows and software from four different instrument vendors [42]. Moreover, by applying inter-laboratory comparisons, the choice of data processing and applied libraries (either in-house produced or external sourced libraries) can be validated too. HRMS performance is dependent upon how the data are analyzed, and minor changes to scoring and analysis settings can produce significant changes in method performance. Compounds of interest are identified by comparing measured mass, retention time, isotope pattern and/or product ion spectrum of a database and/or reference material. Defining how well-observed data matches a reference is critically important for the diagnostic performance of a method. Setting error limits too wide can lead to FP, setting error limits too narrow could result in FN. These critical criteria should be established empirically and validated in real patient samples. The database applied and the criteria for their evaluation should be communicated to the final customer or in published manuscripts.

4.2 Quality Assurance and Certification

While validation of a newly developed method is important to guarantee its "fitness for purpose," an active quality control program is essential to guarantee the quality of results on a daily routine basis. In-house quality control (IQC) guarantees that an analytical run is acceptable, and no problems have occurred during sample preparation or due to malfunction of the instrument. In addition, proficiency testing (PT) will give information on how the laboratory functions in comparison with others, thus how reproducible the results are using the developed and validated method. Moreover, certification according to the International Organization for Standardization norms (ISO 17025 and/or ISO 15189) is more and more requested by national authorities when dealing with forensic or clinical applications.

Proper quality control starts with choosing the proper IQC, number of samples, and the decision limits for the IQC, thus rejection or acceptance/rejection criteria for a batch. Often it is hard to find commercially available matrix matched IQC, and therefore in-house prepared matrix matched IQC are often used. In addition, these matrix matched IQC must be prepared with a high-quality reference material and stability must be guaranteed. According to the FDA and EMA guidelines, a minimum amount of 5% of the number of unknown samples or 6 IQC in total (3 concentrations ranging from 3 x LOQ to the ULOQ analyzed in duplicate), whichever is greater, should be analyzed for evaluation of batches of samples in quantitative MS methods [51, 52]. The criterion for acceptance is 67%, should fall within 15% of the nominal values. Again, no specific rules are established for qualitative HRMS methods. For evaluation of a batch of samples following the targeted workflow in HRMS analysis it would be of interest to evaluate the selectivity via a blank sample, then IQCs with a range of compounds should be injected and evaluated using the whole data processing workflow. All compounds should be detected according to the specified identification criteria. Moreover, one of the IQC samples could be spiked at LOD or LOI concentration in order to evaluate the sensitivity of the system during the run of the batch. In addition to an IQC control for the batch, the addition of a compound that is not found in routine samples (*e.g.*, a deuterated analogue) to each sample, can serve as a kind of internal standard (IS). This IS, if properly evaluated, can indicate problems for a specific sample: loss due to the sample preparation step or the injection system; it will not be used to quantify the samples, but will indicate if something went wrong with that specific sample. The AAFS Standards Board has proposed a general standard for a QC program in forensic toxicology laboratories [66]. In this standard, the proposed QC strategy for qualitative confirmation assays includes a negative control, a positive control, and one process control sample that challenges the LOD (QC concentration that is not higher than three times the LOD or LOI concentration). What is often neglected, for quantitative as well as qualitative analysis, is the performance control of the equipment before a batch is injected. The GTFCh Scientific Committee Quality Control describes evaluation criteria for gas chromatographic as well as liquid chromatographic methods [67]. For LC-methods, there is a test-mix consisting of specific compounds that are chosen wisely so that the obtained chromatograms can indicate if there are issues concerning flow velocity of the mobile phase, the injected volume of the sample, and the dead time and chromatographic separation capability. The retention time of the peaks and the half width value and symmetry of the peaks should be monitored.

Obtaining adequate PT schemes can be difficult, certainly for NPS detection. Fabresse et al. describe participation in a PT provided by the Society of hair testing containing known and unknown substances for inter-laboratory evaluation. Unfortunately, no NPS were included in this PT yet [25]. The French society of analytical toxicology (Société Française de toxicologie analytique; SFTA) has a urinary PT scheme in which various NPS are added (*e.g.*, the following NPS were involved in 2020 SFTA scheme: 2C-B, 3-MMC, 5F-MDMB-PICA, isopropylphenidate, Flualprazolam, 6-APB, 3-Fluoro-Phenmetrazine, 4F-MDMB-BINACA, Alpha PHP, and Ephylone). This type of PT can be of interest to follow up the targeted as well as untargeted HRMS workflows used in the laboratory and to evaluate the whole procedure. If the laboratory cannot subscribe to an existing PT, an independent analyst or QA-responsible can prepare a mix with unknown compounds to which the whole HRMS procedure is followed. Here, new emerging NPS can be added.

5 Conclusion and Future Perspectives

The gold standard for the detection of NPS or other (upcoming) compounds of interest in forensic or clinical toxicology is HRMS. There are numerous technical choices to make, each with their pros and cons. The targeted workflows are established and are easily applicable in routine today. Nontargeted workflows for unknowns are readily available but are rather time-consuming and require experienced personnel. Therefore, these are more suitable for special cases rather than routine; however, they give the toxicologists the necessary tool to look into a case file in more detail whenever symptoms of a patient or an intoxication in a forensic case cannot be explained by the data obtained from targeted MS methods.

Developments in new workflows and new software in order to improve the simplicity of the procedure and the time-constraints will have an enormous impact on the further development of HRMS in a toxicological setting. In addition, the scientific community should address guidelines concerning identification criteria and validation as well as quality assurance looking into the issues of HRMS specifically. Also, the quality control of the databases should be discussed. The improvement of instrument performance and better database standardization will facilitate the robustness of the databases and will promote the overall usability of HRMS in toxicological applications. As more laboratories consider HRMS for comprehensive drug screening, a consistent approach to data analysis and compound identification will be even more crucial. Carefully selecting, validating, and publishing the positivity criteria used in HRMS data analysis is a step in this direction.

Finally, the use of HRMS techniques for NPS detection will increase enormously in the next decade, not only for analytical toxicological screening purposes but also as a tool to look into the metabolism and toxicity of NPS via its application in metabolomics studies. These studies investigate the change of all detectable small molecules in a biological system after use of NPS and will have an enormous impact on the field of toxicology and the way a toxicologist will work in the future [68, 69].

References

1. Council Decision 2005/387/JHA of 10 may 2005 on the information exchange, risk assessment and control of new psychoactive substances. Off J L 127, 20/5/2005, 32–37. Available at http://eurlex.europa.eu/LexUriServ/LexUriServ.do?urie=CELEX:32005D0387:EN/HTML
2. UNODC (United Nations Office on Drugs and Crime) (2017) Early warning advisory on new psychoactive substances: legal responses. Available at https://www.unodc.org/LSS/page/NPS/LegalResponses
3. European Monitoring Centre for Drugs and Drug Addiction (EMCDDA) (2020) European Drug Report. Trends and development 2020. Available at https://www.emcdda.europa.eu/system/files/publications/13236/TDAT20001ENN_web.pdf

4. Australian Criminal Intelligence Commission. Illicit drug data report 2015–2016. Available at www.acic.gov.au
5. UNODC (United Nations Office on Drugs and Crime) World Drug Report 2018 (Vienna: UNODC). Available at wdr.unodc.org
6. Royal Society of Chemistry. High-resolution mass spectrometry. Available at http://www.rsc.org/publishing/journals/prospect/ontology
7. Allen DR, McWhinney BC (2019) Quadrupole time-of-flight mass spectrometry: a paradigm shift in toxicology screening applications. Clin Biochem Rev 40:135–146
8. Mogollon NGS, Quiroz-Moreno CD, Prata PS, de Almeida JR, Cevallos AS, Torres-Guiérrez R, Augusto F (2018) New advances in toxicological forensic analysis using mass spectrometry techniques. J Anal Methods Chem:ID4142527
9. Meyer MR, Maurer HH (2016) Review: LC coupled to low- and high-resolution mass spectrometry for new psychoactive substance screening in biological matrices-where do we stand today? Anal Chim Acta 927:13–20
10. Maurer HH, Meyer MR (2016) High-resolution mass spectrometry in toxicology: current status and future perspectives. Arch Toxicol 403:1203–1220
11. Polettini A, Gottardo R, Pascalie JP, Tagliaro F (2008) Implementation and performance evaluation of a database of chemical formulas for the screening of pharmaco/toxicologically relevant compounds in biological samples using electrospray ionization-time-of-flight mass spectrometry. Anal Chem 80:3050–3057
12. Joye T, Widmer C, Mégevand RM, Longère S, Augsburger M, Thomas A (2020) High-throughput qualitative and quantitative drug checking by MALDI HRMS. Front. 8 article 695
13. Sisco E, Thomas P (2021) Forbes forensic applications of DART-MS: a review of recent literature. Forensic Chem 22:100294
14. Gwak S, Almirall JR (2015) Rapid screening of 35 new psychoactive substances by ion mobility spectrometry (IMS) and direct analysis in real time (DART) coupled to quadrupole time-of-flight mass spectrometry (QTOF-MS). Drug Test Anal 7:884–893
15. Pasin D, Caxley A, Bidney S, Fu S (2017) Current applications of high-resolution mass spectrometry for the analysis of new psychoactive substances: a critical review. Anal Bioanal Chem 409:5821–5836
16. Meira VL, de Oliveira AS, Cohen LSA, de Bhering C, de Oliveira KM, de Siqueira DS, de Oliveira MAM, Aquino Neto FR, Vanini G (2021) Chemical and statistical analyses of blotter paper matrix drugs seized in the State of Rio de Janeiro. Forensic Sci Int 318:110588
17. Belsey S, Flanagan RJ (2020) Analytical detection of novel stimulants by immunoassay and liquid chromatography-high resolution mass spectrometry: case studies on ethylphenidate and mephedrone. J Anal Toxicol 45:521–528
18. Scheidweiler KB, Jarvis MJY, Huestis MA (2015) Nontargeted SWATH acquisition for identifying 47 synthetic cannabinoid metabolites in human urine by liquid chromatography-high-resolution tandem mass spectrometry. Anal Bioanal Chem 407:883–897
19. Sundström M, Pelander A, Angerer V, Hutter M, Kneisel S, Ojenperä I (2013) A high-sensitivity ultra-high performance liquid chromatography/high-resolution time-of-flight mass spectrometry (UHPLC-HR-TOFMS) method for screening synthetic cannabinoids and other drugs of abuse in urine. Anal Bioanal Chem 405:8463–8474
20. Skov-Skov Bergh M, Bogen IL, Wohlfarth A, Wilson SR, Oiestad AML (2019) Distinguishing between cyclopropylfentanyl and crotonylfentanyl by methods commonly available in the forensic laboratory. Ther Drug Monit 41:519–527
21. Bidny S, Gago K, Chung P, Albertyn D, Pasin D (2017) Simultaneous Screening and quantification of basic, neutral and acidic drugs in blood using UPLC-QTOF-MS. J Anal Toxicol 41:181–195
22. Mollerup CB, Dalsgaard PW, Mardal M, Linnet K (2017) Targeted and non-targeted drug screening in whole blood by UHPLC-TOF-MS with data-independent acquisition. Drug Test Anal 9:1052–1061
23. Salomone A, Di Corcia D, Negri P, Kolia M, Amante E, Gerace E, Vincenti M (2020) Targeted and untargeted detection of fentanyl analogues and their metabolites in hair by means of UHPLC-QTOF-HRMS. Anal Bioanal Chem 413:225–233
24. Vincenti F, Montesano C, Cellucci L, Gregori A, Fanti F, Compagnone D, Curini R, Sergi M (2019) Combination of pressurized liquid extraction with dispersive liquid liquid micro extraction for the determination of sixty drugs of abuse in hair. J Chrom A 1605:360348
25. Fabresse N, Larabi IA, Stratton T, Mistrik R, Pfau G, de la Grandmaison GL, Etting I, Delyle SG, Alvarez J-C (2019) Development of a sensitive untargeted liquid chromatography-high resolution mass spectrometry screening devoted to hair analysis

through a shared MS2 spectra database: a step toward early detection of new psychoactive substances. Drug Test Anal 11:697–708

26. Wille SMR, Richeval C, Nachon-Phanithavong M, Gaulier JM, Di Fazio V, Humbert L, Samyn N, Allorge D (2018) Prevalence of new psychoactive substances and prescription drugs in the Belgian driving under the influence of drugs population. Drug Test Anal 10:539–547
27. Richeval C, Dumestre-Toulet V, Wiart J-F, Vanhoye X, Humbert L, Nachon-Phanithavond M, Allorge D, Gaulier J-M (2019) New psychoactive substances in oral fluid of drivers around a music festival in south-west France in 2017. Forensic Sci Int 297:265–269
28. Diamanti K, Aalizadeh R, Alygizakis N, Galani A, Mardal M, Thomaidis NS (2019) Wide-scope target and suspect screening methodologies to investigate the occurrence of new psychoactive substances in influent wastewater from Athens. Sci Total Environ 685:1058–1065
29. Bade R, Tscharke BJ, White JM, Grant S, Mueller JF, O'Brien J, Thomas KV, Gerber C (2019) LC-HRMS suspect screening to show spatial patterns of New Psychoactive substances use in Australia. Sci Total Environ 650:2181–2187
30. Salgueiro-Gonzalez N, Castiglioni S, Gracia-Lor E, Bijlsma L, Celma A, Bagnati R, Hernandez F, Zuccato E (2019) Flexible high resolution-mass spectrometry approach for screening new psychoactive substances in urban wastewater. Sci Total Environ 689:679–690
31. McCullough BJ, Patel K, Francis R, Cain P, Douce D, Whyatt K, Bajic S, Lumley N, Hopley C (2020) Atmospheric solids analysis probe coupled to a portable mass spectrometer for rapid identification of bulk drug seizures. JASMS 31:386–393
32. Yanini A, Esteve-Turrillas FA, de la Guardia M, Armenta S (2018) Ion mobility spectrometry and high resolution mass-spectrometry as methodologies for rapid identification of the last generation of new psychoactive substances. J Chroma A 1574:91–100
33. Losacco GL, Ismail O, Pezzatti J, González-Ruiz V, Boccard J, Rudaz S, Veuthey JL, Guillarme D (2020) Applicability of supercritical fluid chromatography-mass spectrometry to metabolomics. II-Assessment of a comprehensive library of metabolites and evaluation of biological matrices. J Chrom A 1620(461021)
34. Guo J, Huan T (2020) Comparison of full-scan, data-dependent, and data-independent acquisition modes in liquid chromatography-mass spectrometry based untargeted metabolomics. Anal Chem 92:8072–8080
35. Vincenti F, Montesano C, Di Ottavio F, Gregori A, Compagnone D, Sergi M, Dorrestein P (2020) Molecular networking: a useful tool for the identification of new psychoactive substances in seizures by LC-HRMS. Front 8: 572952
36. Allard S, Allard PM, Morel I, Gicquel T (2019) Application of a molecular networking approach for clinical and forensic toxicology exemplified in three cases involving 3-MeO-PCP, doxylamine, and chlormequat. Drug Test Anal 11:669–677
37. Law KP, Lim YP (2013) Recent advances in mass spectrometry: data independent analysis and hyper reaction monitoring. Expert Rev Proteomics 10:551–566
38. Klingber J, Cawley A, Shimmon R, Fu S (2019) Collision-induced dissociation studies of synthetic opioids for non-targeted analysis. Front. https://doi.org/10.3389/fchem.2019.00331
39. Richeval C, Gaulier JM, Romeuf L, Allorge D, Gaillard Y (2019) Case report: relevance of metabolite identification to detect new synthetic opioid intoxications illustrated by U-47700. Int J Legal Med 133:133–142
40. Colby JM, Thoren KL, Lynch KL (2018) Suspect screening using LC-QqTOF is useful tool for detecting drugs in biological samples. J Anal Tox 42:207–213
41. Colby JM, Thoren KL, Lynch KL (2017) Optimization and validation of high-resolution mass spectrometry data analysis parameters. J Anal Toxicol 41:1–5
42. Mardal M, Andreasen MF, Mollerup CB, Stockham P, Telving R, Thomaidis NS, Diamanti KS, Linnet K, Dalsgaard PW (2019) HighResNPS.com: an online crowd-sourced HR-MS database for suspect and non-targeted screening of new psychoactive substances. J Anal Toxicol 43:520–527
43. Davidsen A, Mardal M, Linnet K, Dalsgaard PW (2020) How to perform spectrum-based LC-HR-MS screening for more than 1000 NPS with HighResNPS consensus fragment ions. Plos one. https://doi.org/10.1371/journal.pone.0242224
44. Damanti K, Aalizadeh R, Alygizakis N, Galani A, Mardal M, Thomaidis NS (2019) Wide-scope target and suspect screening methodologies to investigate the occurrence of new psychoactive substances in influent wastewater from Athens. Sci Total Environ 685:1058–1065

45. Allen DR, Warnholtz C, McWhinney BC (2021) Investigation and resolution of interferences in the LC-QTOF-MS detection of 4-MePPP. J Anal Toxicol. https://doi.org/10.1093/jat/bkaa199
46. Little JL, Williams AJ, Pshenichnov A, Tkachenko V (2012) Identification of "Known Unknowns" utilizing accurate mass data and ChemSpider. J Mass Spectrom 23:179–185
47. Gundersen POM, Broecker S, Slordal L, Spigset O, Josefsson M (2020) Retrospective screening of synthetic cannabinoids, synthetic opioids and designer benzodiazepines in data files from forensic post mortem samples analysed by UHPLC-Q-TOF-MS form 2014 to 2018. Forensic Sci Int 311:110274
48. Partridge E, Trobbiani S, Stockham P, Charlwood C, Kostakis C (2018) A case study involving U-47700, diclazepam and flubromazepam-Application of retrospective analysis of HRMS data. Drug Test Anal 10: 651–662
49. Thoren KL, Colby JM, Shugarts SB, Wu AHB, Lynch K (2016) Comparison of information-dependent acquisition of a tandem quadrupole TOF vs a triple quadrupole linear ion trap mass spectrometer for broad-spectrum drug screening. Clin Chem 62:170–178
50. Elmiger MP, Poetzch M, Steuer AE, Kraemer T (2017) Assessment of simpler calibration models in the development and validation of a fast post-mortem multi-analyte LC-QTOF quantitation method in whole blood with simultaneous screening capabilities using SWATH acquisition. Anal Bioanal Chem 409: 6495–6508
51. US Department of Health and Human Services, Food and Drug Administration, Center for Drug Evaluation and Research, Center for Veterinary Medicine (2018) Guidance for industry-bioanalytical method validation. Available at: http://www.fda.gov/downloads/Drugs/Guidance/ucm070107.pdf
52. Committee for medicinal products for human use (CHMP), European Medicines Agency (EMA) (2019) Guideline on validation of bioanalytical methods. Available at http://www.ema.europa.eu/documents/scientific-guideline/guideline-bioanalytical-method-validation_en.pdf
53. AAFS standards Board (2019) Standard practices for method validation in forensic toxicology, ANSI/ASB Standard 036, 1st edn. Available at www.asbstandardsboard.org
54. Scientific Working Group for Forensic Toxicology (SWGTOX) (2013) Standard practices for method validation in forensic toxicology. J Anal Toxicol 37:452–474
55. GTFCh (2009) The validation guidelines of the German speaking Gesellschaft für Toxikologie und Forensische Chemie. Available at https://www.gtfch.org/cms/index.php/en/guidelines
56. Peters FT, Maurer HH (2002) Bioanalytical method validation and its implications for forensic and clinical toxicology-a review. Accreditation Qual Assur 7:441–449
57. Peters FT, Drummer OH, Musshoff F (2007) Validation of new methods. Forensic Sci Int 165:216–224
58. Wille SMR, Peters FT, Di Fazio V, Samyn N (2011) Practical aspects concerning validation and quality control for forensic and clinical bioanalytical quantitative methods. Accreditation Qual Assur 16:279–292
59. Wille SMR, Coucke W, De Baere T, Peters FT (2017) Update of standard practices for new method validation in forensic toxicology. Curr Pharm Des 23:5442–5454
60. Alladio E, Amante E, Bozzolino C, Seganti F, Salomone A, Vincenti M, Desharnais B (2020) Experimental and statistical protocol for the effective validation of chromatographic analytical methods. MethodsX 7:100919
61. Desharnais B, Lajoie MJ, Laquerre J, Mireault P, Skinner CDA (2020) Threshold LC-MS/MS method for 92 analytes in oral fluid collection with the Quantisal® device. Forensic Sci Int 317:110506
62. Alladio E, Amante E, Bozzolino C, Seganti F, Salomone A, Vincenti M, Desharnais B (2020) Effective validation of chromatographic analytical methods: the illustrative case of androgenic steroids. Talanta 215:120867
63. Lemyre FC, Desharnais B, Laquerre J, Morel M-A, Côté C, Mireault P, Skinner CD (2020) Qualitative threshold method validation and uncertainty evaluation: a theoretical framework and application to a 40 analytes liquid chromatography-tandem mass spectrometry method. Drug Test Anal 12:1287–1297
64. Lowes S, Lelacheur R, Shoup R, Garofolo F, Dumont I, Martize S, Zimmer J, Cruz Caturla M, Couerbe P, Awaiye K et al (2014) Recommendations on incurred sample stability (ISS) by GCC. Bioanalysis 6:2385–2390
65. Fast DM, Kelley M, Viswanathan CT, O'Shaughnessy J, King SP, Chaudhary A, Weiner R, DeStefano AJ, Tang D (2009) Workshop report and follow-up_AAPS workshop on current topics in GLP bioanalysis: assay reproducibility for incurred

samples_Implications of Crystal city recommendations. AAPS J 11:238–241

66. ASB AAFS Standards Board (2020) Standard for a quality control program in forensic toxicology laboratories. ASB Standard 054, First Edition. Available at www.asbstandardsboard.org
67. GTFCh Scientific Committee Quality Control. Guideline for quality control in forensic-toxicological analysis. Available at https://www.gtfch.org/cms/images/stories/files/Guidelines-for-quality-control-in-forensic-toxicological-analyses-GTFCh-20090601-20180125.pdf
68. Steuer AE, Daelin D, Boxler MI, Eisenbeiss L, Holze F, Vizeli P, Czerwinska J, Dargan PI, Abbate V, Liechti ME, Kraemer T (2020) Comparative untargeted metabolomics analysis of the psychostimulants 3,4-methylenedioxy-methamphetamine (MDMA), amphetamine, and the novel psychoactive substance mephedrone after controlled drug administration to humans. Metabolites 10:0306
69. Manier SK, Wagmann L, Flockerzi V, Meyer MR (2020) Toxicometabolomics of the new psychoactive substances α-PBP and α-PEP studied in HepaRG cell incubates by means of untargeted metabolomics revealed unexpected amino acid adducts. Arch Toxicol 94:2047–2059

Chapter 10

High-Resolution Mass Spectrometry in Identification of New Analytes & Metabolites

Svante Vikingsson, Henrik Gréen, and Robert Kronstrand

Abstract

Numerous forensic institutes including the National Board of Forensic Medicine in Sweden have been using high-resolution mass spectrometry to identify new psychoactive substances (NPS) and/or their metabolites for more than a decade. Based on these experiences, current practices in generating, processing, and understanding data are presented in this chapter. Due to the rapidly changing market, searching for NPS should begin before reference materials are available. Adding masses to the screening library is one tool, but it is also important to consider the case history and circumstances when identifying new compounds. Similarly, lacking dosing studies, NPS metabolites are often identified using a combination of in vitro model systems and authentic case samples. This chapter provides advice, tips, and tricks on how to design this type of studies. During both screening and metabolite identification, it is important to critically evaluate the data and consider alternative explanations.

Key words New Psychoactive Substance (NPS), In vitro, Metabolism, Hepatocytes, Human liver microsomes, Screening, Synthetic cannabinoid receptor agonist (SCRA), Fentanyl analog

1 Introduction

In most laboratories, high-resolution mass spectrometry (HRMS) is mainly used for screening purposes. However, due to the nature of the instruments and data acquisition they can also be used to search your data for new drugs and analytes showing up in your case work. It can also be used to identify metabolites that could be important markers of drug intake, especially in urine. Although this work is being done in numerous toxicology and research laboratories around the world, the workflow is complex and not easily self-learned. Investigators need to have a systematic approach to be successful and achieve results in an efficient manner.

This chapter describes how the National Board of Forensic Medicine in Sweden has implemented HRMS for identifying new psychoactive substances (NPS) in case work; HRMS approaches for investigating NPS metabolites using in vitro model systems and

Marta Concheiro and Karl B. Scheidweiler (eds.), *Methods for Novel Psychoactive Substance Analysis*, Methods in Pharmacology and Toxicology, https://doi.org/10.1007/978-1-0716-2605-4_10,

authentic case samples are also presented. In this chapter we will share lessons learned over the last decade of employing HRMS in these areas for assisting investigators looking to establish HRMS for identifying new NPS analytes and their metabolites.

2 Identification of New Analytes

In 1995 the European Monitoring Centre for Drugs and Drug Addiction (EMCDDA) was founded. One purpose was to keep track on the substances used for recreational purposes and that may pose a threat to public health. The same year, the first legal high, as they were called then, surfaced and was detected in Sweden. The ecstasy analog MBDB (N-methyl-1,3-benzodioxolylbutanamine), an alpha-ethyl variant of the, at that time common, recreational drug MDMA was detected in urine samples from people suspected of a petty drug offense [1]. A positive amphetamine immunoassay screen combined with negative confirmatory results prompted the toxicologist to look further into the analytical data and perform additional analysis that revealed the presence of a new compound. At that time, the internet had limited capability; the most comprehensive sources of information on legal highs were printed copies of Dr. Alexander Shulgin's books on phenylethylamines and tryptamines [2, 3]. Armed with these resources, the toxicologist set out to find the needle in the haystack. Common screening methods for drugs in forensic cases were based on immunoassays, complemented with chromatographic techniques using either full scan mass spectrometry (MS), detection with nitrogen phosphorous detectors (NPD) or ultra-violet spectrophotometry [4–7].

Legal highs, as the name suggests, are designed to introduce small changes to a chemical structure that evades legislation but retain receptor activity producing a legal alternative to an established drug. Positional isomers and isobars have the same molecular weight whereas other small changes in molecular structure will change the molecular weight but may keep generic fragments found in MS analysis. An example is MDMA, mephedrone (4-methyl-methcathinone) and methedrone (4-methoxy-methcathinone). In Sweden, mephedrone was commonly encountered during 2008 and 2009 but after its scheduling as a narcotic drug in 2009 an analog, methedrone emerged on the drug market and became more prevalent [8]. The close similarity to the already known and scheduled MDMA revealed its presence in initial cases. Methedrone with the same molecular formula, $C_{11}H_{15}NO_2$, as MDMA in combination with very similar chromatographic properties showed up as a peak in a hair drug screen [8, 9]. Confirmatory analyses were negative for MDMA but circumstances around the death drove additional analysis with mass spectrometry. Methedrone and MDMA also share fragments in their electron impact (EI) mass spectra during gas chromatography-mass spectrometry analysis with common base peak at *m/z* 58 during underivatized analysis and a common fragment at *m/z* 135 when derivatized [8, 10]. Therefore, mass spectrometric screening methods may be directed towards backbone structure fragments that may be common for a range of substituted compounds [11].

In addition to mass spectrometric methods, other detection techniques have provided both sensitive and comprehensive methodology [5–7]. Many drugs contain a nitrogen atom such that NPD is an excellent screening technique allowing detection of new, so far unknown, analytes. However, they can also reveal anomalies, as in the case of Krypton, an NPS emerging in 2009. Krypton was a preparation of the herbal Kratom fortified with the active tramadol metabolite O-desmethyl-tramadol. The normal triplicate of analytes seen in the NPD screening after tramadol intake is the detection of the parent compound tramadol, followed by two of its metabolites, O-desmethyltramadol and N-desmethyltramadol, with the metabolites at considerably lower concentrations. However, in a series of cases only O-desmethyltramadol was detected. This anomaly cued the toxicologist to further investigate the presence of mitragynine because case information suggested that the decedents had ingested Krypton bought on the internet [12]. Thus, case history and findings at the scene can be of paramount importance to the progress of the toxicological investigation. Herbal drugs have presented additional complications. During the early 2010s herbal drug mixtures with synthetic cannabinoid receptor agonists increased in popularity. Laboratories, including ours, struggled with detecting synthetic cannabinoids since these new drugs did not cross-react with existing immunoassays nor did they have similar structures to tetrahydrocannabinol or its metabolites and would not be identified via existing cannabis methods. Case history became important for the toxicologists to aim their investigation towards these synthetic cannabinoids. Since the metabolisms of these drugs in humans were unknown, and metabolite reference standards unavailable, at first it seemed easier to direct methods to parent compounds and the analysis of synthetic cannabinoids in blood, rather than urine [13]. This revealed that the blood concentrations of these drugs were even lower than those traditionally encountered for THC and metabolites increasing the expectations of laboratories to analyze for these drugs. Immunochemical methods directed to the analysis of synthetic cannabinoids in urine were developed, but soon became obsolete as antibody cross-reactivities did not detect most new synthetic cannabinoids that were continuously emerging [14–16]. As the metabolism of certain synthetic cannabinoids was characterized, targeted chromatographic methods using tandem mass spectrometry were developed and validated [17–20]. Also, chromatographic methods needed constant updates and the workload associated with adding new analytes to a traditional quantitative LC-MS/MS or GC-MS method made toxicologists look to other solutions. One of these was HRMS. In the late 2000s, our laboratory had begun exploring HRMS as a screening tool for drugs and medications in autopsy cases. Using single stage HRMS acquisition, data was collected and then matched with a database comprising of compounds with acquired HRMS spectra and retention times on that particular chromatography [21]. For screening purposes, the HRMS only data worked well but to confirm the presence of a drug, MS/MS data was desired. At that time, identification criteria using HRMS data were not standardized, and the individual laboratory had to explore possible parameters as well as acceptance ranges. The first confirmatory routine HRMS method published from our laboratory was directed towards synthetic cannabinoids [22]. Using a PCDL (Personal Compound Database and Library) analytes were identified in a manual stepwise procedure using both MS data, MS/MS data, and a concentration threshold for reporting a positive result. MS identification was based on retention time, accurate mass, and the isotopic pattern (mass, abundance, and spacing). Analytes with an accurate mass deviation within ± 5 ppm, a retention time deviation within ± 0.1 min, and a confirmed isotopic pattern were considered presumptive positive if the combined MS score was ≥ 80 (out of 100). In addition, the obtained MS/MS spectrum was matched with accurate mass CID spectra from the PCDL generated at 20 eV, and a score of ≥ 80 out of 100 was considered a positive match. To report a positive finding, peak area ratios were compared with concurrently run threshold calibrators.

Even though a PCDL was used to identify targeted analytes, the data-dependent acquisition using auto MS/MS enabled the detection of unknown analytes. However, it required manually evaluating spectra from unknown peaks. Depending on the case load and human resources, this procedure can be initiated at different stages. For example, when the comparison with the PCDL came out negative but there are still circumstances pointing towards an intoxication with the class of compounds the method targets. Using HRMS data, most data analysis software can generate a list of possible molecular formulas which can be used to search publicly available databases such as Chemspider or Metlin for possible compounds [23]. If a structure seems reasonable and can be explained by the case spectrum the substance can be ordered. However, with NPS there is a high risk that the substance is not yet available as a reference standard. Considering the short lifespan of an NPS, monitoring of the new drug often begins before any reference standard is available. Especially since it is important to identify new NPS trends so that appropriate measures can be taken.

An approach is to enter the available data into the PCDL making sure it is referred to as data from a case. Molecular formula, retention time as well as spectrum can be entered using case data or databases when available [24]. Many times, the local or national crime lab may have seizures of drugs that they have identified and that can be used to qualitatively determine the presence of the drug in biological samples. A minimum requirement would be that the seized drug is accompanied by a certificate of analysis from the crime lab. In some instances, the certificate includes the purity of the drug and might be used for a quantitation, e.g., in a case report [8] but obtaining a commercial reference material that can be used for routine quantitation must be the goal. However, quantitation of NPS has been debated [25]. On the one hand, it is difficult to interpret the concentrations since there are no data for normal or toxic values, on the other hand, if analysis is kept qualitative there will never be reference values available for a quantitative interpretation. Suspected fatal intoxications may be a case category where interpretation is important. One example is the emergence of fentanyl analogs and other μ-opioid agonists, which caused and are causing a high number of deaths. To get a scientific foundation for these interpretations, the community should strive for quantifying at least several cases from the living or from post-mortem cases not related to intoxication and compare those to cases of suspected fatal intoxication [26, 27]. Quantification also helps to set relevant screening thresholds in both chromatographic and immunochemical methods [28, 29].

In summary, several new NPS have been found when scientists and toxicologists explored positive screening results with negative verifications, and also showed that it is important to explore the case circumstances, not only in post-mortem cases but also in samples from living individuals. The interpretation of these drugs will also require quantification of at least some type of compounds and cases. Therefore, continued awareness, curiosity, and ingenuity will always be important. In the next section, the focus is on drug metabolism and how metabolites can be elucidated and subsequently used when screening, confirming, and interpreting toxicological findings.

3 Metabolite Studies

While blood is the preferred matrix to measure impairment, urine is most valuable to verify drug exposure or use, as for many compounds the detection windows are longer and metabolite concentrations higher in urine [30]. In fact, for some NPS, including several synthetic cannabinoids, it is not uncommon to find urine specimens where a metabolite can be detected but no parent [31] (note: when reading these type of studies pay special attention to how the cases were selected – were the samples selected based on the parent being found in the urine sample, the corresponding blood sample, or by screening – the results will be quite different). This creates a need to study the metabolism of NPS, and many such studies have been published, mainly for synthetic cannabinoids [27, 31–44], but also for fentanyl analogs [45–49], synthetic cathinones/stimulants [41, 50, 51] and designer benzodiazepines [52–56]. In this section various aspects of conducting metabolite studies will be presented, hopefully providing guidance to both researchers planning to conduct studies and toxicologists who use metabolism data to better understand their casework.

3.1 Model Systems

To study the pharmacokinetics of a drug, including metabolites, the ideal study design is to administer drugs to volunteers in a clinical study setting, but due to the lack of information about NPS, and the risk of adverse events, only a couple of such studies have ever been conducted [37, 51]. In addition, self-administration studies, where the researchers themselves ingest the drug, have provided important information to our field [34, 36, 52]. However, such studies do carry risks and are not legal in most countries, encouraging researchers to find other ways to study NPS metabolism. These types of studies can be summarized in three different categories, discussed below; authentic urine specimens from case work, in vitro model systems, and animal studies.

Authentic urine specimens are attractive and frequently used to identify metabolites [27, 31, 35, 39, 45, 47–49, 52–56], as the end goal of most research is to provide methods to analyze the metabolites in such samples. However, as very little is known about the purity and amount of drug used, when the drug was used, or the other drugs that were coadministered, misinterpretations can occur. Each specimen only provides a snapshot that may, or may not, be relevant to other specimens, as illustrated by the large inter-specimen variability [39].

An alternative strategy offering more control is to use in vitro models such as hepatocytes, human liver microsomes (HLM), or the S9 fraction. Cryopreserved hepatocytes [27, 32, 38, 46, 48, 49, 53–55], which are donated liver cells pooled from several donors, are commonly used for metabolite identification but another potential source is to use immortalized cell lines such as HepaRG [40, 50]. While the intact cell model of the hepatocytes offers many advantages (such as a more complete metabolism), it is also dependent on liquid nitrogen storage capability and/or tissue culture capabilities. A cheaper alternative is the more stable HLMs which consist of cell membranes from liver cells, which are being used in numerous NPS metabolism studies [31, 33–35, 39, 42–44, 53, 55]. HLMs are limited to membrane bound enzymes (such as the CYP proteins) and are dependent on the addition of cofactors to the incubate [57]. A less used, but interesting alternative is the S9 fraction, which contains both HLMs and cytosol, allowing assessment of additional metabolic pathways [40, 44]. Other organs might also be involved in the metabolism of drugs. For example, esterases are highly active in plasma, and in vitro model systems for plasma and lung tissue have also been used [42].

Most in vitro model systems focus on liver metabolism and cannot account for effects such as the accumulation of polar metabolites in urine. To better study such effect researchers have turned to animal models, including mice [53, 55], rats [44], pigs [33, 44], and zebrafish [40], for their metabolite studies. While this provides a whole-body system including both absorption and elimination, the major drawback is that the enzyme systems differ somewhat between species which changes the metabolites formed, especially with regard to positional isomers and their relative abundance.

It should also be noted that easily scalable model systems such as fungi [41] can be used to produce NPS metabolites. These systems are even more different from humans with regard to the metabolites produced and seem more suitable as a tool to produce quantities large enough to allow isolation and structure determination by, for example, NMR spectroscopy.

3.2 Experimental Design for In Vitro Experiments

The typical experimental design of an in vitro experiment is that the test drug is mixed with the model system (typically HLM, hepatocytes, or S9-fraction) and required cofactors. The reaction is then stopped after a predetermined time, often by the addition of an organic solvent to denature and deactivate the enzymes, thereby stopping the reactions (Fig. 1).

If the experiment is correctly designed, one of the biggest advantages of using an in vitro model is that the conditions are controlled, making it easier to identify which peaks are related to the matrix and which are related to the drug tested. This is accomplished by including several controls. The negative control usually used is a specimen incubated just as the test samples but without drug. No peaks identified in this control can be related to the drug.

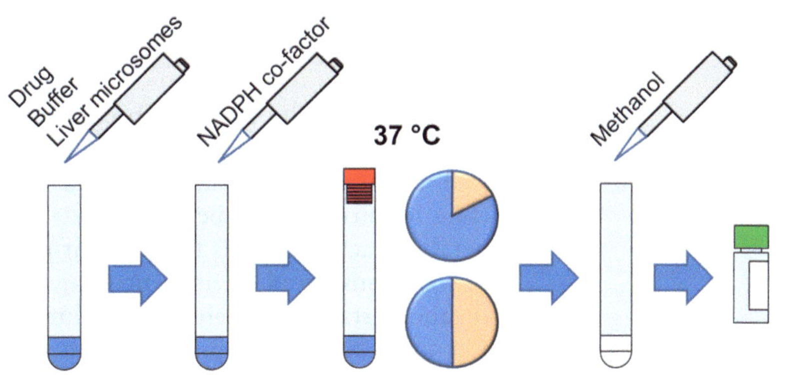

Fig. 1 General design of in vitro metabolite identification

Similarly, a starting point control can be included to identify if any metabolites were present in the starting material, which is especially common for nor-metabolites of fentanyls (as encountered by the authors). However, peaks should still be evaluated if they are substantially larger in the incubated samples, as a compound can be both a degradant in the drug standard and a metabolite. Some metabolic reactions are very fast, so even a couple of seconds of incubation might result in the formation of metabolites. To avoid this, it is recommended to stop the reaction (i.e., by adding solvent) before adding the test drug.

A third common control used for HLM experiments is a control without NADPH, which is necessary for the function of the CYP enzymes. If a metabolite is still formed in this control, it is not formed by NADPH-dependent enzymes such as the CYP enzymes.

Finally, it is important to include a positive control, which is a drug with well-known metabolites incubated together with the test samples. If the expected metabolites are identified, you know that the incubations and sample preparation worked.

We recommend that all samples and controls are incubated in at least duplicate. This allows for the collection of a full data set even if a sample is contaminated or lost during sample preparation. In addition, the duplicates can be used to simplify data processing as discussed below.

Most in vitro studies also include different time points, which serves two purposes. First, different time points can sometimes reveal information about which metabolites are formed first. Second, as the rate of metabolism can vary dramatically between different NPS, using an early, an expected, and a late time point can increase the chances of obtaining a wide variety of metabolites. Time points can be obtained either by removing a specified amount from an incubation tube at different times, or by incubating in different tubes or 96-well plates (Fig. 2). The former has the advantage of being simpler to set up and typically require less sample volume, but if that tube is contaminated it will affect all the different time points and frequent sampling might cool the tube down, affecting the rate of metabolite formation. At the very least, if duplicates are included, they should be incubated in different tubes.

As the viability and metabolic capacity of the model systems decrease over time, excessively long incubations should not be used. As a rough guide, HLM can be used for up to an hour, while hepatocytes can be used up to 6 h (note that a linearity in metabolic activity needs to be confirmed if kinetics is to be evaluated or considered). Different drugs have shown very different metabolic stability which could influence the selection of incubation times. On a group level, synthetic cannabinoids often form metabolites rapidly in vitro [43], while many designer benzodiazepines are very stable [58]. That said, there are also large variations

Plate 1 – 0 h controls

	1	2	3	4	5	6	7	8	9	10	11	12
A			Pos	#1	#2	#3	#4	#5	#6	#7	#8	#9
B			Pos	#1	#2	#3	#4	#5	#6	#7	#8	#9
C												
D												
E												
F												
G												
H												

Plate 3 – 3 h incubation

	1	2	3	4	5	6	7	8	9	10	11	12
A												
B												
C												
D												
E				#1	#2	#3	#4	#5	#6	#7	#8	#9
F				#1	#2	#3	#4	#5	#6	#7	#8	#9
G												
H												

Plate 1 – 1 h incubation

	1	2	3	4	5	6	7	8	9	10	11	12
A												
B												
C				#1	#2	#3	#4	#5	#6	#7	#8	#9
D				#1	#2	#3	#4	#5	#6	#7	#8	#9
E												
F												
G												
H												

Plate 4 – 5 h incubation

	1	2	3	4	5	6	7	8	9	10	11	12
A	D #1	D #5										
B	D #1	D #5										
C	D #2	D #6	D #9									
D	D #2	D #6	D #9									
E	D #3	D #7	Neg									
F	D #3	D #7	Neg									
G	D #4	D #8	Pos	#1	#2	#3	#4	#5	#6	#7	#8	#9
H	D #4	D #8	Pos	#1	#2	#3	#4	#5	#6	#7	#8	#9

Fig. 2 Experiment layout for hepatocyte experiment for 9 test drugs in 96-well plates. Plates can be combined to one injection plate. *Pos* positive control, *Neg* negative control, *D* degradation control

between NPSs from the same group, necessitating each drug to be considered separately.

In addition to the incubation time, the concentration of HLM or hepatocytes, cofactors, and test drug will also affect the obtained results. If high drug concentrations are used, many metabolic systems risk becoming saturated, which will alter the relative formation of various metabolites. Similarly, if the cofactors run out, the metabolic processes will stop. It is important to be aware of these effects, and when conducting clearance measurements, it is important to verify linear kinetics (see below).

When using authentic urine samples or animal models, many of the metabolites will also be conjugated (mainly as glucuronides, but sulfate and glutathione conjugates are also possible). The conjugated metabolites can be analyzed directly, but often suffer from low sensitivity due to poor ionization in positive mode. Another option is to hydrolyze (enzymatically or chemically) the conjugates to their corresponding unconjugated metabolites. If the latter approach is used the degree of conjugation can be estimated by analyzing the sample both before and after hydrolysis. The fraction of unconjugated metabolite can be calculated by dividing the peak area of the unconjugated metabolite before hydrolysis with the peak area after hydrolysis [47].

3.3 Data Acquisition and Chromatography

Many laboratories with a HRMS instrument also have a screening method set up that is specifically designed to analyze a wide range of analytes with varying polarity. While it can be tempting to use that method as is to also analyze the samples for the metabolite study, there are a couple of aspects to consider.

When it comes to sample preparation in metabolite studies, less is more. If you can achieve enough sensitivity with a simple dilute-and-shoot or protein precipitation method, that is preferred. The reason is that metabolites vary a lot in terms of polarity, pKa, and functional groups, and if a more advanced sample preparation, such as SPE or liquid–liquid extraction, is used, some metabolites might be lost, and the relative concentrations affected.

As with many LC-MS methods, chromatographic separation is key. The gradient might need to be extended and tailored to the test drug to achieve efficient separation. We recommend using a 100- or 150-mm column (preferably with sub-2 μm particles) and a gradient spanning 40–50 percentages of organic solvent over 10–15 min. This is important as several positional isomers of a metabolite often are formed. Sometimes you need to work extensively on the chromatography to separate positional isomers. A good example of this is synthetic cannabinoids with an indole moiety and a pentyl side chain, such as JWH-018 [59], see Fig. 3. The gradient should be adjusted so that the test compound elutes towards the end of the chromatographic gradient, as most metabolites are more polar than the parent compound causing them to elute earlier in reversed phase chromatography. However, there are some exceptions, most notably the N-oxides [45, 59–61], that tend to elute slightly after the parent.

Data for metabolite studies can be acquired both in data-dependent mode (by triggering true MSMS spectra) and in data-independent modes (known as e.g., MS^E, all ions or SWATH depending on the vendor) with alternating low and high collision energy scans. While the aim of a screening method is to obtain data of good enough quality to do a library matching, data of higher

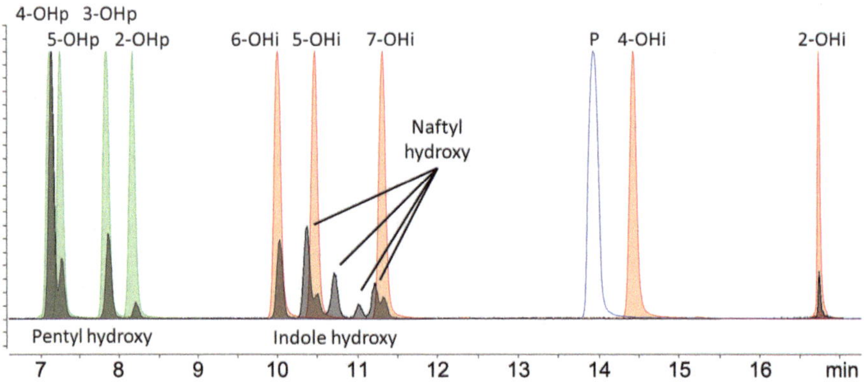

Fig. 3 Separation of monohydroxylated metabolites of JWH-018 after HLM incubation (black). Twelve isobaric peaks were identified of which eight could be matched to reference materials as pentyl chain hydroxylations (green) or indole hydroxylations (red). The remaining four peaks were tentatively identified as naftyl hydroxylations based on MS-spectra. ACE Excel 2 C18-AR, 0.05% formic acid in 10 mM ammonium formate [A] and 0.05% formic acid in methanol [B], gradient from 50–77% B from 0.6–16 min, using data-dependent MS/MS

quality is needed to elucidate metabolites. When using data-dependent acquisition, it might be necessary to increase the number of scans per peak to ensure scans around the apex of the peak and in data-independent acquisition it is important to obtain spectra that are as free from interferences as possible, which might require even further efforts in improving the chromatographic separation. To elucidate structure from a mass spectrum it is important to have as many different fragments as possible and to achieve this it might be necessary to optimize the collision energy. If the instrument allows for it, an approach where the collision energy is ramped during the scan can be considered.

3.4 Metabolite Predictions

The most common way to search for metabolites in QTOF data is to match the peaks in the samples to a list of potential targets, usually as a list of exact masses or molecular formulae but approaches for how to generate that list varies. In general terms, it requires a prediction of which metabolites might potentially be present.

In our laboratory we have successfully applied a list of common metabolic biotransformations, such as hydroxylations, dealkylations, ester hydrolysis, carboxylic acid formation, and glucuronide conjugation (and other phase II biotransformations), as well as combinations thereof, and applied those to the molecular formula of the parent compound. While a description of potential transformations is beyond the scope of this text, the comprehensive summary given by Wen and Nelson [62] is recommended as a good starting point. As a final note, it is also highly recommended to search through the literature for metabolite studies of similar compounds, and to consider the drug from a chemical point of view to ensure that potential biotransformations are covered.

The process of metabolite prediction has been successfully automated and there are several software solutions to predict potential metabolic transformations based on the structure of the parent compound, both connected to vendor-specific metabolite identification workflow solutions (e.g., Waters Metabolynx or Sciex MetabolitePilot) and as standalone applications (e.g. ACD Labs Metasense).

A less common approach is to use metabolomics software to identify peaks that are more common in samples exposed to the parent drug than in control samples [63]. This approach has successfully been applied to Valproic acid [64], synthetic cathinones [50], and cannabidiol [65] but requires either paired samples with and without drugs (such as those from in vitro model systems or animal models) [50, 65] or a substantial number of controls and exposed samples [64].

3.5 Data Analysis and Presentation

After searching for peaks matching a list of molecular masses, there will be a long list of potential matches, containing true matches mixed in with other peaks not connected to the drug. The task is to reduce that list down to a manageable number of analytes to examine in more detail, and there are a few different ways to achieve this. In addition to removing peaks below a minimum area (for example 20000 area units) and with mass difference above the expected ones (for example 5 ppm, keep in mind that high-intensity peaks, e.g. the parent, might show a mass shift in the QTOF due to detector saturation), it can be very useful to remove peaks that are also present in the negative (no parent drug added) control and peaks only found in a single sample (more effective if duplicate samples are used). To limit the data set further, the peak search can be limited to M+H adducts only, as other adducts (such as M +Na) tend to have fragment ions that are very hard to interpret and you might also remove small peaks without any MSMS spectra, since they will not be interpretable. It is key to always document your workflows and decision-making process in detail, to be able to go back and check your data for potential small metabolites that might explain a metabolic pathway.

In-source fragmentation is another source of extra peaks. When passing through the heated source, some analytes, such as hydroxylated metabolites or dihydrodiols, will start to fragment, often producing water loss products. These peaks are sometimes larger than the peak from the unfragmented metabolite but can be identified by the presence of a peak with a mass 18 Da (+H_2O) greater than the water loss product with the same retention time. Keep in mind that ions fragment into smaller molecules, and those fragments cannot contain atoms that were not present in the parent molecule.

Final review of a metabolite includes the matching of the fragmentation pattern to a proposed structure, which can be a very complex process for some analytes such as benzodiazepines [53–55]. Again, a review of fragmentation patterns in LC-MSMS is beyond the scope here but a good starting point is a book chapter on the topic written by Niessen and Correa [66]. In addition, many authors provide very comprehensive justifications for their assignments [32, 48] which can clarify the fragmentation behavior of similar analytes and simplify interpretation of very complex MSMS spectra.

Another good practice is to elucidate the spectra of the parent and then try to identify which fragments are identical in the spectra of the metabolite and for which the mass has changed. For example, it is not likely that the hydroxylation is on a part of the molecule producing the same fragment in both the metabolite and the parent, while it is very likely to be located on the part of the molecule corresponding to a fragment which has increased 16 Da in mass (corresponding to a hydrogen being replaced by a hydroxyl group). Also, a good tip is to start with the most abundant metabolites since their MSMS-spectra often are easier to interpret.

All spectra contain small background ions and sometimes it might contain ions from other compounds eluting at a similar time. (While this has been observed for data-dependent-acquisition, the risk is higher when using data-independent acquisition). It is good practice to ensure that fragments used as the basis of a structure determination are present in spectra from two different samples and be very restrictive in using low-intensity fragments for the determination. Also, be aware that more than one structural analog can elute at the same time, creating mixed spectra indicating multiple sites for the transformation (see chromatographic separation above).

The process of MSMS-spectra elucidation is another part of metabolite identification that is becoming more and more automated, with several computer software providing assistance. Some of these programs offer suggestions on the exact position of a transformation, based on which is most likely to form, but it is recommended to clearly indicate the reason and support of such an assignment.

There is some room for assigning exact structures beyond what can be deduced from the mass spectra alone. In an earlier study from our group [49], we confirmed fentanyl analog positional isomers using reference materials synthesized in-house and noted clear patterns in intensity and retention times, which can be used to suggest the positional isomer of other fentanyl analog metabolites. However, the structure of a metabolite should never be considered confirmed by LC-MS/MS assignments alone. Instead, it should be verified either by NMR or have retention time and MS-spectra matched to a reference material, as there might be surprises. For example, it was shown that β-hydroxylated metabolites of fentanyl and fentanyl analogs in fact show fragmentation indicating that the hydroxylation was located elsewhere [49], most likely leading to an underreporting of this transformation in the literature. Similarly, for synthetic cannabinoids with a pentyl sidechain, the 4- and 5-hydroxylated metabolites are hard to separate and often co-elute (see Fig. 3), making it difficult to ascertain which is the major metabolite. This example also highlights that structural isomers must be excluded to conclusively confirm the structure of a metabolite.

3.6 Writing and Reading Metabolism Studies

When drafting a metabolism manuscript and perhaps even more so when reading articles on the topic, it is important to remember the inherent limitations of these studies. First, the ionization in LC-MS is influenced by the functional groups of the analyte, affecting the observed signal intensity. This means that the most abundant peaks do not necessarily represent the metabolites at the highest concentrations. That said, the most abundant metabolites are also likely to be most abundant in other laboratories, when similar ionization techniques are being used, making them interesting analytical targets.

Second, the results of the studies might not be representative of case samples. For example, a difference between major analytes after hepatocyte [47] and HLM incubations [39] compared to urine specimens has been observed. In addition, as discussed above, there is large inter-specimen variability, especially in authentic urine samples [39, 47], which means that if only a small number of specimens are available, the results might be misinterpreted.

Third, also discussed above, identification by mass spectra elucidation is not an absolute technique and rely on assumptions and scientific judgement. Therefore, when relying on previous literature for elucidation of fragmentation pattern or assignment of structural analogs, it is important to read the original research and understand how the authors came to their conclusions and what evidence support them; especially case selection and experimental design will influence the results.

3.7 Presentation

Data from metabolite studies are often dense and multidimensional, and there seem to be almost as many ways to present it to the reader as there are research groups conducting studies. This section will try to give some advice on how to best present the data to the reader based on our experience, but the authors are aware that many other research groups present their data differently, and do not imply that such presentations are less suitable.

In the methods section include a data processing section detailing which potential metabolites or types of metabolites were included in the search and what the criteria were to include a peak in the final tables or figures. This allows a reader to compare the results to those presented in other studies.

Sometimes the results are presented in a merely qualitative fashion focusing on the presence or absence of different metabolites, as opposed to their relative abundance. Given the limitations regarding abundance and concentrations discussed above, this is a scientifically sound approach, but most readers are interested in both major metabolites, if the metabolism was extensive or not, and (in case of in vitro models) the time course of metabolite formation, all of which motivate some indication of peak abundance.

The identified metabolites are usually presented in three ways: tables, chromatograms, and metabolic schemes. Tables are good for providing details and can include both peak intensity and identifying information such as retention time, mass difference, and identifying fragments. To provide the reader with a quick overview of the major metabolites as well as a sense of the chromatography and separation, an overlaid chromatogram could be considered. A third complementary presentation often used is the metabolic scheme, Fig. 4, which provides the reader with a more complete picture of the metabolism and makes the authors more convinced that they did not miss an important metabolite.

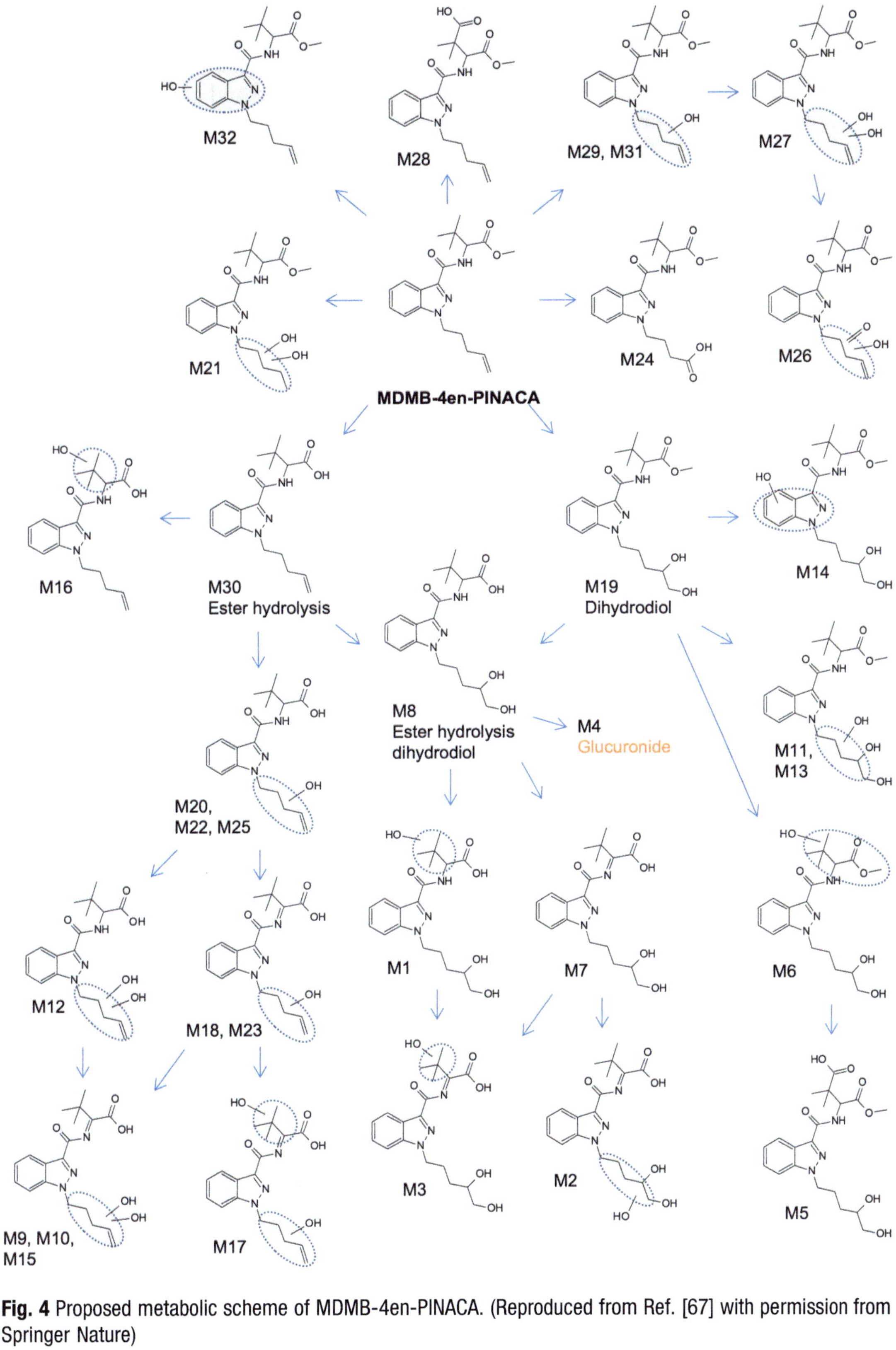

Fig. 4 Proposed metabolic scheme of MDMB-4en-PINACA. (Reproduced from Ref. [67] with permission from Springer Nature)

The rationale and discussion of structural assignments is often provided in the discussion section of the paper. It might be good to include representative mass spectra for each metabolite to allow the reader to make their own interpretations and as a basis for further method development. However, if many metabolites were identified, they can be included as supplemental material.

3.8 Using Results from Metabolite Studies

In this section we have covered how to set up a metabolite experiment, how to interpret the data and how to present it in a paper, but what is data used for and why do we conduct these studies? One major use is to provide information about metabolites so that reference materials can be produced, and the major metabolites be included into analytical methods [49].

Another interesting approach is to run the extracts from metabolite studies using the same analytical conditions as used in screening methods and include the major metabolites in the library [35]. On the one hand, the structure of the metabolites is not confirmed, but on the other hand, presence in samples but not controls, see above, strongly indicates that the metabolites are formed from the parent drug, which make them good analytical targets for screening.

In postmortem toxicology, presence of metabolites can sometimes provide information about how soon after drug intake the individual died. This has been shown for heroin [68], benzodiazepines [69], and buprenorphine [70], but there are also some indications that this could also be true for fentanyl analogs [45].

3.9 Enzyme Kinetics and Identification of Metabolizing Enzymes

When conducting metabolite studies using a model system, there are a couple of special applications that can provide additional information about the pharmacokinetics of the tested drug. One such application is studies of metabolic stability. These studies use HLMs and a similar design as the metabolite studies but focus on the rate of parent elimination. The results can be used to predict in vivo hepatic clearance and half-life [43, 56, 71].

Another application using HLMs is to identify the enzymes responsible for the metabolism, usually with a focus on the CYP enzymes [72]. The identification can be conducted in two ways, either the drug is incubated with microsomes engineered to contain only a single CYP enzyme or an inhibitor specific to a certain CYP enzyme or group of CYP enzymes is added to an HLM incubation. The specific microsomes clearly indicate which enzymes metabolize the parent drug, while the inhibitor studies can provide insight into the relative importance of the enzymes.

4 Discussion and Conclusions

In this chapter we have presented current practices in identifying novel analytes and conducting metabolite studies. We have also tried to share our experiences and provide actionable advice for laboratories starting to work with NPS. However, while we believe that the approaches presented here are fit for purpose, we are also aware that there are many other highly competent research groups and laboratories out there, some of which can be found in the reference list below. While slightly different, their solutions work just as well or perhaps even better than those presented.

The NPS field is vast and constantly changing, making research and screening a resource-intensive effort. To achieve results, a laboratory must maximize the performance given the available resources, which means relying on assumptions about these new drugs. We may assume that a novel synthetic cannabinoid will behave like other synthetic cannabinoids in our sample preparation, that structural analogs will separate on our LC-columns, that certain types of metabolites will be formed from a fentanyl analog, and/or that hydroxylated metabolites will produce fragmentation patterns in a similar way as the parent drug. There is nothing wrong with assuming these things, but we need to remember that we will be surprised or get something wrong sooner or later. We might assume the wrong structure of a metabolite as it has a surprising fragmentation or realize that a novel fentanyl analog has a retention time and mass spectra very similar to a structural analog already in the library. The take home message is to be critical of research findings, consider the assumptions made in the interpretation, and be open to alternative explanations for the results. As usual, we need to take our time and do our due diligence before reporting our findings in forensic casework.

References

1. Kronstrand R (1996) Identification of N-methyl-1-(3,4-methylenedioxyphenyl)-2-butanamine (MBDB) in urine from drug users. J Anal Toxicol 20(6):512–516
2. Shulgin A, Shulgin A (1991) PiHKAL: a chemical love story. Transform Press, Berkeley
3. Shulgin A, Shulgin A (1997) TIHKAL: the Continuation. Transform Press, Berkeley
4. Meyer MR, Peters FT, Maurer HH (2010) Automated mass spectral deconvolution and identification system for GC-MS screening for drugs, poisons, and metabolites in urine. Clin Chem 56(4):575–584
5. Druid H, Holmgren P (1997) A compilation of fatal and control concentrations of drugs in postmortem femoral blood. J Forensic Sci 42(1):79–87
6. Elliott SP, Hale KA (1998) Applications of an HPLC-DAD drug-screening system based on retention indices and UV spectra. J Anal Toxicol 22(4):279–289
7. Bogusz M, Wu M (1991) Standardized HPLC/DAD system, based on retention indices and spectral library, applicable for systematic toxicological screening. J Anal Toxicol 15(4):188–197

8. Wikström M, Thelander G, Nyström I et al (2010) Two fatal intoxications with the new designer drug methedrone (4-methoxymethcathinone). J Anal Toxicol 34(9):594–598
9. Kronstrand R, Nyström I, Strandberg J et al (2004) Screening for drugs of abuse in hair with ion spray LC-MS-MS. Forensic Sci Int 145(2-3):183–190
10. Awad T, Clark CR, DeRuiter J (2006) Chromatographic and mass spectral studies on methoxymethcathinones related to 3,4-methylenedioxymethamphetamine. J Chromatogr Sci 44(3):155–161
11. Pierzynski HG, Neubauer L, Choi C et al (2017) Tips for interpreting GC-MS fragmentation of unknown substituted Fentanyls. Cayman Currents 28:1–3. https://www.caymanchem.com/cms/caymanchem/LiteratureCMS/800181.pdf. Accessed 9 Aug 2021
12. Kronstrand R, Roman M, Thelander G et al (2011) Unintentional fatal intoxications with mitragynine and O-desmethyltramadol from the herbal blend Krypton. J Anal Toxicol 35(4):242–247
13. Kronstrand R, Roman M, Andersson M et al (2013) Toxicological findings of synthetic cannabinoids in recreational users. J Anal Toxicol 37(8):534–541
14. Arntson A, Ofsa B, Lancaster D et al (2013) Validation of a novel immunoassay for the detection of synthetic cannabinoids and metabolites in urine specimens. J Anal Toxicol 37(5):284–290
15. Barnes AJ, Young S, Spinelli E et al (2014) Evaluation of a homogenous enzyme immunoassay for the detection of synthetic cannabinoids in urine. Forensic Sci Int 241:27–34
16. Castaneto MS, Desrosiers NA, Ellefsen K et al (2014) Method validation of the biochip array technology for synthetic cannabinoids detection in urine. Bioanalysis 6(21):2919–2930
17. de Jager AD, Warner JV, Henman M et al (2012) LC-MS/MS method for the quantitation of metabolites of eight commonly-used synthetic cannabinoids in human urine–an Australian perspective. J Chromatogr B Analyt Technol Biomed Life Sci 897:22–31
18. Scheidweiler KB, Huestis MA (2014) Simultaneous quantification of 20 synthetic cannabinoids and 21 metabolites, and semi-quantification of 12 alkyl hydroxy metabolites in human urine by liquid chromatography-tandem mass spectrometry. J Chromatogr A 1327:105–117
19. Freijo TD Jr, Harris SE, Kala SV (2014) A rapid quantitative method for the analysis of synthetic cannabinoids by liquid chromatography-tandem mass spectrometry. J Anal Toxicol 38(8):466–478
20. Castaneto MS, Scheidweiler KB, Gandhi A et al (2015) Quantitative urine confirmatory testing for synthetic cannabinoids in randomly collected urine specimens. Drug Test Anal 7(6): 483–493
21. Roman M, Ström L, Tell H et al (2013) Liquid chromatography/time-of-flight mass spectrometry analysis of postmortem blood samples for targeted toxicological screening. Anal Bioanal Chem 405(12):4107–4125
22. Kronstrand R, Brinkhagen L, Birath-Karlsson C et al (2014) LC-QTOF-MS as a superior strategy to immunoassay for the comprehensive analysis of synthetic cannabinoids in urine. Anal Bioanal Chem 406(15):3599–3609
23. Taylor AM. Accurate mass screening workflows for the analysis of novel psychoactive substances. AB Sciex. https://sciex.com/content/dam/SCIEX/pdf/tech-notes/all/RUO-MKT-02-1884-A_MasterView_TripleTOF_designer_drug.pdf. Accessed 9 Aug 2021
24. Davidsen A, Mardal M, Linnet K et al (2020) How to perform spectrum-based LC-HR-MS screening for more than 1,000 NPS with HighResNPS consensus fragment ions. PLoS One 15(11):e0242224
25. Gerostamoulos D, Elliott S, Walls HC et al (2016) To measure or not to measure? That is the NPS question. J Anal Toxicol 40(4): 318–320
26. Wikström M, Thelander G, Dahlgren M et al (2013) An accidental fatal intoxication with methoxetamine. J Anal Toxicol 37(1):43–46
27. Åstrand A, Vikingsson S, Lindstedt D et al (2018) Metabolism study for CUMYL-4CN-BINACA in human hepatocytes and authentic urine specimens: free cyanide is formed during the main metabolic pathway. Drug Test Anal 10(8):1270–1279
28. Guerrieri D, Rapp E, Roman M et al (2017) Acrylfentanyl: another new psychoactive drug with fatal consequences. Forensic Sci Int 277: e21–e29
29. Guerrieri D, Kjellqvist F, Kronstrand R et al (2019) Validation and cross-reactivity data for Fentanyl analogs with the Immunalysis Fentanyl ELISA. J Anal Toxicol 43(1):18–24
30. Verstraete AG (2004) Detection times of drugs of abuse in blood, urine, and oral fluid. Ther Drug Monit 26(2):200–205

31. Vikingsson S, Gréen H, Brinkhagen L et al (2016) Identification of AB-FUBINACA metabolites in authentic urine samples suitable as urinary markers of drug intake using liquid chromatography quadrupole tandem time of flight mass spectrometry. Drug Test Anal 8(9):950–956
32. Diao X, Wohlfarth A, Pang S et al (2016) High-resolution mass spectrometry for characterizing the metabolism of synthetic Cannabinoid THJ-018 and its 5-Fluoro Analog THJ-2201 after incubation in human hepatocytes. Clin Chem 62(1):157–169
33. Doerr AA, Nordmeier F, Walle N et al (2020) Can a recently developed Pig model be used for in vivo metabolism studies of 7-Azaindole derived synthetic Cannabinoids? A study using 5F-MDMB-P7AICA. J Anal Toxicol 45(6): 593–604
34. Giorgetti A, Mogler L, Haschimi B et al (2020) Detection and phase I metabolism of the 7-azaindole-derived synthetic cannabinoid 5F-AB-P7AICA including a preliminary pharmacokinetic evaluation. Drug Test Anal 12(1): 78–91
35. Haschimi B, Mogler L, Halter S et al (2019) Detection of the recently emerged synthetic cannabinoid 4F-MDMB-BINACA in "legal high" products and human urine specimens. Drug Test Anal 11(9):1377–1386
36. Hutter M, Moosmann B, Kneisel S et al (2013) Characteristics of the designer drug and synthetic cannabinoid receptor agonist AM-2201 regarding its chemistry and metabolism. J Mass Spectrom 48(7):885–894
37. Toennes SW, Geraths A, Pogoda W et al (2018) Excretion of metabolites of the synthetic cannabinoid JWH-018 in urine after controlled inhalation. J Pharm Biomed Anal 150:162–168
38. Truver MT, Watanabe S, Åstrand A et al (2020) 5F-MDMB-PICA metabolite identification and cannabinoid receptor activity. Drug Test Anal 12(1):127–135
39. Vikingsson S, Josefsson M, Gréen H (2015) Identification of AKB-48 and 5F-AKB-48 metabolites in Authentic human urine samples using human liver microsomes and time of flight mass spectrometry. J Anal Toxicol 39(6):426–435
40. Wagmann L, Frankenfeld F, Park YM et al (2020) How to study the metabolism of new psychoactive substances for the purpose of toxicological Screenings-A follow-up study comparing pooled human liver S9, HepaRG Cells, and Zebrafish Larvae. Front Chem 8:539
41. Watanabe S, Kuzhiumparambil U, Fu S (2018) Structural elucidation of metabolites of synthetic Cannabinoid UR-144 by Cunninghamella elegans using nuclear magnetic resonance (NMR) spectroscopy. AAPS J 20(2):42
42. Thomsen R, Nielsen LM, Holm NB et al (2015) Synthetic cannabimimetic agents metabolized by carboxylesterases. Drug Test Anal 7(7):565–576
43. Watanabe S, Wu X, Dahlen J, Konradsson P, Vikingsson S, Kronstrand R, Gréen H (2020) Metabolism of MMB022 and identification of dihydrodiol formation in vitro using synthesized standards. Drug Test Anal 12(10): 1432–1441
44. Walle N, Nordmeier F, Doerr AA et al (2021) Comparison of in vitro and in vivo models for the elucidation of metabolic patterns of 7-azaindole-derived synthetic cannabinoids exemplified using cumyl-5F-P7AICA. Drug Test Anal 13(1):74–90
45. Kronstrand R, Åstrand A, Watanabe S et al (2021) Circumstances, postmortem findings, blood concentrations and metabolism in a series of Methoxyacetylfentanyl related deaths. J Anal Toxicol 45(8):760–771
46. Åstrand A, Töreskog A, Watanabe S et al (2019) Correlations between metabolism and structural elements of the alicyclic fentanyl analogs cyclopropyl fentanyl, cyclobutyl fentanyl, cyclopentyl fentanyl, cyclohexyl fentanyl and 2,2,3,3-tetramethylcyclopropyl fentanyl studied by human hepatocytes and LC-QTOF-MS. Arch Toxicol 93(1):95–106
47. Vikingsson S, Rautio T, Wallgren J et al (2019) LC-QTOF-MS identification of major urinary Cyclopropylfentanyl metabolites using synthesized standards. J Anal Toxicol 43(8):607–614
48. Watanabe S, Vikingsson S, Roman M et al (2017) In vitro and in vivo metabolite identification studies for the new synthetic opioids Acetylfentanyl, Acrylfentanyl, Furanylfentanyl, and 4-Fluoro-Isobutyrylfentanyl. AAPS J 19(4):1102–1122
49. Wallgren J, Vikingsson S, Rautio T et al (2021) Structure elucidation of urinary metabolites of Fentanyl and five Fentanyl analogs using LC-QTOF-MS, hepatocyte incubations and synthesized reference standards. J Anal Toxicol 44(9):993–1003
50. Manier SK, Wagmann L, Flockerzi V et al (2020) Toxicometabolomics of the new psychoactive substances α-PBP and α-PEP studied in HepaRG cell incubates by means of untargeted metabolomics revealed unexpected

amino acid adducts. Arch Toxicol 94(6): 2047–2059
51. Toennes SW, Schneider D, Pogoda W et al (2020) Excretion of 4-fluoroamphetamine and three metabolites in urine after controlled oral ingestion. J Pharm Biomed Anal 179: 113008
52. Huppertz LM, Moosmann B, Auwärter V (2018) Flubromazolam – basic pharmacokinetic evaluation of a highly potent designer benzodiazepine. Drug Test Anal 10(1): 206–211
53. Vikingsson S, Wohlfarth A, Andersson M et al (2017) Identifying metabolites of Meclonazepam by high-resolution mass spectrometry using human liver microsomes, hepatocytes, a mouse model, and authentic urine samples. AAPS J 19(3):736–742
54. Watanabe S, Vikingsson S, Åstrand A et al (2020) Metabolism of the benzodiazepines norflurazepam, flurazepam, fludiazepam and cinolazepam by human hepatocytes using high-resolution mass spectrometry and distinguishing their intake in authentic urine samples. Forensic Toxicol 38:79–94
55. Wohlfarth A, Vikingsson S, Roman M et al (2017) Looking at flubromazolam metabolism from four different angles: metabolite profiling in human liver microsomes, human hepatocytes, mice and authentic human urine samples with liquid chromatography high-resolution mass spectrometry. Forensic Sci Int 274:55–63
56. Noble C, Mardal M, Bjerre Holm N et al (2017) In vitro studies on flubromazolam metabolism and detection of its metabolites in authentic forensic samples. Drug Test Anal 9(8):1182–1191
57. Corning Inc. Mammalian Liver Microsomes – Guidance for Use. https://certs-ecatalog.corning.com/life-sciences/product-descriptions/452161.pdf. Accessed 4 Aug 2021
58. Moosmann B, Auwärter V (2018) Designer Benzodiazepines: another class of new psychoactive substances. Handb Exp Pharmacol 252: 383–410
59. Vikingsson S, Johansson A, Wallgren J et al (2019) S25: Structural identification of metabolites of synthetic cannabinoids JWH-018, AM-2201, THJ-018, THJ-2201 and 5F-AKB-48 using in-house synthesized standards, hepatocytes & LC-QTOF-MS. Presented at the annual meeting of the Society of Forensic Toxicologists (SOFT), San Antonio, TX, 13–18 Oct 2019
60. Cashman JR, Park SB, Yang ZC, Wrighton SA, Jacob P 3rd, Benowitz NL (1992) Metabolism of nicotine by human liver microsomes: stereoselective formation of trans-nicotine N'-oxide. Chem Res Toxicol 5(5):639–646
61. Wohlfarth A, Toepfner N, Hermanns-Clausen-M, Auwärter V (2011) Sensitive quantification of clozapine and its main metabolites norclozapine and clozapine-N-oxide in serum and urine using LC-MS/MS after simple liquid-liquid extraction work-up. Anal Bioanal Chem 400(3):737–746
62. Wen B, Nelson SD (2011) Common biotransformation reactions. In: Lee MS, Zhu M (eds) Mass spectrometry in drug metabolism and disposition: basic principles and applications. Wiley, Hoboken, pp 13–41
63. Steuer AE, Brockbals L, Kraemer T (2019) Metabolomic strategies in biomarker research-new approach for indirect identification of drug consumption and sample manipulation in clinical and forensic toxicology? Front Chem 7: 319
64. Mollerup CB, Rasmussen BS, Johansen SS et al (2019) Retrospective analysis for valproate screening targets with liquid chromatography-high resolution mass spectrometry with positive electrospray ionization: an omics-based approach. Drug Test Anal 11(5):730–738
65. Citti C, Palazzoli F, Licata M et al (2018) Untargeted rat brain metabolomics after oral administration of a single high dose of cannabidiol. J Pharm Biomed Anal 161:1–11
66. Niessen WMA, Correa RAC (2017) Interpretation of MS-MS mass spectra of drugs and pesticides. Wiley, Hoboken
67. Watanabe S, Vikingsson S, Åstrand A et al (2019) Biotransformation of the new synthetic Cannabinoid with an Alkene, MDMB-4en-PINACA, by human hepatocytes, human liver microsomes, and human urine and blood. AAPS J 22(1):13
68. Thaulow CH, Øiestad ÅML, Rogde S et al (2018) Can measurements of heroin metabolites in post-mortem matrices other than peripheral blood indicate if death was rapid or delayed? Forensic Sci Int 290:121–128
69. Forsman M, Nyström I, Roman M et al (2009) Urinary detection times and excretion patterns of flunitrazepam and its metabolites after a single oral dose. J Anal Toxicol 33(8):491–501
70. Kronstrand R, Nyström I, Andersson M et al (2008) Urinary detection times and metabolite/parent compound ratios after a single dose of buprenorphine. J Anal Toxicol 32(8): 586–593

71. Brandon AM, Antonides LH, Riley J et al (2021) A systematic study of the in vitro pharmacokinetics and estimated human in vivo clearance of Indole and Indazole-3-Carboxamide synthetic Cannabinoid receptor agonists detected on the illicit drug market. Molecules 26(5):1396

72. Nielsen LM, Holm NB, Leth-Petersen S et al (2017) Characterization of the hepatic cytochrome P450 enzymes involved in the metabolism of 25I-NBOMe and 25I-NBOH. Drug Test Anal 9(5):671–679

INDEX

Marta Concheiro and Karl B. Scheidweiler (eds.), *Methods for Novel Psychoactive Substance Analysis*, Methods in Pharmacology and Toxicology, https://doi.org/10.1007/978-1-0716-2605-4,

N

O

P

Q

S

T

U

V

MIX
Papier aus verantwortungsvollen Quellen
Paper from responsible sources
FSC® C105338

If you have any concerns about our products,
you can contact us on
ProductSafety@springernature.com

In case Publisher is established outside the EU,
the EU authorized representative is:
Springer Nature Customer Service Center GmbH
Europaplatz 3, 69115 Heidelberg, Germany

Printed by Libri Plureos GmbH
in Hamburg, Germany